Evolutionary Causation

Evolutionary Causation

Biological and Philosophical Reflections

Edited by Tobias Uller and Kevin N. Laland

The MIT Press
Cambridge, Massachusetts
London, England

This book was set in Times Roman by Westchester Publishing Services. Printed and bound in the United States of America.

Library of Congress Cataloging-in-Publication Data
Names: Uller, Tobias, editor. | Laland, Kevin N., editor.
Title: Evolutionary causation : biological and philosophical reflections / edited by Tobias Uller and Kevin N. Laland.
Description: Cambridge, MA : The MIT Press, [2019] | Series: Vienna series in theoretical biology | Includes bibliographical references and index.
Identifiers: LCCN 2018042611 | ISBN 9780262039925 (hardcover : alk. paper)
Subjects: LCSH: Evolution (Biology)—Philosophy. | Causation.
Classification: LCC QH360.5 .E9596 2019 | DDC 576.8—dc23
LC record available at https://lccn.loc.gov/2018042611

10 9 8 7 6 5 4 3 2 1

Contents

Series Foreword

Biology is a leading science in this century. As in all other sciences, progress in biology depends on the interrelations between empirical research, theory building, modeling, and societal context. But whereas molecular and experimental biology have evolved dramatically in recent years, generating a flood of highly detailed data, the integration of these results into useful theoretical frameworks has lagged behind. Driven largely by pragmatic and technical considerations, research in biology continues to be less guided by theory than it may seem.

By promoting the formulation and discussion of new theoretical concepts in the biosciences, this series intends to help fill important gaps in our understanding of some of the major open questions of biology, such as the origin and organization of organismal form, the relationship between development and evolution, and the biological bases of cognition and mind. Theoretical biology has important roots in the experimental tradition of early twentieth-century Vienna. Paul Weiss and Ludwig von Bertalanffy were among the first to use the term *theoretical biology* in its modern sense. In their understanding, the subject was not limited to mathematical formalization, as is often the case today, but extended to the conceptual foundations of biology. It is this commitment to a comprehensive and cross-disciplinary integration of theoretical concepts that the Vienna Series intends to emphasize. Today, theoretical biology has genetic, developmental, and evolutionary components, the central connective themes in modern biology, but it also includes relevant aspects of computational or systems biology and extends to the naturalistic philosophy of sciences.

The Vienna Series grew out of theory-oriented workshops organized by the KLI, an international institute for the advanced study of natural complex systems. The KLI fosters research projects, workshops, book projects, and the journal *Biological Theory*, all devoted to aspects of theoretical biology, with an emphasis on—but not restriction to—integrating the developmental, evolutionary, and cognitive sciences. The series editors welcome suggestions for book projects in these domains.

Gerd B. Müller, Johannes Jäger, Thomas Pradeu, Katrin Schäfer

1 Evolutionary Causation

Tobias Uller and Kevin N. Laland

Introduction

Scientific inference typically relies on establishing causation. This is also the case in evolutionary biology, a discipline charged with providing historical accounts of the properties of living beings, as well as an understanding of the processes that explain the origin of those properties. Familiar phenomena that demand an evolutionary explanation include the fit of form and function (adaptation) and the evolution of reproductive isolation (speciation), but also many others, including the origin of novelty and the organization of biological systems, including genomes and ecosystems. As expected from this diversity of topics, the study of evolution is a quite loosely organized endeavor. Nevertheless, there is broad agreement that an explanation for why organisms have particular features commonly will involve a particular kind of evolutionary process, namely evolution by natural selection.

While the basic principles of evolution by natural selection are simple—variation among individuals, differences in survival and reproduction, and heredity—the biological reality that instantiates these principles is incredibly complex. Furthermore, evolutionary processes encompass causation at different levels of biological organization, from genes to ecosystems, and at different timescales. This complexity makes it necessary to find means to represent biological systems in ways that leave out almost all causal detail, while ensuring that models, observations, and experiments carry explanatory weight. What evolutionary researchers should incorporate into their explanatory accounts, and what they should leave out, is far from a trivial issue.

While biologists are typically content to leave the nature of causation to philosophers, there is scientific value in reflecting on core theoretical concepts. Different views on causation can lead to alternative explanatory frameworks, dictate what counts as legitimate difference makers in evolutionary processes, and shape perspectives on key concepts such as natural selection and adaptation. How causation is understood thus shapes the structure of evolutionary theory, with both historical and contemporary debates in evolutionary biology revolving around the nature of causation (Laland et al. 2011). Yet, these issues

are rarely addressed. This limits the opportunity for constructive exchange and risks con-
fusing outstanding evolutionary problems with semantic disputes, both of which may
hinder scientific progress.

This edited volume brings together biologists and philosophers of science to provide a
comprehensive treatment of evolutionary causation. It is based on the workshop *Cause
and Process in Evolution*, which took place at the KLI in Klosterneuburg in May 2017.
The contributions clarify the nature of causation in the historical and contemporary
representation of evolution, specify alternative perspectives and reveal their underlying
assumptions, and seek ways of thinking about causation that will be helpful to formulate
research programs in evolutionary biology.

Proximate and Ultimate Causation

Whereas biological and philosophical reflections on the nature of causation in living
systems have a long history,[1] contemporary notions of evolutionary causation have been
shaped by the idea that there are two fundamentally different types of causation in biology.
The distinction, brought to prominence in a paper by Ernst Mayr (1961), is that one set
of causes—the "ultimate causes"—provide historical accounts for the existence of particular features in extant organisms, explain the "goal-directedness" of living beings, and
enable predictions about how populations will evolve in the future. The other set—the
"proximate causes"—are not historical and they are considered to explain and predict
biological systems by establishing how different components work, much like a chemist,
physicist, or engineer would establish causation. Under this view, evolutionary biologists
are concerned with ultimate causes, whereas molecular and developmental biologists,
physiologists, and most other biologists are concerned with proximate causes.

Mayr clearly put his finger on an important distinction in biology. An explanation for,
say, why a particular warbler migrates in autumn is different from an explanation for why
warblers migrate at all, rather than staying and enduring winter. Whereas the former
invokes "an immediate set of causes"—such as photoperiod and hormones—that modify
the birds' physiology and behavior, the latter invokes "causes that have a history and that
have been incorporated into the system through many thousands of generations of natural
selection" (Mayr 1961, 1503). More specifically, Mayr lists "ecological causes" and "genetic
causes" as the two categories of ultimate causes. In the case of the warbler, an ecological
cause could be the limited food availability in winter that implies that warblers that do not
migrate would starve. The genetic cause is "a genetic constitution [acquired] in the course
of the evolutionary history of its species which induces it to respond appropriately to the
proper stimuli from the environment" (Mayr 1961, 1503).

Mayr's partition of causation stems from his strong commitment to "population thinking" and variational explanations over "typological thinking" and transformational explanations in evolutionary biology. Sober (1984, 149) made the following helpful analogy to

contrast the two types of explanation. Consider that one wants to explain why all children in a class read at the third-grade level. One account takes the children one at a time and describes how he or she attained that level of reading proficiency. These stories are then aggregated to explain why the class is composed of children that all read at the third-grade level. This is a transformational explanation. It explains by accounting for changes that occur to individuals. But there is an alternative explanatory account that does not rely on aggregation. This is to note that only children who can read at the third-grade level are admitted to the class. This is a variational explanation. In contrast to the transformational account, it explains why the class is composed of children with a particular reading proficiency without referring to any individual causal detail. The variational explanation succeeds if children vary in their level of reading proficiency, allowing their proficiency to make a difference to admissions, and that those who have reached the third-grade level do not lose their ability after they are admitted.

According to Mayr, when evolutionary biologists explain why warblers migrate in autumn, their aim is to explain why the species is composed of individuals that migrate, rather than any other kind of individuals (e.g., those who overwinter). Ecological causes are those that "select" individuals on the basis of their characteristics. Genetic causes ensure that those characteristics vary between individuals before they are selected, and passing on those genes to descendants ensures that the variants, once they have been selected, persist and can accumulate in the population down generations. On this view, mutation and recombination introduce genetic variation and the distribution of these variants changes under natural selection, drift, and migration. From this perspective, there is no room for proximate causes in evolutionary explanations since only genes are inherited, effectively preventing nongenetic developmental causes from becoming evolutionary causes.

Causation and the Role of the Organism in Evolution

The proximate–ultimate distinction has provided conceptual unity for the evolutionary sciences. It has been a core aspect of a shared consensus as to which forms of causality constitute satisfactory evolutionary explanations (i.e., historical and functional accounts) and which do not (i.e., mechanistic and developmental accounts). To this end, the distinction has been used to detect "flawed" evolutionary reasoning—models that, in addition to fitness differences and genetic inheritance, invoke proximate mechanisms to explain the fit of form and function (e.g., Scott-Phillips, Dickins, and West [2011] for human cooperation and Dickins and Rahman [2012] for extra-genetic inheritance).[2] On the above interpretation of the proximate–ultimate distinction, this response appears sound. Since genetic mutations are assumed not to be directed, nor to occur when organisms need them, on this reasoning all of the sustained directionality in evolution comes from fitness differences between genotypes, or natural selection.

While few would contest the value of distinguishing between different types of causal explanation, philosophers and biologists have recognized that this is not the only way to describe causation in living systems (e.g., Calcott 2013; Laland et al. 2013; Watt 2013). Thus, the implementation of the proximate–ultimate distinction, while uniting evolutionary biology, may inadvertently be ruling out certain legitimate classes of evolutionary explanations, and hindering evolutionary biology's ability to draw from adjacent fields (e.g., West-Eberhard 2003; Laland et al. 2011). Particularly contentious has been the implication that developmental processes, which are responsible for bringing phenotypes into being, are considered irrelevant for explaining phenotypic evolution. Several debates within contemporary evolutionary biology concern the extent to which causal effects of environment on organism (e.g., plasticity, epigenetic inheritance) and causal effects of organism on environment (e.g., niche construction) also can be evolutionary causes (Laland et al. 2011, 2015). While the challenges are not new, the exclusion of development in evolutionary explanation is increasingly difficult to reconcile with biological knowledge and how biologists actually go about doing their research.

To illustrate, consider the observation that killer whales have locally adapted diet and specialized hunting techniques (Foote et al. 2016; Hoelzel and Moura 2016). These characters are stably inherited down generations, making phenotypic differences between lineages persist. In fact, sympatric groups with different foraging modes appear to be reproductively isolated (Moura et al. 2014; Moura et al. 2015). Researchers have revealed that the adaptive divergence in diet and hunting technique is not due to genetic differences but to social learning (Foote et al. 2016). It is likely that the origin of novel foraging behavior is due to behavioral innovation through, for example, trial and error rather than genetic mutation (Foote et al. 2016). It thus appears that social learning—traditionally considered a proximate mechanism—contributes to the historical explanation for the adaptive fit between the local environment and phenotype, as well as reproductive isolation of killer whale populations.

That developmental processes like behavioral plasticity and social learning impose persistent adaptive directionality on evolution appears inconsistent with the exclusive role of natural selection in adaptive evolution, as well as the heuristic stance that genetics is the proper mechanistic focus of evolutionary analysis. This inconsistency arises because the explanation for adaptive divergence in killer whales relies, in part, on a transformational explanation[3]. It explains how a particular population becomes composed of fish eaters by referring not only to the fitness differences between individuals with different hunting techniques but also to the acquisition of fish-hunting skills in ontogeny and its spread to other members of the population through social learning. In contrast, a variational explanation for why the population is composed of fish eaters would not invoke environmental influence on how traits originate and are inherited.

It is possible to develop this variational account by reformulating what needs to be explained. What requires an evolutionary explanation, the argument goes, is not why killer whale populations acquired the ability to hunt for seals or salmon, but why they acquired a capacity for social learning. On this account, ecological conditions ("ecological causes") have favored genetic variants ("genetic causes") that enable individuals to adjust their food preferences to local conditions.[4] This avoids a transformational account of evolution by making the adaptive directionality imposed by the organisms themselves a function of long-term natural selection on random genetic variation. This "rescaling" of evolutionary and developmental processes is a very common strategy to protect the notion that there are two fundamentally different kinds of causes in biology. It allows the evolutionary effects of phenotypic plasticity, extra-genetic inheritance, and niche construction to be accommodated into the genetic instantiation of evolution by natural selection (e.g., Scott-Phillips et al. 2014). While this can provide a valid historical explanation, the explanatory shift comes with at least three features that appear problematic.

The first problematic feature is that, to some (but seemingly not all) researchers, the evolutionary explanation simply appears incomplete. Staying with the killer whale example, the traditional account provides only a partial historical analysis, as it explains the killer whales' general reliance on social learning but not why particular populations acquire or maintain particular dietary traditions (killer whale ecotypes are often sympatric). Moreover, the different ecotypes have diverged genetically (Moura et al. 2015) and exhibit morphological adaptations such as population-specific digestive enzymes (Foote et al. 2016). Here the absence of an evolutionary account of why particular populations possess particular feeding habits appears to render incomplete the explanation for the existence and properties of morphological specializations. The latter requires knowledge of the particular dietary traditions of the population and cannot be predicted with knowledge of local ecology.

The second problematic feature is that a strict exclusion of proximate causes in evolutionary explanations appears to confer on genes causal and informational privilege in development. Indeed, when Mayr described genetic causes as ultimate causes, despite that genes exercise their phenotypic effects through development, it reflected his metaphysical view of development as the execution of a genetic program (e.g., Mayr 1961; Mayr 1984). This view was common in the 1960s and continues to be so among contemporary biologists (Moczek 2012), but it has been widely discredited (Oyama 2000; Keller 2010; Griffiths and Stotz 2013). If phenotypes are underdetermined by their genotypes (i.e., if not all adaptive responses are located in the genome as "programs," "blueprints," or "recipes"), it appears conceptually inconsistent to ascribe the adaptive directionality in evolution caused by social learning solely to past selection on genetic variation (e.g., Mesoudi et al. 2013; Jablonka and Lamb 2014).

A third concern is that the justification for the explanatory shift appears to rely on the assumption that variation, differential fitness, and heredity are autonomous processes (Badyaev 2011; Walsh 2015; Uller and Helanterä 2017). For example, it is usually assumed that pattern of selection does not affect the rules of inheritance; inheritance is merely the passing on of whatever alleles were selected. The variation that fuels evolution is considered similarly autonomous. Mutations occur randomly with respect to their consequences for phenotypic variation and fitness, and the acquisition of new variants does not change how variation is transmitted down generations. Process autonomy ensures that all adaptive directionality arising in development in principle can be explained in terms of previous rounds of selection on stably inherited (genetic) variation. But the same logic need not apply when the processes are causally intertwined or entangled (Walsh 2015; Uller and Helanterä 2017). On this account, it does not matter to what phenotype, or how far back in time, we shift our focus; how the principles of evolution by natural selection are instantiated is ever changing, thereby compromising a strictly variational account of adaptation and diversification.

Identifying the consequences of alternative causal representations of evolution is an increasingly pressing task as genome, cell, and developmental biology are producing results that question how well biological systems are represented in gene-centric evolutionary biology (i.e., the framework that motivated the distinction between proximate and ultimate causation). Some argue that evolutionary biologists need to embrace these findings and see them as a means to devise a richer explanation of life's diversity, if evolution is to remain the central guiding principle in biology (e.g., Laland et al. 2015; Sultan 2015). Others point out the traditional framework's ability to generate causal explanation even if some of its core assumptions are relaxed (e.g., Wray et al. 2014; Futuyma 2017). Whereas it remains to be seen if and how evolutionary biology will be transformed by these debates, they will not go away without a concerted effort to resolve conceptual differences or, at the very least, demarcate the fault line(s) of interpretative understanding of the evolutionary process.

Despite the central role of the nature of causation in evolutionary biology, the outstanding issues are rarely addressed. Evolutionary biology textbooks, for instance, hardly ever cover this topic, and such analysis as exists is dominated by philosophers of science. To biologists, at least, the literature on causation in biological systems may appear idiosyncratic and poorly connected to evolutionary theory. This edited volume therefore brings together leading biologists and philosophers of science to focus on the causal structure of evolutionary theory from historical and contemporary perspectives, and thereby endeavor to shed light on current debates.

A Brief Summary of This Book

The opening contribution by Massimo Pigliucci captures the motivation for this volume. Pigliucci outlines four alternatives for the relationship between science and philosophy of science. He makes a case that there can be mutual benefit, and goes on to show—using four case studies of causation—how this benefit can be realized. Pigliucci's analysis emphasizes that, despite their different aims, philosophers and scientists can enrich their respective disciplines through active engagement.

   The remainder of the volume is divided into two parts comprising biological and philosophical reflections on evolutionary causation.

Biological Reflections on Evolutionary Causation

The first section consists of eight chapters that showcase the biological motivations for rethinking evolutionary causation. The chapters provide sufficient biological background on the diverse ways by which development, extra-genetic inheritance, and niche construction challenges our notions of cause and process in evolution. The authors also make new theoretical contributions by detailing how alternative representations of evolutionary causation can help to shed light on a range of evolutionary problems.

   Theoretical population genetics has occupied a central position in evolutionary biology for most of the past century. Within this framework, bias in the introduction of novelty has typically been considered to have little or no influence on the evolutionary outcome. Arlin Stoltzfus explains the logic behind this assertion and finds that it rests upon particular assumptions that became entrenched as the field progressed. In contrast to the received view, he demonstrates how bias in the introduction of variants can have a substantial effect on the evolutionary outcome even in population-genetic models. Stoltzfus's historical account of mutational bias is a reminder that notions of causation and explanation are sensitive to metaphors and analogies, which in turn shape the reception of novel theories.

   The introduction of novelty is of central concern to evolutionary developmental biology, which over the past decades has become an influential discipline within evolutionary biology. Drawing on his experiences as a graduate student and postdoc, Armin Moczek paints a personal picture of this transformation and the promises of developmental studies in evolutionary biology. Exploring how contemporary research sheds light on innovation, diversification, and adaptation, Moczek argues that this body of work provides genuine opportunities to expand our understanding of evolutionary causation.

   One of the most contentious claims regarding the role of development in evolution is that environmentally induced responses can be evolutionarily significant. David Dayan, Melissa Graham, John Baker, and Susan Foster explain the logic behind this perspective, which envisions an evolutionary sequence of events in which phenotypic change leads, and genetic change follows, in evolutionary adaptation. Dayan and colleagues argue that increasing attention to environmental causes of variation in phenotype not only will help to understand phenotypic evolution but that it also can help to resolve several long-standing problems in population genetics. Thus, rather than being in opposition to each other, the genetic and developmental perspectives on evolution can be mutually beneficial.

   Another reason that plastic responses can become evolutionarily consequential is that parental transference of developmental resources or templates influence heredity. Such extra-genetic inheritance is now known to occur through many mechanisms and in many organisms. Sonia Sultan explains how this compromises the logic of treating plasticity as a property of the genotype. The challenge to evolutionary theory is that the treatment of plasticity as a property of the genotype is what maintains the causal privilege of natural selection in evolutionary explanation. Sultan thus points toward a fundamental dilemma facing evolutionary biologists—should they accommodate plasticity and extra-genetic inheritance using idealizations that maintain the contemporary explanatory agenda, or should they seek to ground alternative research programs in constructive views of development and inheritance?

The role of alternative research programs is of central concern to Kevin Laland, John Odling-Smee, and Marcus Feldman. Their survey of the origin of the concept of niche construction, its reception, and the continued debate surrounding its evolutionary implications illustrates how metaphysical and epistemological views shape scientific pursuits. Laland and colleagues single out three conceptual issues of particular importance in this regard: the notion of organismal agency, the concept of development as programmed *versus* constructed, and causal entanglement *versus* causal autonomy of variation, differential fitness, and heredity. They argue that grasping these points is essential to understanding what is new and important about niche construction theory. The message is that, rather than making niche construction fit the current conceptual framework, there is scientific value in actively pursuing alternative points of view.

That organisms co-create the niches in which they develop and are selected is also the theme of the chapter by Renée Duckworth. Building on insights from her work on evolutionary cycles of displacement between western and mountain blue birds, she explains how we can understand better how evolution works by determining how the behavior of biological entities at one scale influence patterns of variation at another scale. In particular, Duckworth is concerned with evolutionary stasis. To be robust at one level is to be dynamically flexible at another level, and Duckworth argues that to understand stasis we must first understand how ecological interactions evolve their robustness. This robustness, in turn, is to be found in the niche-constructing and developmentally plastic responses of the active organism.

The evolution of biological organization is also the focus of the two following chapters. Both are concerned with the evolutionary transition from a population of individuals to a population of collectives. This transition represents a challenge to traditional explanations since it is unclear how fitness benefits at the individual level can result in the formation of collectives that function as their own evolutionary units. Heikki Helanterä and Tobias Uller discuss this problem by applying the Darwinian space framework of philosopher of science Peter Godfrey-Smith (2009) to the evolution of social insect colonies. They conclude that an evolutionary account of how collectives acquire evolutionary individuality requires a focus on the evolution of spatial and functional organization through causal feedbacks between reproductive fitness, plasticity, and niche construction.

Richard Watson and Christoph Thies take this further and ask if plasticity, niche construction, and extra-genetic inheritance are in fact necessary conditions for a transition in evolutionary individuality. Their starting point is that collectives that represent evolutionary units, such as multicellular organisms, must exhibit heritable variation in fitness over and above that of the individual members of the collective. They argue that this is only possible if individuals within collectives exhibit plasticity, extra-genetic inheritance, and niche construction. In fact, Watson and Thies argue that it is these properties of parts that make collectives approximate the causal autonomy of variation, differential fitness, and heredity that the "standard model" assumes. Their conclusion is that the organism plays an active role in explaining transitions in individuality, which may argue in favor of historical explanations that combine variational and transformative accounts.

Philosophical Reflections on Evolutionary Causation

The final section of the book consists of five chapters that analyze evolutionary causation from a philosophical perspective. Denis Walsh begins by drawing attention to what he calls "the paradox of population thinking" in evolutionary biology. The paradox lies in that the causes of individual lives and deaths are what makes the population change, but these causes are precluded from the explanation of how evolution happens. According to Walsh, invoking natural selection or drift as causal explanations is to overpopulate the world with causes. On his account, changes in the composition of trait types in a population is an analytical consequence and higher order effect of the individual causes of evolution. The contention surrounding the role of development in evolution arises from the failure to realize that both "first-level causes" and "higher order effects" explanations are valid.

Jun Otsuka is also concerned with diagnosing the conceptual underpinnings of the debate over proximate causes in evolutionary explanation, in particular the current controversy over the "extended evolutionary synthesis" (EES; Laland et al. 2014). Otsuka concludes that there are different views on both the ontological units of evolution and on the proper methodology for studying their dynamics. Otsuka explains the former using causal graph theory and shows that, in the EES, the evolutionary dynamics arise, not from the genes, but from the entire causal structure of the biological system. Whereas this does not imply that all causes make an equal contribution, it does imply that there is generally no ontological justification to elevate particular causes to evolutionary causes. Otsuka also demonstrates how this ontology is accompanied by a shift in research methodology. He concludes that the challenge now is to demonstrate that this conceptual framework can provide a productive and unified research program.

Arnaud Pocheville turns his attention to the suggestion that causal entanglement of the processes of evolution by natural selection compromises traditional accounts of evolutionary causation. Making the point that entanglement occurs at both levels and timescales, he shows how entanglement is handled by evolutionary theory through "rescaling." In contrast to the interpretation in terms of process (or causal) autonomy, the timescale framework focuses on whether or not one can explain, for example, social learning in terms of imposing constraints on a slower evolutionary process. Pocheville's analysis suggests that this emphasis on local invariants, rather than causes, significantly influences how one should interpret the challenge to evolutionary theory posed by plasticity, extra-genetic inheritance, and niche construction.

Lynn Chiu approaches the entanglement of causes and processes in evolution through an examination of niche construction. In niche construction theory, natural selection and niche construction are two coupled processes that share responsibility for the evolution of adaptive fit between organism and environment. A common strategy is to rescale these two processes, such that natural selection bears responsibility for the direction on evolution imposed by niche construction. Building on Lewontin's insights, Chiu makes the point that organisms do not only physically modify their world but also change how the world is experienced. She argues that this implies that organism and environment can "commingle" in ways that spread selective causes across organism and environment to make niche construction and natural selection inseparable. In these cases, natural selection still explains adaptation, but it is not the same externalist "natural selection" as in the traditional account. Chiu proposes a strategy to identify whether or not it is appropriate to assume that natural selection explanations are externalist.

In the final chapter, Karola Stotz discusses the relationship between causation and information. She points out that biological systems are unique in that they are informed. Whereas the standard interpretation is that this information resides in the genome, Stotz demonstrates how information is acquired, expressed, and stored in development and heredity through a variety of biological processes. This leads her to argue that biological information is a substantive causal factor in both development and evolution. Importantly, the contribution of different causes can be quantified in terms of their specificity, which provides strategies for how information from different sources can be quantified and compared. Stotz suggests that this allows establishing the explanatory relevance of proximate sources of information in evolution.

Conclusion

This collection of essays demonstrates that contemporary debates about evolutionary biology are not solely about new data or novel theoretical findings but also revolve around fundamental conceptual issues (albeit often promoted by new findings). A productive dialogue between biologists and philosophers of science offers the best prospect for resolution of these challenging issues, and we hope this volume makes a contribution toward that resolution.

Acknowledgments

We are grateful to Gerd Müller and Johannes Jäger for hosting the workshop *Cause and Process in Evolution* at the Konrad Lorenz Institute, and to all the speakers and participants for making it a such a stimulating meeting. We are particularly grateful to the authors of this volume, who not only contributed talks at the meeting but also agreed to write up their presentation for publication. The workshop was funded through a grant from the John Templeton Foundation, titled *Putting the Extended Evolutionary Synthesis to the Test* (grant no. 60501). Tobias Uller was supported by a Wallenberg Academy Fellowship from the Knut and Alice Wallenberg Foundation.

Notes

1. It arguably goes back to Aristotle's four causes (Lennox 2001). One historically important debate in biology concerns whether or not biological phenomena can accurately be represented by "mechanistic" causation or if living systems are fundamentally different. This debate is reflected in contemporary disputes over, for example, the reach of reductionism in biology and the nature of agency. Riskin (2016) and Peterson (2016) provide two recent accounts of how this debate played out during different times in history.

2. It is worth noting that transmission genetics is a proximate mechanism. The reasons why this does not appear to compromise a variational account of evolution is explained later in the chapter.

3. Only in part, because there can still be (likely is) selection between individuals on the basis of their characters.

4. An alternative explanatory shift is to consider behavioral variants ("memes") as a population of entities with differential fitness, reproduction, and inheritance (Dawkins 1976).

References

Badyaev, A. V. 2011. "Origin of the Fittest: Link between Emergent Variation and Evolutionary Change as a Critical Question in Evolutionary Biology." *Proceedings of the Royal Society B: Biological Sciences* 278 (1714): 1921–1929. doi: 10.1098/rspb.2011.0548.

Calcott, B. 2013. "Why How and Why Aren't Enough: More Problems with Mayr's Proximate-Ultimate Distinction." *Biology and Philosophy* 28 (5): 767–780. doi: 10.1007/s10539-013-9367-1.

Dawkins, R. C. 1976. *The Selfish Gene.* Oxford: Oxford University Press.

Dickins, T. E., and Q. Rahman. 2012. "The Extended Evolutionary Synthesis and the Role of Soft Inheritance in Evolution." *Proceedings of the Royal Society B: Biological Sciences* 279 (1740): 2913–2921. doi: 10.1098/rspb.2012.0273.

Foote, A. D., N. Vijay, M. C. Avila-Arcos, R. W. Baird, J. W. Durban, M. Fumagalli, R. A. Gibbs, et al. 2016. "Genome-Culture Coevolution Promotes Rapid Divergence of Killer Whale Ecotypes." *Nature Communications* 7. doi: 10.1038/ncomms11693. .

Futuyma, D. J. 2017. "Evolutionary Biology Today and the Call for an Extended Synthesis." *Interface Focus* 7 (5): 20160145. doi: 10.1098/rsfs.2016.0145.

Godfrey-Smith, P. 2009. *Darwinian Populations and Natural Selection.* Oxford: Oxford University Press.

Griffiths, P. E., and K. Stotz. 2013. *Genetics and Philosophy: An Introduction.* Cambridge: Cambridge University Press.

Hoelzel, A. R., and A. E. Moura. 2016. "Killer Whales Differentiating in Geographic Sympatry Facilitated by Divergent Behavioural Traditions." *Heredity* 117 (6): 481–482. doi: 10.1038/hdy.2016.112.

Jablonka, E., and M. J. Lamb. 2014. *Evolution in Four Dimensions: Genetic, Epigenetic, Behavioral and Symbolic Variation in the History of Life.* Cambridge, MA: MIT Press.

Keller, E. F. 2010. *The Mirage of a Space between Nature and Nurture.* Durham, NC: Duke University Press.

Laland, K. N., J. Odling-Smee, W. Hoppitt, and T. Uller. 2013. "More on How and Why: Cause and Effect in Biology Revisited." *Biology and Philosophy* 28 (5): 719–745. doi: 10.1007/s10539-012-9335-1.

Laland, K. N., K. Sterelny, J. Odling-Smee, W. Hoppitt, and T. Uller. 2011. "Cause and Effect in Biology Revisited: Is Mayr's Proximate-Ultimate Dichotomy Still Useful?" *Science* 334 (6062): 1512–1516. doi: 10.1126/science.1210879.

Laland, K. N., T. Uller, M. W. Feldman, K. Sterelny, G. B. Müller, A. Moczek, E. Jablonka, and J. Odling-Smee. 2014. "Does Evolutionary Theory Need a Rethink?—POINT Yes, Urgently." *Nature* 514 (7521): 161–164.

Laland, K. N., T. Uller, M. W. Feldman, K. Sterelny, G. B. Müller, A. Moczek, E. Jablonka, and J. Odling-Smee. 2015. "The Extended Evolutionary Synthesis: Its Structure, Assumptions and Predictions." *Proceedings of the Royal Society B: Biological Sciences* 282 (1813). doi: 10.1098/rspb.2015.1019.

Lennox, J. G. 2001. *Aristotle's Philosophy of Biology. Studies in the Origins of Life Science.* New York: Cambridge University Press.

Mayr, E. 1961. "Cause and Effect in Biology—Kinds of Causes, Predictability, and Teleology Are Viewed by a Practicing Biologist." *Science* 134 (348): 1501–1506. doi: 10.1126/science.134.3489.1501.

Mayr, E. 1984. "The Triumph of the Evolutionary Synthesis." *Times Literary Supplement* 2:1261–1262.

Mesoudi, A., S. Blanchet, A. Charmantier, E. Danchin, L. Fogarty, E. Jablonka, K. N. Laland, et al. 2013. "Is Non-genetic Inheritance Just a Proximate Mechanism? A Corroboration of the Extended Evolutionary Synthesis." *Biology Theory* 7 (3): 189–195. doi: 10.1007/s13752-013-0091-5.

Moczek, A. 2012. "The Nature of Nurture and the Causes of Traits: Toward a Comprehensive Theory of Developmental Evolution." *Integrative and Comparative Biology* 52:E123–E123.

Moura, A. E., J. G. Kenny, R. R. Chaudhuri, M. A. Hughes, R. R. Reisinger, P. J. N. de Bruyn, M. E. Dahlheim, et al. 2015. "Phylogenomics of the Killer Whale Indicates Ecotype Divergence in Sympatry." *Heredity* 114 (1): 48–55. doi: 10.1038/hdy.2014.67.

Moura, A. E., J. G. Kenny, R. Chaudhuri, M. A. Hughes, A. J. Welch, R. R. Reisinger, P. J. N. De Bruyn, et al. 2014. "Population Genomics of the Killer Whale Indicates Ecotype Evolution in Sympatry Involving Both Selection and Drift." *Molecular Ecology* 23 (21): 5179–5192. doi: 10.1111/mec.12929.

Oyama, S. 2000. *The Ontogeny of Information.* Durham, NC: Duke University Press.

Peterson, E. L. 2016. *The Life Organic: The Theoretical Biology Club and the Roots of Epigenetics.* Pittsburgh, PA: University of Pittsburgh Press.

Riskin, J. 2016. *The Restless Clock: A History of the Centuries-long Argument over What Makes Living Things Tick.* Chicago: University of Chicago Press.

Scott-Phillips, T. C., T. E. Dickins, and S. A. West. 2011. "Evolutionary Theory and the Ultimate-Proximate Distinction in the Human Behavioral Sciences." *Perspectives on Psychological Science* 6 (1): 38–47. doi: 10.1177/1745691610393528.

Scott-Phillips, T. C., K. N. Laland, D. M. Shuker, T. E. Dickins, and S. A. West. 2014. "The Niche Construction Perspective: A Critical Appraisal." *Evolution* 68 (5): 1231–1243. doi: 10.1111/evo.12332.

Sober. E 1984. *The Nature of Selection: Evolutionary Theory in Philosophical Focus.* Chicago: University of Chicago Press.

Sultan, S. E. 2015. *Organism and Environment: Ecological Development, Niche Construction and Adaptation.* New York: Oxford University Press.

Uller, T., and H. Helanterä. 2017. "Niche Construction and Conceptual Change in Evolutionary Biology." *British Journal for the Philosophy of Science*, axx050.

Walsh, D. M. 2015. *Organism, Agency and Evolution.* Cambridge: Cambridge University Press.

Watt, W. B. 2013. "Causal Mechanisms of Evolution and the Capacity for Niche Construction." *Biology and Philosophy* 28 (5): 757–766. doi: 10.1007/s10539-012-9353-z.

West-Eberhard, M. J. 2003. *Developmental Plasticity and Evolution.* New York: Oxford University Press.

Wray, G. A., H. E. Hoekstra, D. J. Futuyma, R. E. Lenski, T. F. C. Mackay, D. Schluter, and J. E. Strassmann. 2014. "Does Evolutionary Theory Need a Rethink?—COUNTERPOINT No, All Is Well." *Nature* 514 (7521): 161–164.

2 Causality and the Role of Philosophy of Science

Massimo Pigliucci

Introduction: Science and/vs. Philosophy of Science?

As a scientist (evolutionary biology) and a philosopher (of science), I have long been interested in the relationship between the two disciplines (Pigliucci 2008). This is an issue of importance both in terms of the two academic fields themselves, and in the broader context of how they are perceived by the general public—with further practical implications, obviously, concerning funding of scholarship, faculty positions, and studentships.

While the diatribe between scientific and humanistic disciplines goes back at least to C. P. Snow's ([1960] 2012) famous essay on what he termed "the two cultures," in recent years the noise level has gone up significantly, mostly—though not exclusively—on the side of the scientists. Almost 60 years ago Snow famously complained that humanists, the then-dominant culture on university campuses, would incomprehensibly sneer at the very mention of basic scientific concepts: "Once or twice I have been provoked and have asked the company how many of them could describe the Second Law of Thermodynamics. The response was cold: it was also negative. Yet I was asking something which is about the scientific equivalent of: Have you read a work of Shakespeare's?" (Snow [1960] 2012, 15).

These days the tables are turned, as it is the science departments that hold most of the cash, and therefore the power, in academia. Apparently, and rather curiously, this seems to have engendered a similar sort of open disdain among scientists and science popularizers about the humanities in general, and philosophy in particular. Here are some clear examples, just to set the tone:

• "The insights of philosophers have occasionally benefited physicists, but generally in a negative fashion—by protecting them from the preconceptions of other philosophers. ... I do not mean to deny all value to the philosophy of science, which at its best seems to me a pleasing gloss on the history and discoveries of science." (Weinberg 1993, 166–167)

• "Traditionally these are questions for philosophy, but philosophy is dead. Philosophy has not kept up with modern developments in science, particularly physics." (Hawking and Mlodinow 2010, 5)

• "Philosophy is a field that, unfortunately, reminds me of that old Woody Allen joke, 'those that can't do, teach, and those that can't teach, teach gym.' And the worst part of philosophy is the philosophy of science; the only people, as far as I can tell, that read work by philosophers of science are other philosophers of science." (L. Krauss, in Andersen 2012)

• "Yeah, yeah, exactly, exactly. My concern here is that the philosophers believe they are actually asking deep questions about nature. And to the scientist it's, what are you doing? Why are you concerning yourself with the meaning of meaning?" (N. deGrasse Tyson, in Levine 2014, 20' 19" and following)

And those are only some of the most egregious offenses. Despite this, a good number of scientists do work and publish in philosophy of science, including physicists, biologists, and mathematicians. And there is a large literature in philosophy of science of papers aiming directly at clarifying issues of interest to scientists, from species concepts in biology (Okasha 2002; Pigliucci 2003; Wilkins 2009) to the epistemic status of string theory in fundamental physics (Weingard 2001; Dawid 2009; Johansson and Matsubara 2011).

In this chapter I will use several examples concerning discussions of the nature of causality, particularly as it applies to biological systems, as a conduit to explore how the practices of science and philosophy of science appear to be related, and then to discuss, in a sense, how they ought to be related. Hopefully, this exploration will also result in some degree of clarification of a number of points about the philosophy of causality as it relates to its applications to science. I will begin by briefly presenting four possible models of the relationship between science and philosophy of science. I will then examine four instances of philosophical discussions of causality with respect to their relevance—or lack thereof—to the practice of science. And I will finally attempt to draw some general conclusions in order to move forward both the specific debates about causality and the broader one about the relationship between science and philosophy of science.

Four Models

There are at least four distinct ways in which one could conceive of how science and philosophy of science do, or should, relate to each other:

I) Handmaiden, version 1: Science serves the purpose of achieving philosophical understanding. This, of course, was true for most of the history of philosophy, throughout which "science" did not exist as a separate field of inquiry, but was folded under the heading of natural philosophy. Even as late as Galileo, Newton, and Darwin, scientists thought of themselves as philosophers, with the very term "scientist" being coined by William Whewell (Darwin's mentor, among other things) as late as 1834, in analogy with "artist."

II) Handmaiden, version 2: Philosophy serves the purpose of furthering scientific discoveries. This is the model implicitly adopted by Weinberg, Hawking, Krauss, and deGrasse Tyson in the quotations mentioned above. It explains why—in the same interview—Tyson went on to say: "Philosophy has basically parted ways from the frontier of the physical sciences, when there was a day when they were one and the same.… So, I'm disappointed because there is a lot of brainpower there, that might have otherwise contributed mightily, but today simply does not. It's not that there can't be other philosophical subjects, there is religious philosophy [sic], and ethical philosophy, and political philosophy, plenty of stuff for the philosophers to do, but the frontier of the physical sciences does not appear to be among them." (Levine 2014)

III) Separate Magisteria: Science studies the world, while philosophy (of science) studies science. This has largely, though not exclusively, been the model within philosophy of science since the famous debate between John Stuart Mill and the above-mentioned Whewell on the nature of induction (Forster 2009). It has also characterized the "golden era" of philosophy of science, from logical positivism at the beginning of the twentieth century to Popper, Kuhn, Lakatos, and Feyerabend (Godfrey-Smith 2003), when philosophers of science were preoccupied with producing grand theories of the nature of science.

IV) Overlapping Magisteria: Science studies the world and philosophy of science studies science, but there are mutually beneficial areas of overlap. As I mentioned above, there are a number of these areas that are frequently addressed by both scientists and philosophers, including, but not limited to: species concepts in biology (refs. above); definition and uses of "gene" (Hall 2001; Griffiths and Stotz 2013); discussions of the concept of race (Pigliucci and Kaplan 2003; Hacking 2006); the nature of evolutionary theory (Pigliucci and Müller 2010; Laland et al. 2015), epistemic limits of evolutionary psychology, medical research, neuroscience, and social science (Kaplan 2000; Fine 2017); neuroscience of consciousness (Rose 2006; Block 2007); metaphysical interpretations of quantum mechanics (Barandes and Kagan 2014); desirability of a "post-empirical" science vis-à-vis string theory (refs. above); and several others.

It seems to me fairly clear that the most viable model in contemporary science and philosophy of science is (IV) above. Model (I), as noted, did characterize much of the history of philosophy and of natural philosophy/science, but gradually—beginning with the scientific revolution of the seventeenth and eighteenth centuries—different sciences have successfully emancipated themselves from philosophy. Physics first, then chemistry, biology, and finally social sciences such as economics and psychology. The cognitive study of consciousness is the latest entry in the catalog, gradually but fairly rapidly cutting its umbilical cord from philosophy of mind.

Model (II), as we have seen, is apparently popular among certain scientists, who use it to demonstrate that "obviously" philosophy is dead or irrelevant. But insofar as I can tell, that view of the relationship between the two fields is a fact-free construct of their own minds, with no counterpart in either the historical or the current literature in philosophy of science.

Model (III) does describe what happens most of the time in academic journals: the majority of philosophy of science papers treat science as an external object, to be studied and understood from a distance, so to speak; while the overwhelming majority of science papers simply don't mention philosophy (and take philosophical baggage on board unexamined, as famously observed by Dan Dennett 1996, 21).

Model (IV) has the advantage of accommodating the standard practices described by (III), and yet of leaving enough room for science to directly contribute to some philosophical disputes, as well as for philosophy to help scientists clarify certain issues at the boundaries of their discipline, or when problems are sufficiently underdetermined by the empirical evidence (Stanford 2013) as to accommodate alternative philosophical interpretations.

In the following, I will briefly examine four examples of debates about causality and interpret them as instances of model IV at work. These are: the discussion concerning the distinction between proximate and ultimate causes in biology; the issue of (alleged) top-down causality, in both physics and biology; the nature of functional explanation in biology and the social sciences; and the broad debate over the very nature of causality. I will then attempt to draw some general lessons, hopefully illuminating both my concern about the relationship between science and philosophy, as well as some aspects of the specific scholarship in philosophy on the nature of causality. A fairly large caveat needs to be kept in mind throughout the following, however: I am not attempting a general review of the huge field of the philosophy of causation. That would be unwieldy, redundant with currently available entries in that literature, and especially besides the main point of my analysis, which is focused on using causality as a conduit for a better understanding of the science–philosophy debate. In particular, I will not be discussing causal graphs, which present a semantic approach to causal analysis, where the ontology is implied (Woodward 2016), or interventionist accounts of causality, which are practical, not ontological in nature (Duncan 1976; Morgan and Winship 2007)—characteristics that explain why both of these approaches are especially popular among social scientists and statisticians.

Example I: Proximate vs. Ultimate Causes in Biology

In evolutionary biology, the distinction between so-called proximate and ultimate causes has been common practice since its introduction by Ernst Mayr (1961), one of the most influential figures of twentieth-century biology. Recently, however, the distinction has been under sustained attack in philosophy of science, and for good reasons (Ariew 2003; Amundson 2005; Haig 2013; Laland et al. 2012).

Mayr introduced it in order to explain why developmental biology had famously been left out of the Modern Synthesis of the 1930s and 1940s (Mayr 1993): "The functional biologist is vitally concerned with the operation and interaction of structural elements, from molecules up to organs and whole individuals. His ever-repeated question is 'How?'... [The evolutionary biologist's] basic question is 'Why?'... To find the causes for the existing characteristics, and particularly adaptations, of organisms is [his] main preoccupation" (Mayr 1961, 1502).

A number of years later, he elaborated: "The suggestion that it is the task of the Darwinians to explain development... makes it evident that Ho and Saunders [critics of the Modern Synthesis] are unaware of the important difference between proximate and ultimate causations. ... Expressed in modern terminology, ultimate causations (largely natural selection) are those involved in the assembling of new genetic programmes, and proximate causations those that deal with the decoding of the genetic programme during ontogeny and subsequent life" (Mayr 1984, 1262).

As mentioned above, however, a number of recent authors have critiqued the sharpness of Mayr's clean separation of proximate and ultimate causes. For instance, Amundson (2005, 176) wrote: "In order to achieve a modification in adult form, evolution must modify the embryological processes responsible for that form. Therefore an understanding of evolution requires an understanding of development." Laland et al. (2011, 1512) explain: "In reciprocal processes, ultimate explanations must include an account of the sources of selection (as these are modified by the evolutionary process) as well as the causes of the phenotypes subject to selection."

So far, the two contrasting pictures are on the one hand of ultimate causes acting across generations and proximate causes acting within generations (Mayr), and on the other hand of no meaningful distinction at all to be drawn between the two sorts of causes, in the name of a principle of causal completeness (Mayr's critics, especially Amundson).

Building on these recent re-analyses, Raphael Scholl and I (Scholl and Pigliucci 2014) have proposed to strike a balance between Mayr and his most harsh critics (in line with my more general suggestion that the Extended Evolutionary Synthesis is an expansion, but not a radical revision or rejection, of the MS [Pigliucci and Müller 2010, 3–19]). Our thesis is that it is a mistake to think about the proximate–ultimate distinction in terms of allegedly omitted biological causes, as some of Mayr's critics do. It is much more fruitful to think in terms of the necessary partiality of causal explanations: the issue is not whether certain types of causation (e.g., between genotype and phenotype, or between phenotype and selective environment) exist, but whether these causal paths carry much weight in the explanations we give. For each individual case, we should ask what motivates the foregrounding or backgrounding of some parts of a complete causal account of any given evolutionary transition. The key concern is not causal relevance, in other words, but explanatory salience.

Our approach builds on S. J. Gould's (1985) point that developmental processes sometimes (Most of the time? How frequently?) are not isotropic, as they have classically assumed to be within the Modern Synthesis. When they are, then developmental causes can indeed be backgrounded in favor of natural selection; when they are not, then they need to be foregrounded because they are explanatorily salient.

One well-known case (among many) of explanatory salience of developmental mechanisms is the classic study by Alberch and Gale (1985) of the evolution of foot morphology in amphibians. They present comparisons within and among two orders of amphibians: plethodontid salamanders and anuran frogs. The trait of interest is loss of phalanges in the extremities. Alberch and Gale are able to show that the pattern of phalangeal loss is similar in species belonging to the same order, but different among orders. They further show that different morphologies within the same order can be reproduced experimentally by varying a single developmental parameter: size (i.e., number of cells) in the limb bud, which can be manipulated by the reversible application of colchicine to the developing limb bud. For example, treatment of the limb bud of the salamander *Ambystoma mexicanum* with colchicine results in the loss of various phalanges in such a way as to mimic the normal morphology of the related species, *Hemidactylium scutatum*.

What Alberch and Gale's examples illustrate is a general methodology for demonstrating the need to foreground developmental processes in evolutionary models. What is *not* produced is evidence related to optimality—either evidence that a given feature is non-optimal or that some other structure would be optimal. What *is* produced is evidence that the feature under investigation (adult limb morphology) is a function of a developmental factor (limb bud size), and that the existing adult limb morphology can thus be interpreted as a side-effect of selection on body size alone. In this specific case, then, foregrounding of development becomes useful because developmental variation is not isotropic, and so it needs to be part of the causal story. The pertinent causal claims are supported by experimental data that demonstrate the relevant developmental dynamics. Any additional explanations of the foot morphology by natural selection are not so much disproved, as made redundant, and biologists can meaningfully talk, in these cases, of the evolutionary importance of developmental constraints.

The question of proximate versus ultimate explanations for evolutionary change, then, is empirical, depending on the degree of anisotropy of developmental variation. If Scholl and I are correct in our analysis (and of course, not everyone agrees), which in turn builds on mounting criticism of the original distinction made by Mayr, this is one case in which careful philosophical reflection can help clarify an issue that is very much relevant to the practicing biologist. We have a good instance of philosophy of science helping actual science, instead of studying it from the outside.

Example II: The Question of (Alleged) Top-down Causation

Biologists and other scientists often talk about "top-down" causation, apparently without concern for the fact that fundamental physicists don't recognize any such category of causal interactions, seeing everything from a bottom-up perspective. Philosophers are not in principle wedded to one view or the other, but are obviously very much interested in the nature and manifestations of causality, and are also typically cautious not to unnecessarily multiply entities or levels of description, attempting to keep their ontology as sparse as it is possible given the actual structure of the world (what W. V. O. Quine famously referred to as a "desert ontology": Hylton, 2014).

Craver and Bechtel (2007) provide a splendid example of rigorous philosophical analysis of the problem, arriving at what seems to me a sensible position that they, somewhat paradoxically, term "top-down causation without top-down causes." These authors' concern is that when scientists invoke top-down causation they call upon a concept that is either metaphysically incoherent or implies the existence of "spooky" forces that wholes somehow exert over their components.

Their solution to the problem is to grant that talk of top-down causation intuitively captures a number of observable relationships between the activities of wholes and the behaviors of (some of) their constituent parts. However, they attempt to recast such cases so that one does not actually have to invoke the metaphysically questionable concept of top-down causation. The proposal is to think instead in terms of mechanistic effects that are themselves hybrids of two components: (i) intra-level causal relationships, and (ii) inter-level constitutive relationships. Before I discuss their contribution, I should make clear that my favorable view of their solution to the problem is not universally accepted (for instance, by one of the reviewers of this chapter). That is not a problem for my goal here, though, which is concerned with illustrating the sort of contributions to science that philosophy can make. Whether a particular solution to a specific problem—either philosophical or scientific—turns out to be the right one, or the one that achieves consensus, is another story.

So let us consider a particularly illuminating example among several presented by Craver and Bechtel in the course of their analysis of the issue. They introduce us to Hal, a tennis player, who steps on the court and begins to serve. At the physiological level, the membranes of Hal's cells absorb the glucose that was at that moment circulating in his blood; once transported inside the cell, the glucose is phosphorylated and bound into molecules of hexosediphosphate. A very reasonable description of what's going on is that the fact that Hal, the organism, began to play tennis initiated a cascade of lower-level physiological and molecular effects. These processes started because Hal began to play, and so the scenario appears to be a classical example of top-down causation: the behavior of the whole (Hal) has affected the behavior of the parts (cells and metabolites).

What Craver and Bechtel argue, however, is that a better account of what is going on relies on a combination of intra-level causation and constitutive effects, doing away with the need to invoke top-down causality. Here is their reconstruction of Hal's case:

When Hal started to play tennis, the nerve signals to the muscles caused them to metabolize the available ATP to ADP to provide the energy to contract the muscle cells. The increase in ADP made it available as a receptor for phosphates in high-energy bonds in 1,3-diphosphoglycerate produced at the end of the glycolytic process. This allowed a cascade of reactions earlier in the pathway to proceed, eventually allowing a glucose molecule to take up a phosphate molecule. In this and many similar cases, a change in the activity of the mechanism as a whole just is a change in one or more components of the mechanism which then, through ordinary intra-level causation, causes changes in other components of the mechanism. Hal's playing tennis is in part constituted by activities at neuromuscular junctions, and activities at those junctions cause, in a perfectly straightforward etio-logical sense, changes in the organization and behavior of cellular mechanisms.… Once we have described the mechanism mediating the effect, the drive to speak of this as a case of top-down causation vanishes, although such language might be useful as shorthand. (2007, 559–560)

Craver and Bechtel discuss several other examples like this one, as well as a separate, complementary analysis of bottom-up causation, amounting to the same conclusion: any alleged case of top-down (or bottom-up) causation can always be resolved into the sum of two components: intra-level causation plus constitutive effects.

Craver and Bechtel's analysis represents a second example of philosophy aiding science, but it differs importantly from the first one discussed above: in the case of proximate vs. ultimate causes there was a genuine scientific debate, involving different theoretical positions staked by scientists themselves. The philosophical analysis there helps to resolve that internal debate, by invoking the concept of explanatory relevance to distinguish cases where developmental causes are called for in order to understand the evolution of a structure from cases in which developmental explanations can be backgrounded without significant loss of explanatory power. In this second instance, however, there really isn't much of a discussion or disagreement among scientists who use the concept of top-down causation (and much less among those who use its inverse, bottom-up). And yet, arguably there should be! The philosophical analysis here hints at the possibility that scientists may be sloppy when they engage in *either* top-down *or* bottom-up causal talk, since all non-spooky causation can be construed as being intra-, not inter-level. That said, there are some philosophers who do defend the idea of top-down causation, though I think it is safe to say that they represent a minority position (see, for instance, discussion in Fazekas and Kertész 2010).

Example III: The Nature of Functional Explanations

My third example concerns the rather messy and convoluted history of the debate about the nature of functional explanations in philosophy of science. By contrast with the preceding cases, this is an instance where I think philosophers really ought to take the scientists more seriously, and which shows how long sometimes it can take for a philosophical debate to catch up with standard scientific practice. Even so, the underlying issue—that we need account(s) of how scientists think about functional explanation—is a real one, and one that does have the potential to contribute to clearer thinking on the part of the scientists themselves.

The two classical philosophical proposals for making sense of functional talk in science are from Nagel (1961, 1977) and Cummins (1977, 1983). Very briefly, Nagel's view is that we should understand talk of function on the part of scientists within the broader (and now largely, though not entirely, abandoned) framework of the Deductive-Nomological (D-N) model of explanations (Hempel 1962). For Nagel, the function of a structure *A* embedded in a system *S* with organization *O*, is to enable *S* (in environment *E*) to engage in process *P*. For instance, the function of the vertebrate heart, embedded in a circulatory system of blood and vessels carrying it, is to allow the circulatory system to pump blood, in the typical range of environments encountered by the organism. Nagel dismisses talk of teleology ("the heart is for…") as unnecessary, since the teleological language often used by biologists can be translated into a neutrally descriptive one. This more neutral language can still make reference to criteria explaining why process *P* is there to begin with, such as increased survival and reproduction, the "goals" (so to speak) of all living organisms.

Nagel is sometimes criticized in a way that seems to almost perversely miss the mark. For instance, Wright (1973) objected that the same function (blood circulation) can be performed by artificial structures (synthetic hearts), to which Nagel quite obviously replied that he was concerned with actual natural biological systems, not man-made ones. Wright's "objection" is a good example of modern analytical philosophers going for the clever counter-example at the cost of spectacularly missing what the discussion is actually about. Nevertheless, Nagel's own proposal is marred by its contextualization within the D-N model, which is debatable even with respect to the physical sciences, and certainly falters when it comes to biology. It is interesting to note, however, that mechanistically oriented biologists (especially molecular biologists and some developmental biologists) do talk in the same sort of ahistoric fashion that supporters of the D-N model seek to generalize to all of science.

By contrast, Cummins rejected an approach cast in terms of "goals" of a structure (the heart) within a system (blood circulation), framing things instead as contributions provided by certain structures to the capacity of a containing system. As Cummins (1983, 15) put it: "the natural strategy for answering such a question [about function] is to construct an analysis of S [the system] that explains S's possession of P [the process] by appeal to the properties of S's components and their mode of organization." If we discover that a heart is necessary to the circulatory system in order to pump blood, then *that* is the function of a heart.

Cummins's approach, however, purposely excludes reference to survival and reproduction, a choice that a biologist would find puzzling to say the least. This characteristic of the account also opens it up to the objection that it may identify a large number of "functions" that no biologist would recognize as such. The heart, for instance, also makes noise, and the sketch just given for how to go about recognizing functions would not be able to separate this instance from one of genuine function: the heart is "for" blood circulation, not "for" making noises. The fact that Cummins (1977) responded that there is no objective criterion to distinguish legitimate from doubtful biological functions seems to me to be no reply at all; it is instead an elegant—if likely unwitting—way to concede the point.

Moreover, what happens if an individual instantiation of a component is malfunctioning? Cummins (1977), again, bit the bullet, acknowledging that defective components do not, in fact, have a functional role to perform. Davies (2001, 176) goes so far as to defend Cummins's account on the basis of the alleged fact that "natural traits cannot malfunction," a bizarre statement that would have a biologist justifiably roll her eyes. This example is, again, instructive, because Davies has pushed himself into an untenable position due to his broader commitment to a particular ontology of the natural sciences, an ontology in which common teleological-sounding functional talk in evolutionary theory simply does not fit. As an evolutionary biologist I would say too bad for that particular ontological account of the natural sciences.

We had to wait until Neander's (1991) proposal to get a unified account of function in biology, one that finally takes seriously the theory of evolution: biological functions are the result of natural selection and are therefore inextricably connected to the organism's survival and reproduction. There are, of course, a number of complications, for instance raised by the rather common phenomenon of exaptation (Gould and Vrba 1982), where new functions arise out of the repurposing—so to speak—of old structures. One of the best-known examples is that of dinosaur feathers, which likely initially evolved for thermoregulation, or possibly sexual signaling, but definitely not for flight (Clarke 2013).

Godfrey-Smith (1993) has rejected Neander's approach, however, calling for a pluralist account of function in biology. While I am generally sympathetic to pluralism, I do think philosophers have a tendency to invoke it as a general trump card whenever a discussion becomes difficult, sometimes marring rather than clarifying the issue. Interestingly, Godfrey-Smith's suggestion mirrors the same (misguided) distinction we discussed above between proximate and ultimate causes, since he proposes that a causal account, similar to that of Cummins, is appropriate for physiological investigations that can be carried out independently of historical considerations. My reply would be along similar lines to those articulated in the paper I coauthored with Scholl: while nothing makes sense in biology except in the light of evolution (Dobzhansky 1973), for specific and limited purposes we can background the historical aspect and focus on the causal role. I don't think, however, that this amounts to full-fledged pluralism, as causal roles can be made sense of, more broadly, only by invoking historical explanations as well.

In general, it seems to me that philosophers in this case are taking their sweet time to reach a conclusion that would be obvious to any practicing scientist: functional explanations have no role in the physical sciences; they are framed in terms of teleonomy (not teleology: Monod 1971) in the biological sciences, where the teleonomic agent is natural selection; and retain a fully teleological meaning in the social sciences, where human minds are the source of the teleology. Why it is taking so long to recognize this—and why there is still significant philosophical disagreement about it—is more likely a matter for psychological and sociological analyses than for philosophical debate.

Example IV: The Nature of Causality Itself

Lastly, let us briefly examine the complex literature on the nature of causality itself. This is arguably the furthest away from the interest of the practicing scientist, and yet it underpins pretty much all of the so-called special sciences (i.e., everything but fundamental physics), and possibly science as a whole.

The obligatory starting point is of course David Hume's (1738) idea that the notion of "causation" is a psychological phenomenon arising whenever we see constant conjunction of events, followed by Kant's (1783) famous "awakening" in reaction to Hume, and his proposal that causation is one of the categories of perception with which the human mind comes equipped, and without which it wouldn't be able to make sense of the world. In either case, causation is not something "out there," but rather constructed by the human mind, either from experience (Hume) or innately (Kant).

Modern accounts of causation begin with Bertrand Russell's (1913) theory of causal lines, where he called "a series of events a 'causal line' if, given some of them, we can infer something about the others without having to know anything about the environment" (Russell 1948). Russell was aiming at producing a theory that understood causality in terms of functional relations, making room for the possibility that causal laws—in view of modern quantum mechanics—may turn out not to be deterministic. His approach, incidentally, famously led him to his theory of identity, according to which the identity of an object (including people) over time consists in the fact that the temporal parts of the object all belong to the same causal line. One of the objections raised against Russell is that his theory is actually epistemic (i.e., based on knowledge claims) rather than ontological (i.e., based on claims about what exists). Most modern philosophers think of causation in ontological, not epistemic, terms.

Salmon (1984) articulated an account based on the idea that causal processes transmit "marks," that is, local modifications in structure. Accordingly, spatiotemporally continuous processes capable of transmitting marks are responsible for propagating causal influences. Here too a number of objections have been raised, for instance that mark transmission seems to require a certain degree of uniformity of causal processes over time. If that is the case, then short-lived subatomic particles do not qualify, and yet they do play causal roles in fundamental physics.

The next big contender is the idea that causality consists in the transmission of some conserved quantity, such as momentum or energy, where the transmission is governed by some kind of conservation law. There are two major versions, due to Dowe (1995) and again Salmon (1997). For my purposes here, the differences between these two versions are not important. The general idea, though, does seem to capture a lot of what we want in an account of causality, for instance it neatly explains why certain phenomena, like shadows, don't count as causal processes (they don't transmit any conserved quantity). Then again, there are a number of standing criticisms of both Dowe and Salmon, in particular that large classes of what are normally considered examples of causation do not fit. Consider the fact that I killed my plant by not watering it: there was no exchange of conserved quantities, and yet it seems undeniable that my failure to water the plant is, indeed, what killed it. Dowe (2000) replied that these are not true examples of causation, but rather of a close kin, which he refers to as quasi-causation. Perhaps more worrisome, accounts based on the idea of transmitted quantities do not seem to be adequate in the domain of the social sciences, where it is hard to imagine what sort of conservation law is at play. This brought Cartwright (2004) to suggest that in fact we are dealing with a number of different types of causality, applicable to different domains and phenomena, and which do not seem to have an underlying essence in common. Cartwright is therefore pluralist about causation.

There are several other theories out there, including two major versions of a transference account (Aronson 1971; Fair 1979), which are actually very similar to the conserved quantity accounts just discussed, and a theory of causation as transfer of information (Collier 1999). Notwithstanding all these proposals (or perhaps because of it), there doesn't seem to be a consensus among philosophers on what, in fact, causation is. Then again, we now know on the basis of empirical surveys (Bourget and Chalmers 2014) that philosophers do not settle on unique answers to whatever they are interested in. Instead, they explore the relevant conceptual landscape (Pigliucci 2017) and refine it, eventually identifying a small number of "peaks" corresponding to viable accounts. The choice of a particular peak then may depend just as much on the personality and intellectual preferences of individual philosophers as on the usefulness of a given account within a particular context of application.

Be that as it may, this is an instance where scientists themselves are not going to come up with a theory of causation, because they (implicitly) assume one in the course of their work, and also because it is hard to imagine what sort of experiment may discriminate between different accounts. If this is true, it turns out that causation is a quintessentially philosophical problem. Even so, it is certainly one that requires input from the sciences, perhaps most clearly evident in the case of Dowe's and Salmon's theory of causation as transference of conserved quantities, which they claim is an instance of "empirical analysis," as distinct from "conceptual analysis." This empirical aspect is often presented by opponents of Dowe and Salmon as an objection to the account (I guess because the theory isn't sufficiently philosophical, too contaminated by science!), but that seems to me a very misguided attitude, as it is hard to imagine a lot of interesting philosophical questions that are completely detached from the empirical world. Then again, at the other extreme we find Fair (1979), the proponent of one of the two transference theories, who goes so far as to claim that science has actually "discovered" what causality is (a transfer of energy and/ or momentum), an assertion that seems incongruent with how scientists operate (since many of them talk of causation when there is no discernible transfer of energy or momentum), not to mention with how philosophical questions are articulated and pursued (again, see Pigliucci 2017).

Conclusion: Causality, Philosophy, and Science

I have briefly examined four case studies concerning how philosophers think about causality. In the first instance, that of proximate versus ultimate causes in biology, we arguably have a good example of philosophical analysis that is not only valuable in its own right but also potentially useful to scientists themselves in order to think more clearly about a lively debate within their own field. The second instance, about the alleged existence of top-down (and bottom-up, as distinct from intra-level) causality, is also a case of philosophy being helpful to science, but in a different fashion: opening up a debate that scientists should have but apparently have overlooked. The third case, the nature of functional explanations, conversely, represents a situation where the philosophical debate ought to be more settled than it appears to be, and this is at least in part because philosophers have discounted pretty clear input from the sciences, input that is germane to the way they think about the issue. Finally, we have looked at the very broad problem of the nature of causality itself, one that really seems to constitute a genuine instance of a purely philosophical issue, where it is difficult even to imagine how direct input from the sciences would be helpful.

If we consider all four cases together, and factor in the more general observation that of course there is a great deal of science that goes on without showing up on the radar of philosophers, and vice versa, a lot of philosophy of science that is clearly engaged in studying science from the outside, so to speak, then we have to conclude that the best model to capture the relationship between science and philosophy is the fourth one: overlapping magisteria. Just how large the overlap is depends on the specific type of philosophy of science (e.g., philosophy of the social sciences, philosophy of biology, philosophy of quantum mechanics, and so forth).

One thing ought to be clear: the value of philosophical investigations of science does not hinge on the size of the overlap. Even if model III above were correct, that of a philosophy of science concerned solely with studying and making sense of science itself from the outside, that would still make it a worthy intellectual pursuit, made even more pertinent to society by the huge amount of money and resources spent nowadays on behalf of the sciences. Presumably, we want to have a better understanding of the sort of enterprise that we are spending all this money on. Moreover, let us not forget that a general public increasingly less literate about science is being called upon to make ethically crucial decisions about the societal value of scientific claims concerning climate change, vaccines, and so forth. This is another reason for philosophers to join the high table and contribute their expertise. All of this clearly means that philosophers are, at least occasionally, directly useful to scientists (and, let's not forget, vice versa). The Lawrence Krausses and Stephen Hawkings of the world ought, therefore, to pay attention.

Acknowledgments

I wish to thank Kevin Laland and Tobias Uller for organizing the workshop *Cause and Process in Evolution* at the Konrad Lorenz Institute for Theoretical Biology, and for kindly allowing me the opportunity to present my ideas on the subject. Two anonymous reviewers have been both very generous and constructive in their criticisms of an earlier version of this chapter.

References

Alberch, P., and E. A. Gale. 1985. "A Developmental Analysis of an Evolutionary Trend: Digital Reduction in Amphibians." *Evolution* 39:8–23.

Amundson, R. 2005. *The Changing Role of the Embryo in Evolutionary Thought: Roots of Evo-Devo*. Cambridge: Cambridge University Press.

Andersen, R. 2012. "Has Physics Made Philosophy and Religion Obsolete?" *The Atlantic*, April 23. Accessed April 26, 2017. https://www.theatlantic.com/technology/archive/2012/04/has-physics-made-philosophy-and-religion-obsolete/256203/.

Ariew, A. 2003. "Ernst Mayr's 'Ultimate/Proximate' Distinction Reconsidered and Reconstructed." *Biology and Philosophy* 18:553–565.

Aronson, J. 1971. "On the Grammar of 'Cause.'" *Synthese* 22:414–430.

Barandes, J., and D. Kagan. 2014. "A Synopsis of the Minimal Modal Interpretation of Quantum Theory." *arXiv*. 1405.6754v3.

Block, N. 2007. "Consciousness, Accessibility, and Then Mesh between Psychology and Neuroscience." *Behavioral and Brain Sciences* 30:481–499.

Bourget, D., and D. J. Chalmers. 2014. "What Do Philosophers Believe?" *Philosophical Studies* 170:465–500.

Cartwright, N. 2004. "Causation: One Word, Many Things." *Philosophy of Science* 71:805–819.

Clarke, J. 2013. "Feathers before Flight." *Science* 340:690–692.

Collier, J. 1999. "Causation Is the Transfer of Information." In *Causation and Laws of Nature*, edited by H. Sankey, 215–245. Dordrecht, NL: Kluwer.

Craver, C. F., and W. Bechtel. 2007. "Top-down Causation without Top-down Causes." *Biology and Philosophy* 22:547–563.

Cummins, R. 1977. "Programs in the Explanation of Behavior." *Philosophy of Science* 44:269–287.

Cummins, R. 1983. *The Nature of Psychological Explanation*. Cambridge, MA: MIT Press.

Davies, P. 2001. *Norms of Nature: Naturalism and the Nature of Functions*. Cambridge, MA: MIT Press.

Dawid, R. 2009. "On the Conflicting Assessments of String Theory." *Philosophy of Science* 76:984–996.

Dennett, D. 1996. *Darwin's Dangerous Idea: Evolution and the Meaning of Life*. New York: Simon & Schuster.

Dobzhansky, T. 1973. "Nothing in Biology Makes Sense Except in the Light of Evolution." *The American Biology Teacher* 35:125–129.

Dowe, P. 1995. "Causality and Conserved Quantities: A Reply to Salmon." *Philosophy of Science* 62:321–333.

Dowe, P. 2000. *Physical Causation*. Cambridge: Cambridge University Press.

Duncan, O. D. 1976. "Introduction to Structural Equation Models." *American Journal of Sociology* 82:731–733.

Fair, D. 1979. "Causation and the Flow of Energy." *Erkenntnis* 14:219–250.

Fazekas, P., and G. Kertész. 2010. "Causation at Different Levels: Tracking the Commitments of Mechanistic Explanations." *Biology and Philosophy* 26:365–383.

Fine, C. 2017. *Testosterone Rex: Myths of Sex, Science, and Society*. New York: W. W. Norton & Co.

Forster, M. 2009. "The Debate between Whewell and Mill on the Nature of Scientific Induction." In *Handbook of the History of Logic. X. Inductive Logic*, edited by D.M. Gabbay and J. Woods, 93–115. Amsterdam, NL: Elsevier.

Godfrey-Smith, P. 1993. "Functions: Consensus without Unity." *Pacific Philosophical Quarterly* 74:196–208.

Godfrey-Smith, P. 2003. *Theory and Reality: An Introduction to the Philosophy of Science*. Chicago: University of Chicago Press.

Gould, S. J. 1985. *Ontogeny and Phylogeny*. Cambridge, MA: Harvard University Press.

Gould, S. J., and E. S. Vrba. 1982. "Exaptation—A Missing Term in the Science of Form." *Paleobiology* 8:4–15.

Griffiths, P., and K. Stotz. 2013. *Genetics and Philosophy: An Introduction*. Cambridge: Cambridge University Press.

Hacking, I. 2006. "Genetics, Biosocial Groups, and the Future of Identity." *Daedelus* (Fall):81–95.

Haig, D. 2013. "Proximate and Ultimate Causes: How Come? And What For?" *Biology and Philosophy* 28:781–786.

Hall, B. K. 2001. "The Gene Is Not Dead, Merely Orphaned and Seeking a Home." *Evolution and Development* 3:225–228.

Hawking, S., and L. Mlodinow. 2010. *The Grand Design*. New York: Bantam.

Hempel, C. G. 1962. "Deductive-Nomological vs. Statistical Explanations." *Minnesota Studies in the Philosophy of Science* 3:98–169.

Hume, D. 1738. *A Treatise of Human Nature, and an Enquiry Concerning Human Understanding*. Accessed April 26, 2017. https://www.gutenberg.org/files/4705/4705-h/4705-h.htm.

Hylton, P. 2014. "William van Orman Quine." *Stanford Encyclopedia of Philosophy*. Accessed December 31, 2017. https://plato.stanford.edu/entries/quine/.

Johansson, L. G., and K. Matsubara. 2011. "String Theory and General Methodology: A Mutual Evaluation." *Studies in History and Philosophy of Modern Physics* 42:199–210.

Kant, I. 1783. *Prolegomena to Any Future Metaphysics*. Accessed April 26, 2017. https://archive.org/stream /kantsprolegomena00kantuoft/kantsprolegomena00kantuoft_djvu.txt.

Kaplan, J. M. 2000. *The Limits and Lies of Human Genetic Research: Dangers for Social Policy*. Abingdon-on-Thames, UK: Routledge.

Laland, K. N., J. Odling-Smee, W. Hoppitt, and T. Uller. 2012. "More on How and Why: Cause and Effect in Biology Revisited." *Biology and Philosophy* 28:1–27.

Laland, K. N., K. Sterelny, J. Odling-Smee, W. Hoppitt, and T. Uller. 2011. "Cause and Effect in Biology Revisited: Is Mayr's Proximate-Ultimate Dichotomy Still Useful?" *Science* 334:1512–1516.

Laland, K. N., T. Uller, M. W. Feldman, K. Sterelny, G. B. Müller, A. Moczek, E. Jablonka, and F. J. Odling-Smee. 2015. "The Extended Evolutionary Synthesis: Its Structure, Assumptions and Predictions." *Proceedings of the Royal Society B: Biological Sciences* 282. doi: 10.1098/rspb.2015.1019.

Levine, K. 2014. "Neil deGrass Tyson Returns Again." *The Nerdist Podcast*, March 7. Accessed April 26, 2017. http://nerdist.com/nerdist-podcast-neil-degrasse-tyson-returns-again/.

Mayr, E. 1961. "Cause and Effect in Biology." *Science* 134:1501–1506.

Mayr, E. 1984. "The Triumph of Evolutionary Synthesis." *Times Literary Supplement* 2:1261–1262.

Mayr, E. 1993. "What Was the Evolutionary Synthesis?" *Trends in Ecology and Evolution* 8 (1): 31–34.

Monod, J. 1971. *Chance and Necessity: An Essay on the Natural Philosophy of Modern Biology.* New York: Alfred A. Knopf.

Morgan, S. L., and C. Winship 2007. *Counterfactuals and Causal Inference: Methods and Principles for Social Research.* Cambridge: Cambridge University Press.

Neander, K. 1991. "The Teleological Notion of 'Function.'" *Australasian Journal of Philosophy* 69:454–468.

Nagel, E. 1961. *The Structure of Science.* Indianapolis: Hackett.

Nagel, E. 1977. "Teleology Revisited: Goal-Directed Processes in Biology." *Journal of Philosophy* 74:261–301.

Okasha, S. 2002. "Darwinian Metaphysics: Species and the Question of Essentialism." *Synthese* 131:191–213.

Pigliucci, M. 2003. "Species as Family Resemblance Concepts: The (Dis-)solution of the Species Problem?" *BioEssays* 25:596–602.

Pigliucci, M. 2008. "The Borderlands between Science and Philosophy: An Introduction." *Quarterly Review of Biology* 83:7–15.

Pigliucci, M. 2017. *The Nature of Philosophy: How Philosophy Makes Progress and Why It Matters.* New York: Footnotes to Plato.

Pigliucci, M., and J. Kaplan. 2003. "On the Concept of Biological Race and Its Applicability to Humans." *Philosophy of Science* 70:1161–1172.

Pigliucci, M., and G. Müller. 2010. *Evolution—The Extended Synthesis.* Cambridge, MA: MIT Press.

Rose, D. 2006. *Consciousness: Philosophical, Psychological, and Neural Theories.* Oxford: Oxford University Press.

Russell, B. 1913. "On the Notion of Cause." *Proceedings of the Aristotelian Society* 13:1–26.

Russell, B. 1948. *Human Knowledge.* New York: Simon & Schuster.

Salmon, W. 1984. *Scientific Explanation and the Causal Structure of the World.* Princeton, NJ: Princeton University Press.

Salmon, W. 1997. "Causality and Explanation: A Reply to Two Critiques." *Philosophy of Science* 64:461–477.

Scholl, R., and M. Pigliucci. 2014. "The Proximate-Ultimate Distinction and Evolutionary Developmental Biology: Causal Irrelevance versus Explanatory Abstraction." *Biology and Philosophy* 30:653–670.

Snow, C. P. 1960/2012. *The Two Cultures and the Scientific Revolution.* Cambridge: Cambridge University Press.

Stanford, K. 2013. "Underdetermination of Scientific Theory." *Stanford Encyclopedia of Philosophy.* Accessed April 26, 2017. https://plato.stanford.edu/entries/scientific-underdetermination/.

Weinberg, S. 1993. *Dreams of a Final Theory: The Scientist's Search for the Ultimate Laws of Nature.* New York: Pantheon.

Weingard, R. 2001. "A Philosopher Looks at String Theory." In *Physics Meets Philosophy at the Planck Scale,* edited by C. Callender and N. Huggett, 138–151. Cambridge: Cambridge University Press.

Wilkins, J. 2009. *Species: The History of the Idea.* Berkeley: University of California Press.

Woodward, J. 2016. "Causation and Manipulability." *Stanford Encyclopedia of Philosophy.* Accessed April 26, 2017. https://plato.stanford.edu/entries/causation-mani/and references therein.

Wright, L. 1973. "Functions." *Philosophical Review* 82:139–168.

3 Understanding Bias in the Introduction of Variation as an Evolutionary Cause

Arlin Stoltzfus

Most evolutionary geneticists would agree that the major problems of the field have been solved. We understand both the nature of the mutational processes that generate novel genetic variants and the populational processes which cause them to change in frequency over time—most importantly, natural selection and random genetic drift, respectively ... we will never again come up with concepts as fundamental as those formulated by the "founding fathers" of population genetics. (Charlesworth 1996)

Understanding Biases in the Introduction of Variation

Climbing Mount Probable

Imagine, as an analogy for evolution, a climber operating on a rugged mountain landscape (figure 3.1A). A human climber would scout a path to the highest peak and plan accordingly, but an analogy for evolution must disallow foresight and planning; therefore, let us imagine a blind robotic climber. The climber will move by a two-step proposal–acceptance mechanism. In the "proposal" or "introduction" step, the robotic climber reaches out with one of its limbs to sample a point of leverage, some nearby hand-hold or foot-hold. Each time this happens, there is some probability of a second "acceptance" step, in which the climber commits to the point of leverage, shifting its center of mass.

Biasing the second step, such that relatively higher points of leverage have relatively higher probabilities of acceptance, causes the climber to ascend, resulting in a mechanism, not just for moving, but for climbing.

What happens if a bias is imposed on the proposal step? Imagine that the robotic climber (perhaps by virtue of longer or more active limbs on one side) samples more points on the left than on the right during the proposal step. Because the probability of proposal is greater on the left, the joint probability of proposal and acceptance is greater (on average), so the trajectory of the climber will be biased, not just upward, but to the left as well. If the landscape is rough (as in figure 3.1A), the climber will tend to get stuck on a local peak that is upward and to the left of its starting point. If the landscape is a smooth hemisphere, the climber will spiral up and to the left, ascending to the single summit.

Figure 3.1
(A) The rugged mountain landscape of Aiguille Verte, Massif du Mont Blanc (photo credit: see Acknowledgments). (B) A case of two possible beneficial mutations. Mutations to alternative points in a state-space are shown as arrows projected onto two dimensions, with fitness in the vertical dimension, and with mutation rate represented by line thickness. Two of the mutations lead to more beneficial states, one going to the left, and the other going to the right.

A Population-Genetic Model

The metaphor of "climbing mount probable" helps us conceptualize the influence of a bias in the introduction of novelty in evolution—unless, of course, the metaphor is misleading. To find out, we must ask whether we can make it work in terms of population genetics.

In attempting to establish any such principle, one begins with the simplest case. In the simplest case, the climber has access to only two different moves, one going up and to the left, and the other going up and to the right. If the same path is favored by biases in both the proposal step and in the acceptance step, then that path is obviously favored. The nontrivial question is what happens when the leftward outcome is more strongly favored by the proposal step, and the rightward outcome more strongly favored by acceptance. That is, suppose that we have some set of possible mutations (possible moves) illustrated in figure 3.1B. Each mutation goes from the current state to some other state (e.g., from one DNA sequence to another), and each one is characterized by a mutation rate u and a selection coefficient s. At some point in time, there are two beneficial mutations (black arrows), with the leftward mutation having a higher mutation rate $u_2 > u_1$, but a lower fitness benefit $s_2 < s_1$.

Let B represent the bias in mutation favoring the leftward move, $B = u_2/u_1$, and let K represent the bias in selection coefficients favoring the rightward move, $K = s_1/s_2$. How often will evolution climb to the left, instead of the right? How much will B influence the outcome? How will the outcome depend on K, or on the strength of selection?

To simulate evolution with this model, we will start with a pure population that may evolve by mutation and fixation to either the left (type 2) or the right (type 1), considering a range of population size N from 100 to a million. We will vary B from 1 to 1000 by setting $u_2 = 10^{-5}$ per generation, and allowing the other mutation rate to vary; meanwhile, $s_1 = 0.02$ and $s_2 = 0.01$, so that $K = 2$. Under these conditions, the time to fixation varies inversely with N, on the order of 10^3 to 10^5 generations, because in all but the largest populations, most of the time is spent waiting for a beneficial allele to arise and reach sufficient numbers that stochastic loss becomes unlikely.

The results of simulating the model (hundreds of times for each set of conditions) are shown for a range of population sizes in figure 3.2A, with the bias in outcomes—the ratio of left to right outcomes—as a function of mutation bias B (for details, see Yampolsky and Stoltzfus 2001). All of the lines are going up: the bias in outcomes increases with the bias in mutation. When a line goes above $y = 1$, the mutationally favored outcome prevails. For $N = 100$ or $N = 1000$, the bias in outcomes is approximately B/K (dashed gray line), the outcome expected from origin-fixation dynamics (explained below). That is, under certain conditions, an exact relationship is expected between mutation bias and the outcome of evolution. For larger population sizes, we depart from the origin-fixation regime, but mutation bias continues to influence the outcome of evolution.

Let us consider the behavior of this model in further detail (see Yampolsky and Stoltzfus 2001). What about the strength of selection? Is the behavior above simply a matter of selection not being strong enough? Actually, $s = 0.02$ or $s = 0.01$ is considered to be "strong selection" given a population size of $N = 1000$ or higher. But let us demonstrate this point more clearly. In figure 3.2B below, the bias in outcomes is shown as a function of the higher selection coefficient (s_1) over a 200-fold range from $s_1 = 0.001$ to $s_1 = 0.2$. The lines are flat, indicating that the biasing effect of mutation is not some kind of "opposing forces" contest with selection. We have increased the magnitude of s, increasing the force of selection, with no effect. If effect (output) does not increase with force (input), then the "force" analogy fails.

Likewise, we can ask whether the biasing effect is dependent on high mutation rates. Figure 3.2C shows the bias in outcomes for four different values of B as a function of the rate of mutation (the higher rate, u_2), which decreases from right to left (for $N = 1000$ and

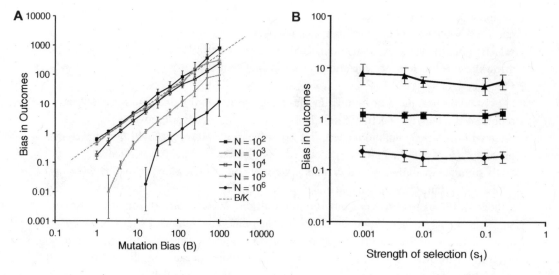

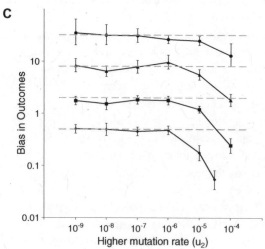

Figure 3.2
Results of simulations from Yampolsky and Stoltzfus (2001). (A) Effect of population size. For each population size indicated on the right, evolution was simulated over a 1,000-fold range of mutation bias, with hundreds of replicates for each set of parameters. (B) Effect of the strength of selection, which here varies 200-fold for $B = 16$ (triangles), $B = 4$ (squares), and $B = 1$ (diamonds), with $N = 1000$ and other parameters as before. (C) Effect of the strength of mutation. From top to bottom, the lines here are for B values of 64, 16, 4, and 1, where $K = 2$, thus B/K (dashed gray lines) is 32, 8, 2, and 0.5. As mutation rate decreases going from right to left, the bias in outcomes approaches B/K. Error bars, 95 percent confidence intervals.

other conditions the same as above). As the mutation rate decreases, the bias in outcomes converges on B/K (dashed gray lines) for each set of conditions. Again, this violates the "force" theory: larger forces must yield larger effects, otherwise the concept is inapplicable.

The behavior above is readily understood from origin-fixation models, which represent evolutionary change as a simple two-step process of (1) the introduction of a new allele by mutation, followed by (2) its fixation or loss (McCandlish and Stoltzfus 2014). The most familiar version of this formula is $K=4Nus$, where $2Nu$ is the rate of mutational introduction of beneficial alleles with selection coefficient s, and $2s$ is the probability of fixation for such alleles. For the case of neutral alleles, the probability of fixation is $1/(2N)$, so the origin-fixation rate of neutral evolution is $K=2Nu/(2N)=u$.

Origin-fixation models have distinctive implications regarding the effect of biases in the introduction of new alleles (McCandlish and Stoltzfus 2014). For example, consider a model in which a population is currently fixed at allele i and can mutate to a variety of other alleles. Using u for a mutation rate and p for a probability of fixation (dependent on a selection coefficient s and a population size N), the odds that the population will next become fixed at allele j rather than allele k are given by Eqn. 1:

$$\frac{P_{ij}}{P_{ik}} = \frac{2Nu_{ij}\pi(s_{ij},N)}{2Nu_{ik}\pi(s_{ik},N)} = \frac{u_{ij}}{u_{ik}} \times \frac{\pi(s_{ij},N)}{\pi(s_{ik},N)} \tag{1}$$

This can be described as the product of two factors: a ratio of mutation rates, and a ratio of fixation probabilities. For the case of beneficial alleles, we can use the approximate probability of fixation $2s$ (Haldane 1927), and this reduces to Eqn. 2 (Yampolsky and Stoltzfus 2001):

$$\frac{P_{ij}}{P_{ik}} = \frac{u_{ij}}{u_{ik}} \times \frac{s_{ij}}{s_{ik}} \tag{2}$$

The B/K ratio mentioned above is simply an instance of this relation, that is, $B=u_{ij}/u_{ik}$ and $1/K=s_{ij}/s_{ik}$. Thus, not only does a bias in mutation have a direct, proportional effect on the odds of one outcome relative to another, this effect is neither weaker nor stronger, but exactly the same as, a bias in fixation of the same magnitude. That is, doubling the mutation rate to an allele or doubling its probability of fixation both have the same effect on the odds that it is the next allele to be fixed.

The Sushi Conveyor and the Buffet

The behavior described in the previous section is not universal. In some regimes of population genetics, biases in mutation are completely ineffectual. To understand why, let us consider another analogy, comparing two regimes of population genetics with two styles of self-service restaurant—the buffet and the sushi conveyor (figure 3.3). In the former, we

Figure 3.3
Two styles of self-service restaurant, (A) the sushi conveyor and (B) the buffet (photo credits: see Acknowledgments).

begin with a practically inexhaustible abundance of static choices in full view, and fill our plate with the desired amount of each dish, often choosing many different items; in the latter, we iteratively make a yes-or-no choice of the chef's latest creation as it passes by, typically accepting only a few choices and rejecting the vast majority. In either case, we exercise choice, and we may end up with a satisfying meal.

The thinking of molecular evolutionists frequently corresponds to the sushi conveyor model, which illustrates a proposal-acceptance process like the one described in "Climbing Mount Probable." We choose (we select), but we don't control what is offered or when: instead, we accept or reject each dish that passes by our table. Though we decide whether each dish is right for us, initiative and creativity belong largely to the chef.

By contrast, the architects of the Modern Synthesis were committed to the buffet view, in the form of a totipotent "gene pool" with sufficient variation to respond to any challenge (addressed in more detail below). Just as the staff who tend the buffet will keep it stocked with a variety of choices sufficient to satisfy every customer, the gene pool is said to *maintain* abundant variation, so that selection never has to wait for a new mutation. Adaptation happens when the customer gets hungry and proceeds to select a platter of food from the abundance of choices, each one ready at hand, choosing just the right amount of each ingredient to make a well-balanced meal.

A bias in the choices offered to customers (analogous to a bias in variation) will have different effects in the two regimes. Let us suppose that the buffet has five apple pies and one cherry pie. This quantitative bias makes no difference. A rational customer who prefers cherry pie is unaffected by relative amounts and will choose a slice of cherry pie every time. The only kinds of biases in supply that are relevant at the buffet are absolute constraints— the complete lack of some possible dish—that is, the customer who prefers cherry pie will end up with apple pie only if there are no cherry pies.

But at the sushi conveyor, the effect of a bias will be different. Let us suppose that the dishes of sushi on the conveyor show a 5 to 1 ratio of salmon to tuna. Even a customer who would prefer tuna in a side-by-side comparison may eat salmon more often, because a side-by-side comparison simply is not part of the process.

How does this difference in analogies map to a difference in regimes of evolutionary genetics? We have already seen sushi conveyor dynamics (figure 3.2). The question is whether the buffet regime exists, and if so, what are its defining characteristics? The key condition in a hypothetical buffet regime would be that the variants relevant to the outcome of evolution are abundantly present in the initial gene pool.

Therefore, let us consider what is the effect of initial variation in the model presented earlier. The result, shown in figure 3.4, is simple and compelling. Here we have simulated the model exactly as before. The upper line repeats the results (from figure 3.2) of evolution from a pure starting population, and the lower lines represent cases in which both alternative types are present in the initial population at non-zero frequencies (Yampolsky and

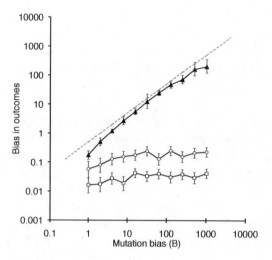

Figure 3.4
Effect of initial variation in the Yampolsky–Stoltzfus model. The upper series (closed triangles) represents the same results for $N = 1000$ shown in figure 3.2A. In the lower two series, the initial population contains both alternative types at a low frequency of either 0.005 (open circles) or 0.01 (open squares). Error bars, 95 percent confidence intervals. Results from Yampolsky and Stoltzfus (2001).

Stoltzfus 2001). Even when the initial frequencies of the alternatives are just 0.5 percent, mutation bias has essentially no effect.

Why is selection so much more powerful in the buffet regime? The 2-fold difference between $s_1 = 0.02$ and $s_2 = 0.01$ corresponds to a 2-fold preference for type 1 in the origin-fixation regime, given that the behavior of selection in this regime follows the probability of fixation, $2s$. In the buffet regime, the impact of exactly the same fitness difference is far greater: if we have two alternative alleles already in a population and they have escaped random loss, selection nearly always establishes the more fit alternative, even if it starts out at a lower frequency.

Why is mutation bias so important in the sushi conveyor regime? Figure 3.4 shows that a bias in outcomes does not always follow from a bias in mutation: the bias B increases over 3 orders of magnitude, but sometimes this has no effect. If mutation bias is strong but there is no effect on outcomes, i.e., if the putative cause is present but the expected effect is not, this suggests that mutation bias *per se* is not the proper cause. Instead, the bias operating so effectively in the origin-fixation regime is a bias in the *introduction process*. In the buffet regime, there is no bias in the introduction process, because *there is no introduction process*—all relevant alleles are present already.

Note that the above results are from simulations. The behavior of this model does not have obvious mathematical solutions except in two limiting cases. In the limiting case as $u \to 0$ with N fixed (where $u = u_1 + u_2$), we may consider an absorbing Markov chain whose behavior is readily understood with origin-fixation dynamics, yielding Eqn 2. In the limiting case as $N \to \infty$ with u fixed, i.e., a deterministic case of shifting frequencies with all types present immediately, we may consider a set of coupled differential equations whose behavior dictates that the fittest type asymptotically approaches a frequency near 1. This behavior corresponds to the buffet regime. In the intermediate regime where concurrent mutation is possible (as per Weissman et al. 2009; Desai and Fisher 2007), the problem has not been solved, yet the simulation results in the preceding section indicate clearly that biases in mutation are important.

Distinctive Implications

The results above reveal a kind of cause–effect relationship that seems very fundamental but is absent from standard treatments of evolutionary causation. Stated more precisely, it is the cause–effect relationship by which biases in the introduction of variation have a difference-making power characterized by the following statements:

· **It applies when fixations are selective.** Hypotheses in which mutation bias shapes features, though not allowed in the original Modern Synthesis (OMS), quickly became accepted by molecular evolutionists as implications of neutrality, under the assumption that mutation bias can be effective only when selection is absent. By contrast, biases in the introduction of variation do not require neutrality, leading to the novel prediction of mutation-biased adaptation.

· **It poses a directional, quantitative dependence.** In conventional thinking, variation is only a material cause, a source of substance but not form, and so the obvious questions to ask about variation concern the amount of raw material available. By contrast, the cause–effect relationship proposed here identifies quantifiable *directions* of variation that may influence directions of evolution and proposes idealized conditions under which the effect of a mutational bias of magnitude B is a B-fold bias on evolution.

· **It is regime-dependent.** The maximal effect just mentioned is fully realized in the sushi conveyor regime but is negligible in the buffet regime, where the variants relevant to the outcome of evolution are abundantly present. By contrast, conventional thinking assigns *fixed roles* to mutation and selection, independent of population-genetic regime.

· **It depends on the rareness of mutations.** In this kind of causation, the influence of biases in the generation of variation emerges as $uN \to 0$. By contrast, conventional arguments based on the force of mutation assume that mutation rates must be large in magnitude for mutational effects to be important in evolution.

· **It establishes a condition of parity with selection.** Under ideal conditions, the effect of a bias in the introduction of new alleles is the same in magnitude as that of a bias in fixation probabilities of equal magnitude. By contrast, conventional thinking treats variation as a different *kind* of cause that is passive and subordinate to selection.

• **It establishes a condition for composition and decomposition of causes.** The proposed causal relationship allows that biases in mutation and biases in fitness effects may, under ideal conditions (Eqn 1), both act like vectors that combine to determine the resultant direction of evolution. This provides a basis for combining or disentangling causes.

Such a cause–effect relationship is not described in the canonical works of the Modern Synthesis. It is not found, to my knowledge, in present-day textbooks of evolution. In fact, as will be described further below, a famous argument from Fisher and Haldane says that biases in variation cannot influence the course of evolution. Various examples could be given in which the *absence* of this cause–effect relationship is conspicuous. For example, if the authors of a famous "developmental constraints" paper (Maynard Smith et al. 1985) had understood this relationship, surely they would have invoked it to resolve the contentious issue of how developmental biases in variation were supposed to act as evolutionary causes.

Some Evidence

In the field of molecular evolution, hypotheses that invoke mutational causes for patterns are common (Stoltzfus and Yampolsky 2009). The effects of mutation bias appear to be pervasive in molecular evolution. Nevertheless, when this is interpreted as an aspect of neutral evolution, it is superficially consistent with conventional "forces" thinking: mutation pressure can be effectual when the opposing pressure of selection is absent.

The distinctive implication of biases in the introduction of variation is that their effectiveness, unlike that of mutation pressure, does not depend on neutrality. Thus, in seeking evidence for this kind of causation, we are particularly interested in cases where beneficial changes are happening. Available cases include instances in which mutation bias affects the outcome of experimental adaptation (Cunningham et al. 1997; Rokyta et al. 2005; Meyer et al. 2012; Couce, Rodríguez-Rojas, and Blázquez 2015; Sackman et al. 2017), the argument of Galen et al. (2015) for the role of a CpG mutational hotspot in altitude adaptation of Andean house wrens (see also Stoltzfus and McCandlish 2015), and a recent meta-analysis of parallel adaptive changes (Stoltzfus and McCandlish 2017).

Cunningham et al. (1997) adapted bacteriophage T7 in the presence of the mutagen nitrosoguanidine. Deletions evolved 9 times and nonsense codons 11 times, sometimes with the same change occurring multiple times. All the nonsense mutations were GC-to-AT changes, which is the kind of mutation favored by the mutagen. Couce et al. (2015) carried out experimental adaptation in *E. coli* with replicate cultures from wild-type, mutH, and mutT parents, the latter two being "mutator" strains with elevated rates of mutation and different mutation spectra. When these strains were subjected to increasing concentrations of the antibiotic cefotaxime, the resulting adaptive changes reflected the differences in mutational spectrum between lines: resistant cultures from the mutT parent tended to adapt by a small set of A:T→C:G transversions, while resistant cultures from the <u>mutH</u> parent tended to adapt by another small set of G:C→A:T and A:T→G:C transitions.

Meyer et al. (2012) reported changes in the J gene of bacteriophage Lambda in 48 replicate cultures of *E. coli*, half of which had acquired the ability to utilize the OmpF receptor as an attachment site. The complete set of 241 differences (all non-synonymous) from the parental J gene in 48 replicates is shown in figure 3.5. Among these non-synonymous mutations, 22 are found at least twice (asterisks), including 16 transitions observed in adapted strains 181 times, and 6 transversions observed 42 times. Thus the transition:transversion ratio is $16/6 = 2.7$ when we count paths (i.e., columns in figure 3.5), and $181/42 = 4.3$ when we count events (cells in figure 3.5). This is far above the null expectation of 0.5 under an absence of mutation bias.

To explore the role of mutational biases in parallel adaptation more systematically, Stoltzfus and McCandlish (2017) gathered data for a set of experimental cases such as Meyer et al. (2012), and for a comparable set of natural cases of parallel adaptation. They used these data to test for an effect of transition:transversion bias, a widespread kind of mutation bias. In the dataset of 389 parallel events along 63 paths from experimental studies of evolution, they find a highly significant tendency—from 4-fold to 7-fold in excess of null expectations—for adaptive changes to occur by transition mutations rather than transversion mutations. For the dataset of natural cases of parallel adaptation consisting of 231 parallel events along 55 paths, they found a bias of 2-fold to 3-fold over null expectations, which was statistically significant for both paths and events. They conclude that parallel adaptation takes place by nucleotide substitutions that are favored by mutation, noting that the size of this effect is not a small shift, but a substantial effect of 2-fold or more.

This analysis was made possible by prior work that used empirical data on the effects of mutations to show that transitions and transversions do not differ importantly in their distributions of fitness effects (Stoltzfus and Norris 2016). This was important to establish because the lore in molecular evolution held that amino acid changes tend to occur by transitions because they are more conservative in their effects (Keller, Bensasson, and Nichols 2007; Rosenberg, Subramanian, and Kumar 2003; Wakeley 1996). Thus, taking the work of Stoltzfus and McCandlish (2017) at face value, a several-fold effect previously attributed to selection is now best understood as being entirely or almost entirely due to mutation.

More generally, I take it that the way to establish that a kind of cause is important in evolution is to show that it has quantitatively large effect-sizes in regard to features that evolutionary biologists consider important or interesting. An effect-size of 2-fold is large. In a data-rich field such as molecular evolution, enormous numbers of papers are published every year based on far smaller effect-sizes. The subjectivity of "important or interesting" cannot be avoided, but may be exploited, as in the above example, by stealing explanatory

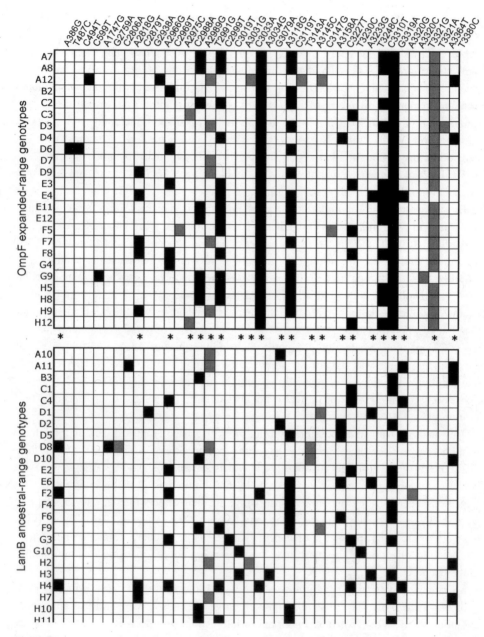

Figure 3.5
Parallel evolutionary changes (asterisks) in the Lambda tail tip gene (after figure 1 of Meyer et al. 2012). Black (transition) and gray (transversion) boxes indicate specific nucleotide mutations (columns) found among 48 replicates (rows).

power from selection: if some evolutionary phenomenon (such as transition:transversion bias) was important enough to elicit a selective explanation, then an alternative mutational explanation is also important.

Note that this evidence does not address all of the distinctive implications outlined above. Specifically, it addresses the claim that an effect of mutation bias is possible under non-neutral conditions. With more data, it will be possible to address a graduated relationship between cause and effect (given that the magnitude of mutation bias varies with taxonomic context), the composition of causes, and the role of population-genetic regime.

Three Larger Implications

The Potential for Directional Trends

The model above addresses what theoreticians would call a 1-step adaptive walk. Is it possible for a sustained bias in the introduction of new variation to cause a sustained trend over many steps? To address this question, one must consider a multistep adaptive walk on some larger landscape.

Precisely this issue is addressed by Stoltzfus (2006), using a model of a protein-coding gene, based on an abstract NK fitness model to represent interactivity of amino acids in a protein. The question addressed in the model is whether a mutational bias toward G and C nucleotides could result in a biased protein composition, even when all of the changes are beneficial. This question is relevant to interpreting observed differences in protein composition that correlate with GC content (Singer and Hickey 2000): organisms with high-GC synonymous sites and inter-genic regions also have proteins enriched for amino acids encoded by high-GC codons (Gly, Ala, Arg, Pro) and depleted for those encoded by low-GC codons (Phe, Tyr, Met, Ile, Asn, Lys). Typically such differences are attributed to mutational biases operating in the context of neutral evolution. However, we can now understand the potential for mutation biases to operate in non-neutral evolution.

The general result of simulating evolution of a protein-coding gene via beneficial changes, subject to biases in the introduction of variants, is to confirm the intuitions of the metaphor of Climbing Mount Probable, showing that it is not misleading in its implications. On a perfectly smooth landscape, the trajectory of evolution is deflected initially by mutation bias, but the ultimate destination—the optimal sequence—is unaffected. On rough landscapes, a mutation-biased composition evolves along a trajectory to a local peak. The rougher the landscape, the shorter the trajectory and the greater the per-step effect of mutation bias. In the model used by Stoltzfus (2006), a modest mutational bias toward GC (or AT), with a magnitude consistent with mutational biases inferred from models of genome composition evolution, can cause a progressive shift in protein composition of a magnitude consistent with observed biases in protein composition.

A

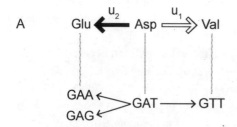

B

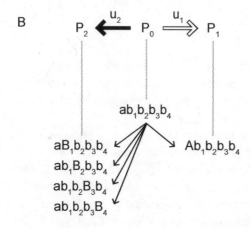

C

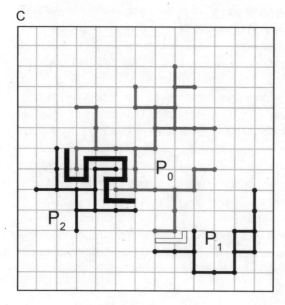

The Causal Efficacy of Developmental Biases

The novel implications of biases in the introduction process apply to some biases that are properly developmental, in the sense of reflecting the mapping of genotypes to phenotypes. The simplest and most complete such mapping is the genetic code, which represents the developmental process of translation, by which an amino acid emerges as the expression of a triplet codon. Like other developmental processes, translation is a complex, multistep process. Like other developmental processes, there is not a 1:1 mapping of phenotypes to genotypes: most amino acid phenotypes can be encoded in multiple ways; some changes are phenotypically silent. Unlike many other developmental processes, translation results in discrete phenotypes. More distinctly, translation results in discrete phenotypes in such a highly canalized way that the developmental realization of the modal phenotype occurs over 99.9 percent of the time, that is, less than 1 in a 1,000 amino acids is the wrong one.

Though translation isn't the typical case of development discussed in the evolutionary literature, its differences from the typical cases of discrete phenotypic variants are a matter of degree. Many developmental processes result in discrete phenotypes, for example, limb development in chordates results in an autopod with a discrete set of digits, with a modal number of 5 for mammals. Cranial development is so strongly canalized that nearly every mammal is born with precisely one head and not two, three, or more. Yet, rare cases indicate that double-headedness is developmentally possible (typically with duplication of the neck and upper torso), with a chance of occurrence less than that of a translation error.

Consider a single codon encoding an amino acid as an evolving system. The system begins with the Aspartate phenotype, encoded by GAT, as in figure 3.6A. The developmental expression of a genotype is indicated by the dotted gray line. From this starting point, there are twice as many mutational paths to genotypes expressed as the Glutamate phenotype than expressed as Valine. Assuming (for the sake of example) that all individual mutation rates are the same, this means that, in a population of Asp-encoding genotypes, there will be a 2-fold bias in the introduction of alternative Glu phenotypes, relative to Val phenotypes.

The asymmetry described above is analogous to a form of developmental bias invoked repeatedly in the evo devo literature, by which some phenotypes are more likely to arise by mutation than others (e.g., Alberch and Gale 1985; Emlen 2000). Figure 3.6B gives

Figure 3.6
Biases induced by features or processes above the DNA level. (A) A 2-fold bias in mutational paths induced by the genetic code. (B) A 4-fold bias in mutational paths induced by an asymmetric mapping of genotypes to phenotypes. One mutation ($a \rightarrow \underline{A}$) at the A locus leads to P_1, and four different mutations at loci B_1, B_2, B_3, and B_4 lead to P_2. (C) Differential accessibility of phenotypic states induced by the structure of connected neutral networks in sequence space (after Fontana 2002). Only three networks are shown, each connecting RNA sequences with the same fold (phenotype). Over time, a sequence traverses a network neutrally, and the chance of evolving to a different network (phenotype) depends on its proximity. Because the boundary (thick solid line) between P_0 and P_2 is several times longer than between P_0 and P_1, P_0 is several-fold more likely to evolve to P_2.

this concept a precise interpretation, in which, given the starting phenotype P_0, there are four times as many ways to mutate to P_2 as to P_1. In the language of genetics, the mutants that generate P_2 are phenocopies of each other. This kind of bias differs formally from a DNA-level bias in that it cannot be defined in terms of DNA chemistry, but must make reference to development or phenotypes.

The evolutionary impact of such biases can be understood by mapping them to the Yampolsky–Stoltzfus model. We could define u_2 to cover mutations from Asp to Glu, whether they are to GAA or GAG, and u_1 to cover mutations to Val. Or we could define u_2 to cover all four mutations from P_0 to P_2, and u_1 to cover mutations to P_1. Then, all of the same implications would apply, and in particular, the 2-fold or 4-fold bias will exert a 2-fold or 4-fold bias (respectively) on the outcome of evolution in the origin-fixation regime.

An Interpretation of Structuralism

Classic thinking in evolutionary biology includes a preference (often implicit) for explanations that are externalist, functionalist, and mechanistic: the explanations have to do with external rather than internal causes; they explain form by relating it to function; and they are seen to rest on Aristotelean material and efficient causes rather than formal causes. Yet, there has always been a structuralist tradition distinct from (and often in conflict with) the mainstream neo-Darwinian view, in which it is considered important to give explanations of the forms of things in terms of structural principles, for example, relating the form of a mushroom to the form of a mushroom cloud. One would associate structuralism historically with authors such as D'Arcy Thompson, and in more recent times, with authors such as Goodwin (1994), Kauffman (1993), and Fontana (2002).

Contemporary structuralists often make arguments relating the chances of evolving a particular form to its relative accessibility or density in state-space (e.g., Kauffman 1993; Fontana 2002), and these arguments can be understood as implications of biases in the introduction of variation. To understand how, let us begin with an argument from molecular evolution about the rough correspondence observed between the frequency with which an amino acid is found in proteins, and the number of codons assigned to it in the genetic code (which ranges from 1 to 6). Originally, King and Jukes (1969) pointed to this correlation as an argument for neutral evolution, but King (1971) later stepped back from this position, arguing that the same correlation could arise under a stochastic view with selective allele fixation, explaining this position as follows:

Suppose that, at a given time, there are several possible amino acid substitutions that might improve a protein, and among these are changes to serine and to methionine; serine, with its six codons, has roughly six times the probability of becoming fixed in evolution. Once this has occurred, the mutation to methionine may no longer be advantageous. (King 1971, 89)

Amino acids with more codons occupy a greater volume of sequence space. By the same token, they have more mutational arrows pointed at them: they are more likely to be proposed, and therefore more likely to evolve, other things being equal. This is analogous to the argument of Garson, Wang, and Sarkar (2003) in "How development may direct evolution," where there is a discrete map between genotypes and phenotypes, and the chance of evolving is a function of the number of genotypes assigned to a phenotype.

The resulting tendency for the frequency of an amino acid (phenotype) to correspond to its number of codons (genotypes) is not caused by natural selection, even if all changes are beneficial: instead, this pattern arises from the way in which the space of possibilities is explored by the mutation process. This is why, in the arguments of Kauffman (1993), the "ensemble properties" whose emergence Kauffman attributes to "self-organization" still arise when differential selection is removed.

A related idea concerns the role of the local accessibility of genotypes with a given alternative phenotype. In the RNA-based models of Fontana (2002), sequences that have the same fold (phenotype) may be represented by a network as in figure 3.6C (after Fontana 2002). An RNA under selection to maintain function may diffuse neutrally within the network for its current fold; the chances of evolving a different fold (phenotype) depend on the accessibility of the new fold over the entire network. For the networks shown in figure 3.6C, a sequence with phenotype P_0 is more likely to evolve to P_2 than P_1, because a random step out of the P_0 network is more likely to yield P_2 than P_1.

This is not the same as an effect of the number of genotypes assigned to an alternative phenotype: P_1 and P_2 both have exactly 13 genotypes (sequences). Instead, it is an effect of the number of alternative genotypes *within the mutational horizon* (the set of 1-mutant neighbors) of the network for the current phenotype.

On this basis, we can see that a key concern of structuralism—understanding the relative chances for the evolution of various forms by considering their distribution and local accessibility in genetic state-space—relies on the action of biases in the introduction of variation. This means that, in ideal cases, we can take advantage of the theory provided above, which allows one to combine biases in the introduction of variants and biases in their fixation within a common framework. This is important because it represents a genuine framework of dual causation that supports quantitative links between causes and effects. By contrast, in the history of evolutionary thought, the dual causation of evolution by internal and external causes is typically addressed by first declaring a fundamental asymmetry by which the dominant cause governs the subordinate cause, or an asymmetry by which one kind of cause is nuanced and quantitative, whereas the other is a matter of yes or no.

A Contrast of Theories

One's immediate reaction to the suggestion that there are fundamental but previously unrecognized implications of biases in mutation may be one of bafflement. Aren't these merely straightforward implications of discrete events of mutation—a kind of event that scientists have known for over a century? How could such a fundamental idea escape recognition for 80 years after theoreticians began in earnest to explore the evolutionary implications of genetics in the 1920s?

Actually, the history of evolutionary thought provides multiple reasons to believe that, through the latter half of the twentieth century, a well-informed evolutionary biologist would be strongly disposed *against* recognizing or appreciating the concept of biases in the introduction of variation. The suspicion of internal causes in the dominant neo-Darwinian culture ran so deep that every internalist idea, no matter how reasonable, was treated as an appeal to vitalism. As explained below, the hypothesis that variational biases could influence evolution was considered by the founders of theoretical population genetics—and was specifically rejected as impossible. Until roughly 1969, theoreticians did not construct models of evolutionary change using an introduction process, because the traditional approach assumed that evolution begins with all relevant alleles already present. This gene-pool assumption was part of a neo-Darwinian theory by which evolution has no direct dependence on the chance occurrence of new mutations. This combination of views favored a view of causation in terms of mass-action forces, whereas individual events of mutation that introduce novelty were labeled as proximate causes acting at the wrong level to be important in evolution.

Exploring these barriers helps to clarify the significance, for evolutionary thinking, of recognizing biases in the introduction of variation as potentially important causes of evolutionary bias.

The "Opposing Pressures" Argument

In the mid-1980s, evolutionary developmental biologists proposed a reappraisal of the role of development in evolution via the concept of developmental "constraints." Some uses of this term corresponded to the previous idea of "orthogenesis," meaning the generation of variation in a developmentally preferred direction, rather than indiscriminate "ambigenesis" in all directions (Popov 2009). The workshop article on developmental constraints by Maynard Smith and colleagues (1985) became a touchstone, cited over 1,000 times today.

What became of this idea? Years later, critics such as Reeve and Sherman (1993) could point to this seminal paper and complain that there was "no delineation of a mechanism," only a description of the idea of developmental biases in variation. That is, Maynard Smith et al. (1985) did not actually offer a mechanistic cause. They wrote that

Since the classic work of Fisher (1930b) and Haldane (1932) established the weakness of directional mutation as compared to selection, it has been generally held that directional bias in variation will not produce evolutionary change in the face of opposing selection. This position deserves reexamination.

However, the issue was not re-examined until the work of Yampolsky and Stoltzfus (2001).

Fisher and Haldane showed that, starting with a population of A1 individuals, an inequality of forward and backward mutation rates favoring allele A2 will not result in fixation of A2 if this is opposed by selection. Instead, given that mutation rates are small in comparison to selection coefficients, a deleterious allele will persist at a low level reflecting a balance between the opposing pressures of mutation and selection. In the simplest haploid case, the "mutation-selection balance" is approximated by $f = u/s$, where u is the forward rate of mutation to A2 (the backward rate hardly matters) and s is the selective disadvantage of A2.

This kind of equation makes it easy to depict mutation and selection as opposing forces, with selection pushing down the frequency of A2 with magnitude s, and mutation pushing the frequency up with magnitude u. Given that selection coefficients on the order of 10^{-2} or 10^{-3} were thought to be common, whereas mutation rates were never so large, a wide-ranging conclusion seemed warranted:

For mutations to dominate the trend of evolution it is thus necessary to postulate mutation rates immensely greater than those which are known to occur, and of an order of magnitude which, in general, would be incompatible with particulate inheritance. ... The whole group of theories which ascribe to hypothetical physiological mechanisms, controlling the occurrence of mutations, a power of directing the course of evolution, must be set aside. (Ch. 1 of Fisher 1930b)

By treating mutation and selection as opposing pressures on allele frequencies, Fisher and Haldane reduced the vexing issue of internal orienting factors to *a simple matter of size*. No difficult empirical analysis of evolutionary history was necessary to evaluate different hypotheses for the causes of trends. No experiments were needed. Instead, the entire issue could be solved by extrapolating from a few simple facts and an equation. The argument was breathtaking in its scope and simplicity: population genetics tells us that, because mutation rates are small, internal orienting factors acting via mutation are ruled out, *completely*.

Provine (1978) identifies this as one of the vital contributions of population genetics theory to the success of the OMS (original Modern Synthesis), on the grounds that it appeared to eliminate the rival theories of orthogenesis and Mendelian-mutationism. Indeed, the architects of the OMS repeatedly denied mutational causes based on this argument (Huxley 1942, 56; Simpson 1953, 114; Mayr 1960, 355; Mayr 1959, 7–9; Ford 1971, 391).

This is how the "opposing pressures" argument, to the effect that mutation is a weak force unable to overcome selection or influence the direction of evolution, emerged from

Fisher (1930b, a) and Haldane (1932, 1933), as well as Wright (1931). More recent sources also cite this argument directly (e.g., Gould 2002, 510), or draw on the argument that mutation is a weak force (e.g., see Section 5.3 of Freeman and Herron 1998; Sober 1987, 105–118; Maynard Smith et al. 1985, 282). This even happens in the literature of molecular evolution (e.g., Li, Luo, and Wu 1985, 54). An interesting example is from a white paper on evolution, science, and society endorsed by various professional societies (Futuyma et al. 1998). One section discusses accomplishments of mathematical population genetics, claiming "it is possible to say confidently that natural selection exerts so much stronger a force than mutation on many phenotypic characters that the direction and rate of evolution is ordinarily driven by selection even though mutation is ultimately necessary for any evolution to occur."

Population Genetics without the Introduction Process

For Fisher and Haldane to reject a role for mutation bias without considering its effect on the introduction process may seem absurd to us today. This absurdity is a measure of how far evolutionary thinking has drifted from the OMS. As noted above, the architects of the OMS were committed to a buffet view in which all the variants relevant to the outcome of evolution are present initially. The nature of the OMS view is explored in the next section: our focus in this section is on its impact on modeling.

When the alleles relevant to the outcome of evolution are present initially, the influence of mutation is merely to shift the relative amounts of the alleles. These effects are minor and typically negligible compared to the frequency-shifting effects of selection and even drift, because mutation rates are so small. That is, mutation is a weak pressure because mutation rates are small. As Lewontin (1974) noted, "There is virtually no qualitative or gross quantitative conclusion about the genetic structure of populations in deterministic theory that is sensitive to small values of migration, or any that depends on mutation rates" (267).

As a result, mutation was often simply left out of models. Classic works such as Lewontin and Kojima (1960) don't include mutation because its effects are trivial. The treatment of the mathematical foundations of population genetics by Edwards (1977) includes hundreds of equations, none with a term for mutation: the word "mutation" occurs only on page 3, in the sentence, "All genes will be assumed stable, and mutation will not be taken into account."

Thus, there was no theory relating the rate of evolution to the rate of mutational introduction. The need for such a theory became obvious in the 1960s, when protein sequence comparisons suggested that evolution was a recurring process of mutation and fixation. It was in this context that origin-fixation models emerged in 1969 (McCandlish and Stoltzfus 2014).

However, for decades after their first appearance, origin-fixation models remained associated with molecular evolution, and were used primarily to address the fate of neutral or slightly deleterious mutations (McCandlish and Stoltzfus 2014). Outside of molecular evolution, theoretical population genetics still relied on the gene-pool assumption. In the 1990s, theoreticians began to notice this restriction; for example, Yedid and Bell (2002) write:

In the short term, natural selection merely sorts the variation already present in a population, whereas in the longer term genotypes quite different from any that were initially present evolve through the cumulation of new mutations. The first process is described by the mathematical theory of population genetics. However, this theory begins by defining a fixed set of genotypes and cannot provide a satisfactory analysis of the second process because it does not permit any genuinely new type to arise. (p. 810)

Likewise, Hartl and Taubes (1998) write:

Almost every theoretical model in population genetics can be classified into one of two major types. In one type of model, mutations with stipulated selective effects are assumed to be present in the population as an initial condition. ... The second major type of models does allow mutations to occur at random intervals of time, but the mutations are assumed to be selectively neutral or nearly neutral. (p. 525)

Eshel and Feldman (2001) make a similar distinction:

We call short-term evolution the process by which natural selection, combined with reproduction (including recombination in the multilocus context), changes the relative frequencies among a fixed set of genotypes, resulting in a stable equilibrium, a cycle, or even chaotic behavior. Long-term evolution is the process of trial and error whereby the mutations that occur are tested, and if successful, invade the population, renewing the process of short-term evolution toward a new stable equilibrium, cycle, or state of chaos. (p. 182)

They argue that,

Since the time of Fisher, an implicit working assumption in the quantitative study of evolutionary dynamics is that qualitative laws governing long-term evolution can be extrapolated from results obtained for the short-term process. We maintain that this extrapolation is not accurate. The two processes are qualitatively different from each other. (p. 163)

Thus, due to the way that prevailing views shaped approaches to modeling, theoreticians were not prepared to recognize a role for biases in the introduction of variation. A half-century after theoretical population genetics emerged as a discipline, the introduction process began to appear as an unnamed technical feature of certain types of models, though not as a feature of contending theories (which were focused instead on selection, drift, recombination, sex, and so on). It took several more decades for the argument to emerge that recognizing the introduction process is not a minor technical detail, but a major theoretical innovation that challenges previous thinking about how evolution works, and opens new avenues for theoretical and empirical research.

Structure and Implications of the OMS

In the early years of the twentieth century, Johannsen's experiments with beans showed that selection can be effective at sorting out existing varieties, but it does not create new types from masses of environmental fluctuations. That experiment spelled the end of Darwinism among those who embraced genetics (Gayon 1998). In the mutationist view that emerged among early geneticists such as Morgan (1916), "Evolution has taken place by the incorporation into the race of those mutations that are beneficial to the life and reproduction of the organism" (p. 194).

Given this view, one might imagine that the introduction of a new mutation would be a key event that provides initiative for evolutionary change, and thus influences dynamics as well as direction. Following Shull (1936), one might suppose that "If mutations are the material of evolution, as geneticists are convinced they are, it is obvious that evolution may be directed in two general ways: (1) by the occurrence of mutations of certain types, not of others, and (2) by the differential survival of these mutations or their differential spread through the population" (p. 122), and we might suppose that a new allele "produced twice by mutation has twice as good a chance to survive as if produced only once" (p. 140).

We might think this way today, but the architects of the OMS emphatically did not. In their view, Johannsen's experiments showed nothing. Instead, the true nature of evolution was revealed in Castle's experiments with the hooded rat, because unlike Johannsen, "Castle had been able to produce new types by selection" (Provine 1971, 114). Castle was able to shift coat-color from mottled to nearly all black, or nearly all white, in less than 20 generations of selection, not enough time for new mutations to play any important role—that is, *selection can create new types without mutation.*

The genetic interpretation of this result was that selection simultaneously shifts the frequencies of available alleles at many loci, leveraging recombination to combine many small effects in one direction (Provine 1971). Thus, recombination (not mutation) is the proximate source of new genetic variation every generation. Evolution, rather than being a process of the mutational introduction and reproductive sorting of variation, is envisioned as a process of shifting allele frequencies in the gene pool. Even though this mode of change requires abundant standing variation, it prevails in nature (they argued), because natural populations have a "gene pool" that "maintains" variation. Thus, the maintenance of variation in the gene pool, logically necessary to prop up the Castle experiment as the paradigm of evolution, became a major theme, what Gillespie (1998) called "The Great Obsession" of population genetics.

Thus, in the OMS, the term "gene pool" is not merely descriptive but evokes the theory that natural populations *maintain* abundant genetic variation; for example, Stebbins (1966) writes that "a large 'gene pool' of genetically controlled variation exists at all times" (p. 12). This theory, proposed by Chetverikov and popularized by Dobzhansky, holds that various features of genetics and population genetics—including recessivity, chromosome assortment, crossing over, sexual mixis, frequency-dependent selection, and heterosis—come together to create a dynamic in which variation is soaked up like a "sponge" and "maintained." In the OMS, this gene-pool dynamic ensures that evolution is always a multifactorial process in which selection never waits for a new mutation but can shift the population to a completely new state based on readily available variation.

This gene-pool theory was used to argue against the early geneticists in regard to (1) the source of initiative in evolution (mutation or selection), and (2) rates of evolution, which do not depend on mutation rates. For example, Stebbins (1959) writes:

Second, mutation neither directs evolution, as the early mutationists believed, nor even serves as the immediate source of variability on which selection may act. It is, rather, a reserve or potential source of variability which serves to replenish the gene pool as it becomes depleted through the action of selection. ... The factual evidence in support of these postulates, drawn from a wide variety of animals and plants, is now so extensive and firmly based upon observation and experiment that we who are familiar with it cannot imagine the appearance of new facts which will either overthrow any of them or seriously limit their validity. (p. 305)

Stebbins shows an irreversible commitment to denying any direct role of mutation in evolution. An episode of evolution begins when the environment changes, bringing on selection of an abundance of small-effect variation in the gene pool so that, in the words of Mayr (1963, 613), "mutation merely supplies the gene pool with genetic variation; it is selection that induces evolutionary change." As Mayr (1963, 101) explains at greater length,

In the early days of genetics it was believed that evolutionary trends are directed by mutation, or, as Dobzhansky (1959) recently phrased this view, "that evolution is due to occasional lucky mutants which happen to be useful rather than harmful." In contrast, it is held by contemporary geneticists that mutation pressure as such is of small immediate evolutionary consequence in sexual organisms, in view of the relatively far greater contribution of recombination and gene flow to the production of new genotypes and of the overwhelming role of selection in determining the change in the genetic composition of populations from generation to generation.

The gene-pool theory has a particular implication about rates of evolution. If an event of mutation is never the token cause that initiates evolutionary change, then the rate of evolution will not depend in any strong way on the rate of mutation; for example, Stebbins (1966, 29) writes:

Mutations are rarely if ever the direct source of variation upon which evolutionary change is based. Instead, they replenish the supply of variability in the gene pool which is constantly being reduced by selective elimination of unfavorable variants. Because in any one generation the amount of variation contributed to a population by mutation is tiny compared to that brought about by recombination of pre-existing genetic differences, even a doubling or trebling of the mutation rate will have very little effect upon the amount of genetic variability available to the action of natural selection. *Consequently, we should not expect to find any relationship between rate of mutation and rate of evolution. There is no evidence that such a relationship exists.* [my emphasis]

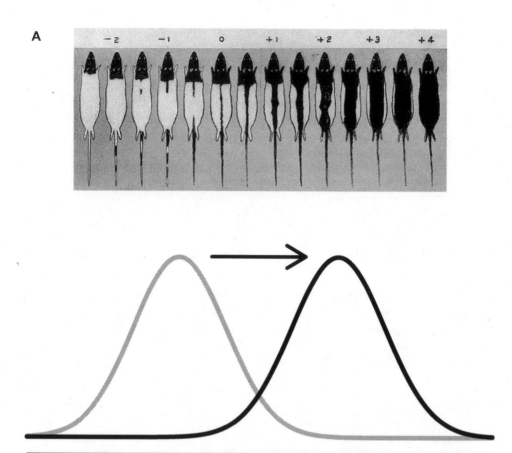

Figure 3.7
The forces view of the OMS. In the OMS, evolution happens when the environment changes, and (A) selection shifts the phenotypic value, sometimes shifting it so far as to result in a new type. (B) At the genetic level, this shift is a multifactorial process, dependent on small-effect variation at many loci. Change is not based on mutation-fixation events but quantitative shifts in frequency of alleles maintained in the gene pool. Therefore, (C) we can understand the process as taking place in the topological interior of an allele-frequency space (i.e., the interior of the cube, but not its edges or surfaces). In this interior space, the trajectory of an evolving system can be understood with the theory of forces. However, the forces theory becomes inadequate when movement occurs outside of this interior space.

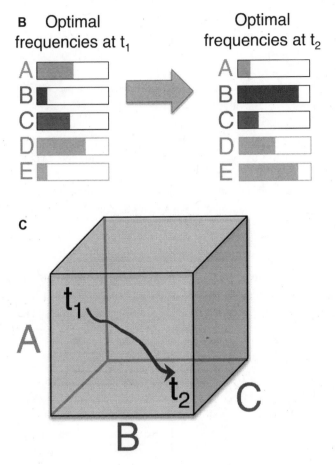

B Optimal
frequencies at t_1

Optimal
frequencies at t_2

A
B
C
D
E

A
B
C
D
E

c

t_1

A

t_2

C

B

Figure 3.7
(continued)

Likewise, in their 1970s textbook, Dobzhansky et al. (1977, 72) write:

The large number of variants arising in each generation by mutation represents only a small fraction of the total amount of genetic variability present in natural populations. ... It follows that rates of evolution are not likely to be closely correlated with rates of mutation. ... Even if mutation rates would increase by a factor of 10, newly induced mutations would represent only a very small fraction of the variation present at any one time in populations of outcrossing, sexually reproducing organisms.

The Forces Theory

The structure of the OMS encourages a particular conception of evolutionary causes as frequency-shifting forces. Evolution at the phenotypic level is envisioned as a smooth shift to a new adaptive optimum, as in figure 3.7A, with a rate and direction that depend entirely on the rate and direction of the environmental changes to which selection is responding. At the genetic level, evolution occurs by shifting the frequencies of alleles already present in the population. This is a multifactorial process, that is, an episode of adaptation is a shift in frequencies of alleles at many loci simultaneously (figure 3.7B).

If so, then evolution can be understood as a process that takes place in the *topological interior* of an allele-frequency space (e.g., the interior of the cube in figure 3.7C, but not its edges or surfaces); evolution is more specifically a selection-driven shift from a previous optimum to a new multi-locus optimum (e.g., from t_1 to t_2 in figure 3.7C). Forces are conceived as processes that can displace an evolving system in this interior space. Thus the forces jointly determine the rate and direction of movement. As noted by Fisher and Haldane, the forces are not all equal in strength. Selection is typically the most powerful, and mutation the weakest. For alleles with intermediate frequencies, even drift will have a larger impact than mutation (except in very large populations).

For instance, consider just two of the dimensions in figure 3.7C. We will say that the position from 0 (origin) to 1 (top) in the A dimension is the frequency of allele A2, as distinct from A1, and in the B dimension, likewise for B2 and B1. We can imagine a small shift from the center of this area {0.500, 0.500} to {0.506, 0.506}. In a population of 1,000, that would mean adding 6 more A2 and 6 more B2 individuals. We can imagine such a shift happening by drift, selection, or mutation, although in practice, a shift of ~ 1 percent is too much for mutation to accomplish in a generation: it is indeed a weak force.

We could change the details of this example in many ways and the statement would still be true, and this is the whole point of the force analogy. A force is a force because it has a generic ability to shift a frequency f to $f + \delta$. The forces all have a common currency of causation, which is allele-frequency shifts. By way of this common currency, they have a kind of comparability and, in ideal cases, are combinable, that is, we can combine the shift caused by one force with the shift caused by another. The mutation-selection balance equation would be a simple example of combining two forces using their common currency of causation.

For our purposes, the most important feature of the forces theory is that it fails to include the novelty-introducing role of mutation, which renders it inadequate once we step outside the buffet regime assumed by the OMS. The unique novelty-introducing role of mutation— but not selection, drift, or recombination—is represented by its ability to move the population from $x = 0$ to $x = 1/N$, or from $x = 1$ to $x = 1 - (1/N)$. The shift from {0.500, 0.500} to {0.506, 0.506} is mathematically identical to the shift from {0.000, 0.000} to {0.006, 0.006}, but it is not evolutionarily identical: the latter shift absolutely requires the involvement of mutation. Mutation allows the system to jump off of an axis into the interior where the forces of selection and drift operate. Yet, this role is not covered by the "forces" formalism.

That is, the introduction process draws on a different currency of causation inaccessible to selection or drift. This is why the forces theory fails to provide correct reasoning for the behavior of the Yampolsky–Stoltzfus model in the sushi conveyor regime, or even in regimes that are close to it. We described these failures earlier: neither increasing the magnitude of selection coefficients (which increases the force of selection), nor decreasing the magnitude of mutation rates (which decreases the force of mutation), causes selective preferences to overpower mutational ones.

The limitations of the forces view are also revealed in how scientists construct hypotheses and explanations. In molecular evolution, we are regularly faced with comparative evidence suggesting that molecular features, particularly aggregate features like codon usage or proteome composition, are systematically shaped by mutation biases. By the logic of opposing forces, this must mean that the force of selection is effectively absent. Accordingly, for decades, molecular evolutionists habitually treated mutational explanations as references to neutral models (e.g., Sueoka 1988; Gillespie 1991; Gu, Hewett-Emmett, and Li 1998; Knight, Freeland, and Landweber 2001; Lafay et al. 1999; Wolfe 1991).

This interpretation literally suggests that the pressure of recurrent mutation is driving alleles to fixation in the absence of opposing selection. Although this is the correct way to apply the forces theory, the resulting model is not very reasonable, because fixation by recurrent mutation is too slow. The more likely basis for mutation-biased neutral evolution would be that fixation happens by drift, while the mutational bias resides in the introduction process. That is, once we recognize two kinds of causes with two different currencies—introduction and fixation—this immediately suggests a different interpretation in which the mutation bias is applied only to the introduction process. From this point, it is a small step to consider that the fixation process might be either drift or selection, leading to the novel prediction of mutation-biased adaptation.

Conclusion

The notion that a fundamental cause–effect relationship could emerge 80 years after the origins of theoretical population genetics must be considered surprising, as suggested by the epigraph at the head of this chapter (Charlesworth 1996). One eminent theoretician argued that explaining the nature of the results of Yampolsky and Stoltzfus (2001) was not a worthwhile activity because we have "matured past the phase when discussions of basic principles are useful for professionals. We all believe in mutation, selection, and drift, I hope" (see Stoltzfus 2012). Lynch (2007) cites Yampolsky and Stoltzfus (2001) along with a handful of other sources going back to Charles Darwin, explaining that "the notion that mutation pressure can be a driving force in evolution is not new."

A theory is new if scientists have not considered it before, either because it was never introduced, or because it was introduced and then ignored. The question of whether some set of past statements about mutation or variation constitutes a prior description of the kind of cause–effect relationship described here is a matter of applying the duck test: if it looks like a duck, swims like a duck, and quacks like a duck, it's a duck. Either the distinctive implications of biases in the introduction process have been previously shown to arise from a consideration of basic principles of population genetics, or they have not.

One's disbelief at the belated discovery of this kind of cause–effect relationship must begin to fade rapidly when the historical development of evolutionary thought is considered. In fact, (1) the founders of theoretical population genetics considered the possible role of biases in variation as an influence on the course of evolution, (2) they reached the conclusion that such an effect was impossible, and (3) this conclusion became a foundation of the OMS, cited until contemporary times. A role for internal tendencies of variation was rejected by leading thinkers *precisely on the grounds of lacking a population-genetic mechanism*. Indeed, the idea was ridiculed, for example, as when Simpson (1967) argued that selection is the only plausible cause of directional changes in the fossil record, dismissing "the vagueness of inherent tendencies, vital urges, or cosmic goals, without known mechanism" (159).

Decades of results from molecular studies of evolution have conditioned us to accept the mutationist idea that the timing and character of evolutionary change may depend on the timing and character of mutations. Because of this, scientists today may find the evolutionary role of biases in the introduction process to be intuitively obvious. However, the foundations of contemporary evolutionary thought were laid by scientists who explicitly rejected mutationist thinking as an apostasy, and who were committed to the doctrine that variation plays only the role of a material cause, a source of substance only, never a source of form or direction.

From this context, we can begin to see that evolutionary reasoning about causes and explanations is not merely a matter of mathematics, and is not merely a matter of following obvious implications of basic principles that everyone accepts.

Other factors are at work to make the recognition and appreciation of causes non-obvious. We have addressed two of them already. One factor is the role of high-level conjectural theories asserting that evolution works in some ways and not in others, and in particular, the role of neo-Darwinism in shaping the OMS theory of evolutionary genetics. To understand evolutionary reasoning requires not merely a knowledge of mathematics but an awareness of the substantive conjectural theories that have shaped prevailing mathematical approaches to evolution—for example, the gene-pool theory. One cannot understand the strengths and weaknesses of contemporary evolutionary thinking if one does not understand how it relies on the OMS conjecture that evolution operates in the buffet regime of abundant preexisting variation (Stoltzfus 2017).

A second factor, which requires more explication, is the role of verbal theories of causation and explanation. Even if we attempt to set aside the OMS and other broad conjectures, we must recognize that the conceptual tools we use to negotiate issues of causation and explanation include verbal theories that draw on metaphors and analogies (e.g., pressures, raw materials). Mathematics itself is not a language of causation but a notation for relating quantities, along with a set of rules for manipulating these relations. The "equals" sign has no direction, thus it cannot possibly represent an arrow of causation. There is no necessary causal translation of $a = bc$, or the mathematically equivalent statements $b = a/c$ or $a/b = c$.

Thus, the rules of mathematics do not compel any particular interpretation of the equations of population genetics. For example, little in the breeder's equation ($\Delta z = h^2 S$) differentiates variation and selection: they are two factors multiplied together to yield a product. If we choose to call the product "the response to selection," rather than something else, it is because we have situated this equation within some conceptual framework that extends beyond mathematics. Indeed, it is natural to refer to the left-hand side as "the response to selection" because, in practice, selection has nearly always been the experimentally manipulated variable, even though it is possible in principle to manipulate the heritable component of available variation in a population, in which case we might call the left-hand side "the response to variation."

After Lande and Arnold (1983) generalized the breeder's equation to multiple dimensions as $\Delta Z = G\beta$ (where β is a vector of selection differentials, G is a matrix of variances and covariances quantifying standing variation, and ΔZ is a vector of expected changes in trait values), quantitative geneticists began to understand that, in the words of Steppan, Phillips, and Houle (2002), "Together with natural selection (the adaptive landscape) it [the G matrix] determines the direction and rate of evolution." This verbal theory of dual causation, in which the *rate and direction* of evolution are *jointly* attributed to selection and standing variation, was novel: it does not correspond to the previous verbal theories in which the rate and direction of evolution *are governed by selection* (while variation merely supplies raw materials), even though quantitative genetics is the branch of mathematical theory that most closely follows neo-Darwinian assumptions.

That is, Darwin's verbal theory led to Fisher's mathematical theory, and then further mathematical developments along with empirical results (e.g., Schluter 1996; McGuigan 2006) led to conflicts with the original verbal theory, resulting in a new verbal theory that aligns better with mathematical models and empirical results.

This kind of reform, in which verbal theories of causation are replaced or modified so as to achieve better alignment with mathematical models and empirical results, has not yet occurred in regard to the introduction process, and particularly in regard to quantitative biases in the generation of variants in discrete spaces (which are not the same thing as asymmetries in standing variation of continuous traits represented by the G matrix). The architects of the OMS were explicit that events of mutation are not evolutionary causes but proximate causes that act at the wrong level to be evolutionary causes. The emergence of origin-fixation models in 1969 posed a threat to this view, yet the threat did not materialize: it did not induce a shift in conceptualizations of causation. Neither those who explicitly invoked the need to address long-term dynamics dependent on new mutations, nor those who recognized a renaissance in thinking about adaptation, called for explicit recognition of a point process of introduction absent from the forces theory.

The practice of scientific reasoning is not a form of magic. A complete understanding of the evolutionary implications of mutation does not blossom forth in one's mind simply by uttering the word "mutation." Instead, our understanding of causal factors must be constructed. In practice, it is constructed in a context that includes not only equations and facts but low-level folk theories of causation (e.g., pressures, raw materials) and high-level conjectures about how causes work together to account for larger phenomena. Recognizing the introduction process as a causal process in evolution, and then recognizing the consequences of biases in the introduction process, makes variation-biased evolution intelligible in a way that led to the prediction of mutation-biased adaptation (now verified), and which yields new insights into the potential role of development and self-organization in evolution.

To summarize, the novelty of scientific theories is not judged by whether they are considered intuitively obvious with benefit of hindsight, or whether they require a complex mathematical derivation. To recognize a causal role of biases in the introduction process is novel because this kind of causation has distinctive behavior and implications. Generic references to mutation pressure, contingency, chance, constraints, and so on, simply are not the same thing as references to biases in the introduction process, because *the former concepts were never previously understood to exhibit the implications that the latter concept can be demonstrated to exhibit.* Generations of theoreticians who invoked the term "mutation" failed to comprehend the implications of biases in the introduction process: again, such comprehension does not emerge magically by uttering the word "mutation," but requires a modeling framework and a conceptualization of causation.

Mutational and developmental biases in the introduction of variation appear to be important determinants of the outcome of evolution, representing a novel kind of causation absent from the Modern Synthesis. The extent of their importance, and particularly their importance as predictable factors acting at the level of phenotypic evolution, is a major unresolved issue in evolutionary biology.

Acknowledgments

The author thanks David McCandlish, Yogi Jaeger, and an anonymous reviewer for helpful comments. The identification of any specific commercial products is for the purpose of specifying a protocol and does not imply a recommendation or endorsement by the National Institute of Standards and Technology. Figure 1A is a photo by John Johnston, cropped and converted to grayscale from the original (https://www.flickr.com/photos /89324753@N00/79046791/), available via CC-BY (https://creativecommons.org/licenses /by/2.0/). In Figure 3.3, the sushi conveyor photo is by NipponBill from the original (https://en.wikipedia.org/wiki/Conveyor_belt_sushi#/media/File:Sushi_conyeyor_chain _1.jpg), and the salad buffet photo is by Heider Ribeiro, from the original (https://commons .wikimedia.org/wiki/File:Salad_bar-02.jpg), both converted to grayscale and both available via CC-BY-SA license (https://creativecommons.org/licenses/by-sa/3.0/).

References

Alberch, P., and E. A. Gale. 1985. "A Developmental Analysis of an Evolutionary Trend: Digital Reduction in Amphibians." *Evolution* 39 (1): 8–23.

Charlesworth, B. 1996. "The Good Fairy Godmother of Evolutionary Genetics." *Current Biology* 6 (3): 220.

Couce, A., A. Rodríguez-Rojas, and J. Blázquez. 2015. "Bypass of Genetic Constraints during Mutator Evolution to Antibiotic Resistance." *Proceedings of the Royal Society of London B: Biological Sciences* 282 (1804).

Cunningham, C. W., K. Jeng, J. Husti, M. Badgett, I. J. Molineux, D. M. Hillis, and J. J. Bull. 1997. "Parallel Molecular Evolution of Deletions and Nonsense Mutations in Bacteriophage T7." *Molecular Biology and Evolution* 14 (1): 113–116.

Desai, M. M., and D. S. Fisher. 2007. "Beneficial Mutation Selection Balance and the Effect of Linkage on Positive Selection." *Genetics* 176 (3): 1759–1798. doi: 10.1534/genetics.106.067678.

Dobzhansky, T. 1959. "Variation and Evolution." *Proceedings of the American Philosophical Society* 103:252–263.

Dobzhansky, T., F. J. Ayala, G. L. Stebbins, and J. W. Valentine. 1977. *Evolution*. San Francisco: W. H. Freeman.

Edwards, A. W. F. 1977. *Foundations of Mathematical Genetics*. 2nd ed. New York: Cambridge University Press.

Emlen, D. J. 2000. "Integrating Development with Evolution: A Case Study with Beetle Horns." *BioScience* 50 (5): 403.

Eshel, I., and M. W. Feldman. 2001. "Optimality and Evolutionary Stability under Short-term and Long-term Selection." In *Adaptationism and Optimality*, edited by S. H. Orzack and E. Sober, 161–190. Cambridge: Cambridge University Press.

Fisher, R. A. 1930a. "The Distribution of Gene Ratios for Rare Mutations." *Proceedings of the Royal Society of Edinburgh* 50:205–220.

Fisher, R. A. 1930b. *The Genetical Theory of Natural Selection*. London: Oxford University Press.

Fontana, W. 2002. "Modelling 'evo-devo' with RNA." *Bioessays* 24 (12): 1164–1177. doi: 10.1002/bies.10190.

Ford, E. B. 1971. *Ecological Genetics*. 3rd ed. London: Chapman & Hall.

Freeman, S., and J. C. Herron. 1998. *Evolutionary Analysis*. Upper Saddle River, NJ: Prentice-Hall.

Futuyma, D. J., T. R. Meagher, M. J. Donoghue, C. H. Langley, L. Maxson, A. F. Bennett H. J. Brockmann, et al. 1998. *Evolution, Science and Society: Evolutionary Biology and the National Research Agenda*. White paper. Available at http://www.ugr.es/~jmgreyes/Evolwhite.pdf.

Galen, S. C., C. Natarajan, H. Moriyama, R. E. Weber, A. Fago, P. M. Benham, A. N. Chavez, Z. A. Cheviron, J. F. Storz, and C. C. Witt. 2015. "Contribution of a Mutational Hotspot to Hemoglobin Adaptation in High-Altitude Andean House Wrens." *Proceedings of the National Academy of Sciences USA* 112 (45):13958–13963. doi: 10.1073/pnas.1507300112.

Garson, J., L. Wang, and S. Sarkar. 2003. "How Development May Direct Evolution." *Biology and Philosophy* 18 (2): 353–370.

Gayon, J. 1998. *Darwinism's Struggle for Survival: Heredity and the Hypothesis of Natural Selection*. Cambridge: Cambridge University Press.

Gillespie, J. H. 1991. *The Causes of Molecular Evolution*. In *Oxford Series in Ecology and Evolution*, edited by R. M. May and P. H. Harvey. New York: Oxford University Press.

Gillespie, J. H. 1998. *Population Genetics: A Concise Guide*. Baltimore, MD: Johns Hopkins University Press.

Goodwin, B. 1994. *How The Leopard Changed Its Spots: The Evolution of Complexity*. New York: Charles Scribner's Sons.

Gould, S. J. 2002. *The Structure of Evolutionary Theory*. Cambridge, MA: Harvard University Press.

Gu, X., D. Hewett-Emmett, and W. H. Li. 1998. "Directional Mutational Pressure Affects the Amino Acid Composition and Hydrophobicity of Proteins in Bacteria." *Genetica* 103 (1–6): 383–391.

Haldane, J. B. S. 1927. "A Mathematical Theory of Natural and Artificial Selection. V. Selection and Mutation." *Proceedings of the Cambridge Philosophical Society* 26:220–230.

Haldane, J. B. S. 1932. *The Causes of Evolution*. New York: Longmans, Green and Co.

Haldane, J. B. S. 1933. "The Part Played by Recurrent Mutation in Evolution." *The American Naturalist* 67 (708): 5–19.

Hartl, D. L., and C. H. Taubes. 1998. "Towards a Theory of Evolutionary Adaptation." *Genetica* 103 (1–6): 525–533.

Huxley, J. S. 1942. *Evolution: The Modern Synthesis*. London: George Allen & Unwin.

Kauffman, S. A. 1993. *The Origins of Order: Self-organization and Evolution*. New York: Oxford University Press.

Keller, I., D. Bensasson, and R. A. Nichols. 2007. "Transition-Transversion Bias Is Not Universal: A Counter Example from Grasshopper Pseudogenes." *PLoS Genetics* 3 (2): e22.

King, J. L. 1971. "The Influence of the Genetic Code on Protein Evolution." In *Biochemical Evolution and the Origin of Life*, edited by E. Schoffeniels, 3–13. Viers: North-Holland Publishing Company.

King, J. L., and T. H. Jukes. 1969. "Non-Darwinian Evolution." *Science* 164:788–797.

Knight, R. D., S. J. Freeland, and L. F. Landweber. 2001. "A Simple Model Based on Mutation and Selection Explains Trends in Codon and Amino-Acid Usage and GC Composition within and across Genomes." *Genome Biology* 2 (4): research0010.1—research0010.13.

Lafay, B., A. T. Lloyd, M. J. McLean, K. M. Devine, P. M. Sharp, and K. H. Wolfe. 1999. "Proteome Composition and Codon Usage in Spirochaetes: Species-specific and DNA Strand-specific Mutational Biases." *Nucleic Acids Research* 27 (7): 1642–1649.

Lande, R., and S. J. Arnold. 1983. "The Measurement of Selection on Correlated Characters." *Evolution* 37 (6): 1210–1226.

Lewontin, R. C. 1974. *The Genetic Basis of Evolutionary Change*. New York: Columbia University Press.

Lewontin, R. C., and K. Kojima. 1960. "The Evolutionary Dynamics of Complex Polymorphisms." *Evolution* 14:458–472.

Li, W.-H., C.-C. Luo, and C.-I. Wu. 1985. "Evolution of DNA Sequences." In *Molecular Evolutionary Genetics*, edited by R. J. MacIntyre. New York: Plenum.

Lynch, M. 2007. "The Frailty of Adaptive Hypotheses for the Origins of Organismal Complexity." *Proceedings of the National Academy of Sciences USA* 104 Suppl 1: 8597–8604.

Maynard Smith, J., R. Burian, S. Kauffman, P. Alberch, J. Campbell, B. Goodwin, R. Lande, D. Raup, and L. Wolpert. 1985. "Developmental Constraints and Evolution." *Quarterly Review of Biology* 60 (3): 265–287.

Mayr, E. 1959. "Darwin and the Evolutionary Theory in Biology." In *Evolution and Anthropology: A Centennial Appraisal*, 1–10. Washington, DC: Anthropological Society.

Mayr, E. 1960. "The Emergence of Evolutionary Novelties." In *Evolution after Darwin: The University of Chicago Centennial*, edited by S. Tax and C. Callender, 349–380. Chicago: University of Chicago Press.

Mayr, E. 1963. *Animal Species and Evolution*. Cambridge, MA: Harvard University Press.

McCandlish, D. M., and A. Stoltzfus. 2014. "Modeling Evolution Using the Probability of Fixation: History and Implications." *Quarterly Review of Biology* 89 (3): 225–252.

McGuigan, K. 2006. "Studying Phenotypic Evolution Using Multivariate Quantitative Genetics." *Molecular Ecology* 15 (4): 883–896. doi: 10.1111/j.1365–294X.2006.02809.x.

Meyer, J. R., D. T. Dobias, J. S. Weitz, J. E. Barrick, R. T. Quick, and R. E. Lenski. 2012. "Repeatability and Contingency in the Evolution of a Key Innovation in Phage Lambda." *Science* 335 (6067): 428–432. doi: 10.1126/science.1214449.

Morgan, T. H. 1916. *A Critique of the Theory of Evolution*. Princeton, NJ: Princeton University Press.

Popov, Igor. 2009. "The Problem of Constraints on Variation, from Darwin to the Present." *Ludus Vitalis* 17 (32): 201–220.

Provine, W. B. 1971. *The Origins of Theoretical Population Genetics*. Chicago: University of Chicago Press.

Provine, W. B. 1978. "The Role of Mathematical Population Geneticists in the Evolutionary Synthesis of the 1930s and 1940s." *Studies in History of Biology* 2:167–192.

Reeve, H. K., and P. W. Sherman. 1993. "Adaptation and the Goals of Evolutionary Research." *Quarterly Review of Biology* 68 (1): 1–32.

Rokyta, D. R., P. Joyce, S. B. Caudle, and H. A. Wichman. 2005. "An Empirical Test of the Mutational Landscape Model of Adaptation Using a Single-Stranded DNA Virus." *Nature Genetics* 37 (4): 441–444.

Rosenberg, M. S., S. Subramanian, and S. Kumar. 2003. "Patterns of Transitional Mutation Biases within and among Mammalian Genomes." *Molecular Biology and Evolution* 20 (6): 988–993.

Sackman, A. M., L. W. McGee, A. J. Morrison, J. Pierce, J. Anisman, H. Hamilton, S. Sanderbeck, C. Newman, and D. R. Rokyta. 2017. "Mutation-Driven Parallel Evolution during Viral Adaptation." *Molecular Biology and Evolution*. doi: 10.1093/molbev/msx257.

Schluter, D. 1996. "Adaptive Radiation along Genetic Lines of Least Resistance." *Evolution* 50 (5): 1766–1774.

Shull, A. F. 1936. *Evolution*. New York: McGraw-Hill.

Simpson, G. G. 1953. *The Major Features of Evolution*. New York: Simon & Schuster.

Simpson, G. G. 1967. *The Meaning of Evolution*. 2nd ed. New Haven, CT: Yale University Press.

Singer, G. A., and D. A. Hickey. 2000. "Nucleotide Bias Causes a Genomewide Bias in the Amino Acid Composition of Proteins." *Molecular Biology and Evolution* 17 (11): 1581–1588.

Sober, E. 1987. "What Is Adaptationism?" In *The Latest on the Best*, edited by J. Dupré, 105–118. Cambridge, MA: MIT Press.

Stebbins, G. L. 1959. "The Synthetic Approach to Problems of Organic Evolution." *Cold Spring Harbor Symposium of Quantitative Biology* 24:305–311.

Stebbins, G. L. 1966. *Processes of Organic Evolution*. Englewood Cliffs, NJ: Prentice-Hall.

Steppan, S. J., P. C. Phillips, and D. Houle. 2002. "Comparative Quantitative Genetics: Evolution of the G Matrix." *Trends in Ecology and Evolution* 17 (7): 320–327.

Stoltzfus, A. 2006. "Mutation-Biased Adaptation in a Protein NK Model." *Molecular Biology and Evolution* 23 (10): 1852–1862.

Stoltzfus, A. 2012. "Constructive Neutral Evolution: Exploring Evolutionary Theory's Curious Disconnect." *Biology Direct* 7 (1): 35. doi: 10.1186/1745-6150-7-35.

Stoltzfus, A. 2017. "Why We Don't Want Another 'Synthesis.'" *Biology Direct* 12 (1): 23. doi: 10.1186/s13062-017-0194-1.

Stoltzfus, A., and D. M. McCandlish. 2015. "Mutation-Biased Adaptation in Andean House Wrens." *Proceedings of the National Academy of Sciences USA* 112 (45): 13753–13754. doi: 10.1073/pnas.1518490112.

Stoltzfus, A., and D. M. McCandlish. 2017. "Mutational Biases Influence Parallel Adaptation." *Molecular Biology and Evolution*. doi: 10.1093/molbev/msx180.

Stoltzfus, A., and R. W. Norris. 2016. "On the Causes of Evolutionary Transition: Transversion Bias." *Molecular Biology and Evolution* 33 (3): 595–602. doi: 10.1093/molbev/msv274.

Stoltzfus, A., and L. Y. Yampolsky. 2009. "Climbing Mount Probable: Mutation as a Cause of Nonrandomness in Evolution." *Journal of Heredity* 100 (5): 637–647.

Sueoka, N. 1988. "Directional Mutation Pressure and Neutral Molecular Evolution." *Proceedings of the National Academy of Sciences USA* 85 (8): 2653–2657.

Wakeley, J. 1996. "The Excess of Transitions among Nucleotide Substitutions: New Methods of Estimating Transition Bias Underscore Its Significance." *TREE* 11 (4): 158–162.

Weissman, D. B., M. M. Desai, D. S. Fisher, and M. W. Feldman. 2009. "The Rate at which Asexual Populations Cross Fitness Valleys." *Theoretical Population Biology* 75 (4): 286–300. doi: 10.1016/j.tpb.2009.02.006.

Wolfe, K. H. 1991. "Mammalian DNA Replication: Mutation Biases and the Mutation Rate." *Journal of Theoretical Biology* 149 (4): 441–451.

Wright, S. 1931. "Evolution in Mendelian Populations." *Genetics* 16:97.

Yampolsky, L. Y., and A. Stoltzfus. 2001. "Bias in the Introduction of Variation as an Orienting Factor in Evolution." *Evolution and Development* 3 (2): 73–83.

Yedid, G., and G. Bell. 2002. "Macroevolution Simulated with Autonomously Replicating Computer Programs." *Nature* 420 (6917): 810–812.

4 The Shape of Things to Come: *Evo Devo* Perspectives on Causes and Consequences in Evolution

Armin P. Moczek

Introduction

When I was a young and innocent postdoctoral researcher hunting for the elusive tenure track position, I would begin my seminars by briefly highlighting the major contributions made by alternative frameworks in conceptualizing what matters in directing evolution. Showing a slide of Darwin's finches, I would emphasize the role of adaptive evolution and the view of organisms as Swiss army knives—accumulations of gadgets, each with a specific function honed over time. Showing a drawing of a prehistoric small mammal gnawing on a dinosaur carcass, I would highlight the role of chance and accidents in the diversification of life on earth. And lastly I would show a drawing of a bird embryo squeezed to its limits within an egg and emphasize the role of developmental constraints in determining where and where not evolution may be allowed to go. I would close this introduction with three major conclusions: First, all three of these perspectives have been incredibly illuminating. Second, they are not mutually exclusive. Third, they are all roughly equally useless when it comes to understanding the origins of novelty in evolution because selection cannot select for traits that do not yet exist, accidents can only sort among preexisting variation, and constraints only limit options, but by themselves do not create new ones (Moczek 2008). Instead I would posit that how the origin of novelty can be integrated within a framework of descent with modification, how novel complex traits may originate from within the confines of ancestral variation, the baby steps of innovation needed to eventually yield the first limb, wing, eye, feather, photic organ, and so on all remain remarkably poorly understood in spite of over 150 years of vibrant evolutionary biology since the publication of the *Origin of Species* (Darwin 1859). And then I would say that my research program addresses these shortcomings by integrating the role of development into our understanding of speed and direction in organismal evolution, in particular in the origins of novelty, and that I will finally resolve this long-standing, foundational challenge to evolutionary biology. And that you really should strongly consider hiring me.

About 15 years have passed since and it is appropriate that I look back and assess where we stand—the larger field of evolutionary developmental biology (*evo devo*) in general, and my own research program as one of its many representatives—with respect to our abilities to contribute meaningfully to our current understanding of the evolutionary process, in particular with respect to challenges for which previous approaches and schools of thought have struggled to find resolution, such as innovation and the origins of novelty. Specifically, in this chapter I will begin by reviewing a few key terms and concepts, and the creative tensions between them, that will be critical for subsequent discussions. Secondly, drawing from the work of others as well as my own, I will then highlight examples that illustrate how, on what levels of biological organization, and on what level of causation, an understanding of how organisms build themselves enriches our understanding of how and why they evolve the way they do. Lastly, I will discuss some of the challenges that remain, and in particular conceptual challenges *evo devo* is now itself encountering, and opportunities for their resolution. Let us set the stage, however, with a brief discussion of some of the conceptual constellations that led to the birth of *evo devo* as a discipline, and which motivate many of its practitioners, including myself.

What Can an Evolutionary Biologist Possibly Learn from Studying Development?

This was the question I would hear during every single meeting I had with my dissertation committee while in graduate school. It was always posed by the same faculty member who shall remain unnamed, a highly accomplished evolutionary biologist and population geneticist, who to this day I respect very much, and who posed his question not to tease me: instead, he truly did not understand why anyone interested in understanding evolution's paths would bother learning about how development works. In this of course he was and is not alone—it is a mindset that scientists like myself encounter to this day. *Evo devo*'s contributions to our understanding of the evolutionary process are often either considered modest at best, or alternatively, thought of as not really evolutionary in nature. So how did we get to this point?

Evolutionary biology as a discipline first emerged in the first half of the twentieth century by integrating natural selection and Mendelian inheritance into the then-coalescing framework of population genetics (Mayr 1982). In the decades that followed, evolutionary biology continued to expand and mature into a highly sophisticated and successful framework able to address a broad range of biological phenomena. In the process, several key concepts and dichotomies became deeply engrained in how we conceptualize organismal evolution in research and how it is taught in our courses: most importantly, we grew to understand phenotypes as rooted in genes and genomes, and as long as phenotypic variation could be associated with genetic variation in some way, this enabled the opportunity to conceptualize, and thereby equate, *phenotypic* evolution as a change in the *genotypic* composition of a population over time (Laland et al. 2015). Doing so removed the need to understand exactly how genotypic information and variation manifested in phenotypes and phenotypic variation. As a by-product, this then removed any need to understand how organisms are built during development from understanding how they evolve over time. The resulting quantitative framework enabled important advances in our understanding of the nature of diverse evolutionary phenomena, though a subset of challenges stubbornly resisted resolution, including as already noted the questions surrounding the origins of novel complex traits and the corresponding major transitions and radiations they enabled. Given their entrenchment in deep time, and the lack of phenotypic variation accessible to quantitative and population-genetic approaches, standard evolutionary biology struggled to generate satisfactory answers and had stopped trying by the time I had entered graduate school. The resulting disconnect between what I considered to be among the most important questions in evolution, and the tools I was being taught in my upper-level graduate evolution classes, caused me to look elsewhere, and the then rapidly transforming field of *evo devo* very quickly revealed to me novel ways to both conceptualize, and empirically interrogate, the nature of innovation in evolution.

What Is an Evolutionary Novelty?

Prior to the advent of *evo devo*, Ernst Mayr (1960) defined novelty as "any newly acquired structure or property that permits the assumption of a new function," which parallels corresponding statements as far back as Lamarck and Darwin, holds intuitive appeal, yet runs into trouble when we try to use it to derive hypotheses regarding how novelties might originate: selection can only act on traits that already exist, but if they already exist in some shape or form they are no longer exactly novel. Something else is needed to account for the initiation of novelty independent of future functionality. A second definition was proposed by Müller (1990), who defined novelty as "a qualitatively new structure with a discontinuous origin, marking a relatively abrupt deviation from the ancestral condition." This definition remained neutral regarding functionality, but left it up for interpretation where quantitative variation ends and qualitative distinctness begins. How different is novel? This is where a third definition, proposed by Müller and Wagner (1991), stepped in to provide what seemed like an iron-clad cutoff: "A morphological novelty is a structure that is neither homologous to any structure in the ancestral species or homonomous to any other structure in the same organism." Now novelty began where homology ended. But where does homology end? Traditionally, and as taught in the introductory zoology classes I took in Germany in the early 1990s before moving to the United States, classic homology criteria included relative position, intermediate forms, and the all-encompassing special qualities, and they neatly dichotomize traits into those that are homologous and those that are not (Remane 1952). *Evo devo* very swiftly forced a revision of this framework into a far more complex, nuanced, and layered understanding of homology, for two major reasons. First, it forever rejected the notion that the extraordinary phenotypic diversity that exists on the level of organisms must somehow be paralleled by a corresponding diversity in genetic and developmental mechanisms. Instead, researchers now recognize that the developmental genetic underpinnings of phenotypic diversity are remarkably conserved, and that highly divergent organisms rely on much of the same developmental mechanisms to instruct the building of very different, and by conventional criteria clearly non-homologous, organs and structures (reviewed in, for example, Shubin, Tabin, and Carroll 2009; Held 2017; and see below for concrete examples). And we had to recognize the existence of the *opposite* constellation as well: traits that by conventional criteria are clearly and unambiguously homologous may form during development in surprisingly different, *non-homologous* ways, a phenomenon we now recognize as developmental systems drift or phenogenetic drift (Weiss and Fullerton 2000; True and Haag 2001). *Evo devo* thus forced a transformation of our understanding of homology away from a neat and discrete black and white to layered set of shades of gray (Wagner 2014). At the same time, as we will see next, it brought us significantly closer to understanding the origins of novelty in ways no previous discipline had been able to achieve.

Novelty and Diversity from the Confines of Ancestral Variation

(a)   Cooption, Parallelism, and the Modular Nature of Development

By discovering the remarkable conservation of developmental building blocks and processes that characterizes phenotypic diversity, *evo devo* forced a view of organismal diversity akin to that of Lego creations: rather than being shaped primarily or solely by adaptive responses to selection pressures, diverse organisms emerged as the modified re-assemblages of the same and seemingly very limited pool of genes, developmental pathways, and morphogenetic processes. Clearly, natural selection remained a leading force in the creation of organism–environment fit, but one that suddenly had to draw from a heavily restricted pool of resources from which to generate diversity. No surprise that many embraced the new *evo devo* findings pouring in during the 1990s as a reflection of overwhelming evidence for *developmental constraint* on what evolution might otherwise be able to accomplish. If diversification seemed heavily constrained, then how anything *novel* could ever emerge in the process was anyone's guess. While I sense that many evolutionary biologists looking in on the discipline of *evo devo* have retained the perspective to this day that the best understanding development can do for an evolutionary biologist is to understand *the limits it imposes on diversification*, *evo devo* itself managed to move on, in large part because of the realization that what may act as a constraint in one context may provide critical opportunities in others, and that diversity and novelty may have evolved not *in spite* of the deep homology of genes, pathways, and processes across phyla, but *because of it*. Two examples will help to illuminate this perspective.

The first concerns the eyes of vertebrates, insects, mollusks, or jellyfish—morphologically distinct structures that arise in disparate embryological contexts (reviewed in Shubin et al.

2009). Eyes are thus assumed to have originated independently many times in different phyla, and traditionally they have been interpreted as representing remarkable cases of convergent evolution. Thanks to comparative developmental studies, however, we now know that even across phyla homologous transcription factors help specify and organize eye formation, while homologous opsin proteins provide the light-absorbing properties needed to convert incoming photons into outgoing action potentials (Oakley 2003). Further, the interneurons involved in processing this visual input—that is, the optic lobe of *Drosophila* and the retina of vertebrates—while only analogous in function, are both specified by homologous transcription factors, both project to visual centers of their respective brains, and both subsequently express yet again homologous transcription factors to aid in further integration of signal processing (Erclik et al. 2008; Shubin et al. 2009). Collectively, these data suggest that rather than being simply the product of convergent evolution, eye diversity across phyla may more appropriately be understood as reflecting parallelisms, enabling the evolution of lineage-specific eyes by utilizing the same ancestral toolbox of patterning mechanisms, visual pigments, cell types, and cellular circuitry.

The second example concerns appendage formation. While the legs of flies and mice, the tube feed of echinoderms, the siphons of ascidians, or the horns of scarab beetles have very little in common in terms of strict homology, they all share that they are outgrowths whose development requires some kind of mechanisms specifying which cells will adopt a distal fate, and which will be proximal, anchoring the outgrowth in the remainder of the body. We now know that to do so, all of them, despite the enormity of phylogenetic distance among them, share the use of the same genetic regulatory cascade including the use of a key transcription factor (e.g., *Distal-less* in insects, *Dlx* in vertebrates) and shared downstream effector genes, a machinery likely already in existence in the earliest bilaterians (Panganiban et al. 1997; Mercader et al. 1999; Moczek et al. 2006; Moczek and Rose 2009). As before, parts of an ancient developmental toolbox became reused over and over again, independently in very different lineages, in this case to enable the formation of various kinds of outgrowths, limbs, and appendages.

Collectively, these and by now hundreds of similar examples illustrate, on one side, the modular nature of developmental processes: a module exists that specifies proximo-distal polarity and will do so regardless of exact context, just like a module exists that can pattern the formation of cell types that can be assembled into light sensors. Secondly, what may have originally been perceived as a constraint, because it seemed there is developmentally only one way to make an eye or a limb, may now be simultaneously recognized as a set of preexisting opportunities in disguise. Put another way, while there may indeed only be one way to specify proximo-distal identity in development, if a lineage ever needed to add that ability to whatever novel context, it was already endowed with a preexisting developmental toolbox ready to do just that. If light perception offered new ecological opportunities, the developmental genetic properties necessary for the specification of some type of light-sensing organ did not need to be evolved *de novo*, instead it already lay in wait in every bilaterian lineage. This creative ability of development to focus evolution along specific productive directions becomes further apparent if we consider the major avenues along which modest developmental modifications enable the modular nature of development to yield remarkable phenotypic diversity and novelty, as discussed next.

(b) Facilitating Developmental Evolution along Four Dimensions: The Power of "the Four Hs"

The modular nature of development allows developmental units, from genetic pathways to morphogenetic processes, to be executed semi-independently from other modules. The four Hs—heterotopy, heterochrony, heterometry, and heterocyberny—all illustrate that no new modules are needed to generate profound diversity via relatively simple changes in the execution of modules relative to each other.

(i) Heterotopy refers to an evolutionary change in the precise location of a developmental event. No *new* modules are needed; instead all that is altered is the ontogenetic location of their actions relative to each other. Executing "old" modules in novel spatial contexts is now recognized as yielding some of the most spectacular examples of innovation in evolution: For example, the recruitment of appendage patterning genes ancestrally used to establish the proximodistal axis of arthropod legs, antennae, and mouthparts into the dorsal head and pronotum of beetles represented key contributions to the origin and subsequent rapid diversification of beetle horns now used in male combat (Moczek et al. 2006; Moczek and Rose 2009). Recruitment of some of the same genes onto the wing surfaces of butterflies facilitated the developmental evolution of wing spots, now used in mate choice and predator avoidance (Brakefield et al. 1996). Heterotopy can change—not just add parts to where they did not exist before; it can also contribute old functions to newly positioned traits: for example, cooption of signaling via the *hedgehog* and *doublesex* pathways now allows beetle horns to adjust their development in a highly nutrition responsive manner, yielding alternative male morphologies with species as a consequence (reviewed in Moczek and Kijimoto 2014; Casasa, Schwab, and Moczek 2017). In all of these cases the Lego analogy of organismal development and evolution is perhaps most apt and obvious: novel traits and functions may arise with ease by placing old building blocks in new places. However, additional routes to novelty exist, perhaps more subtle in nature, but potentially just as consequential.

(ii) Heterochrony refers to a change in the timing of developmental events relative to each other from one generation to the next. Heterochrony can occur on any level of biological organization, from gene regulation to behavior. All heterochronies share, however, that, as with heterotopy, no new events are introduced; instead, all that is altered is the ontogenetic timing of their actions relative to each other. The consequence of heterochrony range from subtle and quantitative to profound and qualitative. For example, bats and kiwis arrive at relatively large and smaller forelimbs than other quadrupeds by advancing and delaying the onset of forelimb growth, respectively (Richardson 1999). The forelimbs of

marsupials also develop earlier compared to placental mammals, enabling the marsupial embryo to climb into the pouch and suckle (Smith 2003; Sears 2004). Heterochrony can change not just the relative sizes of parts but also their number: for example, in the lizard genus *Hemiergis*, the duration of *sonic hedgehog (shh)* activation within the limb bud determines whether the bud gives rise to 3, 4, or 5 digits (Shapiro, Hanken, and Rosenthal 2003).

Heterochrony's consequences, however, can also be truly profound: A dramatic example is the origin of holometabolous development in insects, as seen in butterflies, moths, beetles, or bees, which develop from an embryo to an immature larva, which undergoes several larval-to-larval molts before molting into a pupa, and ultimately an adult. The pupal stage thus effectively decouples the larval from the adult stage and is credited with having allowed larval holometabolous insects to utilize feeding niches otherwise unavailable to other insects, culminating in the dramatic divergence in form, physiology, and behavior between larval and adult stages of extant holometabolous insects. This is in contrast to hemimetabolous development as observed in, for example, grasshoppers or cockroaches, in which embryos molt via a brief pronymph stage into a nymphal stage, which in many ways resembles a miniature and incomplete version of the final adult. Through a series of nymphal-to-nymphal molts animals grow in size, culminating in a final nymphal-to-adult molt (Gullan and Cranston 2014). The most widely held hypothesis regarding the origin of holometabolous development postulates that holometabolous larvae are homologous to hemimetabolous nymphs, and that the origin of holometabolous metamorphosis was made possible through the invention of the pupal stage which, consequently, lacks a homologous counterpart among the Hemimetabola. Recent work challenges this long-standing view and argues that the holometabolous pupa instead arose from a *compacting* of the nymphal stages into a single life stage, thus making pupal and nymphal stages homologous (Truman and Riddiford 1999). The holometabolous larvae in turn arose as an *elaboration* of the hemimetabolous pronymph stage. The pronymphal stage of hemimetabolous insects is a distinct stage directly following the embryo, but it is so brief and ephemeral that it is spent entirely to largely while the animal is still inside the egg. Compelling evidence now exists that supports the hypothesis that the holometabolous larva may indeed have arisen through a "de-embryonization" of the pronymph stage, converting a largely embryonic stage into a free-living larva. The hemimetabolous nymphal stages, in turn, collapsed into what we now recognize as the holometabolous pupa. Consequently, a three-part life cycle already existed prior to the origin of holometabolous development, which instead arose via heterochronic changes in the endocrine regulation of growth and molting. If correct, no origins of a new stage are needed, instead heterochronic modifications of preexisting stages may have sufficed to fuel the single-most important developmental transition in insect evolution.

(iii) Heterometry is commonly defined as an evolutionary change in the amount of a gene product but could also be applied to higher order developmental products, such as hormone titers. Like heterochrony, it seems like a subtle change which in turn should only

have subtle consequences. One of the best examples illustrating the opposite is the developmental evolution of Darwin's finches, via heterometric changes in two genes, bone morphogenetic protein 4 (Bmp4) and calmodulin (cal), which both encode proteins that through different routes promote cell division and thus tissue growth (Abzhanov et al. 2004, 2006). Experimental and modeling work shows that evolved changes in the expression levels of both genes are sufficient to explain the diversity of beak shapes among Darwin's finches, and that experimental induction of a subset of these changes in chick embryos results in matching changes in beak formation (Wu et al. 2004, 2006).

(iv) Heterocyberny, lastly, is a term few seem to use, but it nevertheless illustrates an important concept: an evolutionary change in *governance*, that is a change in the upstream regulation of a conserved downstream process (Gilbert and Epel 2009). Used most broadly, it refers to the process whereby initially environmentally induced traits may over generations become genetically stabilized and incorporated into lineage's norm of reaction. We will return to this broad notion of heterocyberny toward the end of this chapter; for now I want to emphasize that evolutionary changes in upstream regulation can of course also occur on many other levels. As before, no new modules or building blocks need to be introduced, instead both up and down-stream components already exist; all that changes is the nature of interaction between them. Evolutionary developmental geneticists now broadly recognize, for example, the ease with which transcription factors acquire novel targets, even in traits that themselves constitute relatively recent evolutionary inventions: for example, the somatic sex-determination gene *doublesex (dsx)* regulates the relative size and sex-specific expression of evolutionary novel beetle horns, just like it regulates the same features of much more ancient traits, such as genitalia. Yet in horns it acts on a largely non-overlapping repertoire of target genes (compared to genitalia or brains), suggesting that both heterotopic recruitment of novel regulators (such as dsx) and novel target genes into their governance can occur with surprising ease (Ledon-Rettig, Zattara, and Moczek 2017).

In summary, heterotopy, heterochrony, heterometry, and heterocyberny all illustrate that much diversification and innovation may be possible without the need to generate new genes, pathways, or cell fates. Instead, phenotypic diversity emerges through heritable changes channeled along four simple developmental axes—developmental time, developmental location, quantity of developmental product, and nature of regulatory interactions. Lastly, combinations of these four processes operating sequentially or at the same time have the power to further potentiate the developmental degrees of freedom available for rapid evolutionary diversification and innovation.

One may ask then which mechanisms in turn enable developmental processes to be so modular in developmental time, space, upstream regulation and downstream output? The reasons for this derive to a significant degree from the fact that the mechanisms in question are themselves highly modular, and on a variety of levels (reviewed in Carroll, Grenier, and Weatherbee 2004; Gerhart and Kirschner 2010; Gilbert 2013): for example, cellular transduction pathways convert signals external to a cell, such as information on nutrient availability or position, into signals that enter the nucleus and affect gene expression. Cellular transduction of information primarily takes the form of on/off switches, which are in operation all the time during the life of a cell, in all cells and tissues, developmental stages, and in response to a remarkable diversity of external cues, soliciting a corresponding diversity of intracellular responses. Collectively, this diversity of regulatory decisions reliant upon signal transduction pathways is mindboggling, especially when juxtaposed to the comparatively minute number of transduction pathways that facilitate it. All of those share that they are ancient, predating the traits or processes they regulate in extant organisms by at times billions of years. They also share remarkable degrees of conservation across phyla, and an exquisitely fine-tuned, robust, and reliable nature of interactions among their respective component parts. And lastly, they all share that their modular and combinatorial re-use across diverse contexts facilitates precise developmental decision-making, yet without having to evolve a comparable diversity of switch mechanisms (Gerhart and Kirschner 2010). Transcription factors by themselves, too, contribute modularity, most strikingly through their highly combinatorial action in regulating gene expression. Precise combinations of transcription factors are needed to drive gene expression in specific contexts, and subtle changes in the timing or location of a single transcription factor may suffice to generate heterochronic or -topic developmental changes, without resulting in negative developmental consequences in other aspects of phenotype formation elsewhere, and without the need to evolve new factors for new developmental decisions (Carroll et al. 2004). Lastly, cis-regulatory elements, or CREs, are those genomic regions transcription factors bind to regulate gene expression. CREs are themselves highly modular, with different CREs enabling different facets of a given gene's expression, such as expression in specific locations, developmental stages, cell types, and so on. The highly modular nature of CREs then allows each facet to be regulated—and to evolve—semi-independently, again minimizing pleiotropic constraints (Prud'homme, Gompel, and Carroll 2007). As before, by relying on a preexisting and clearly finite arsenal of building blocks, in this case signal transduction pathways, other transcriptional regulators, and the respective DNA binding sites they interact with, developmental systems are able to generate diverse and novel regulatory settings without having to generate novel regulatory machineries.

On different levels of biological organization we thus see how the nature of developmental processes establishes the degrees of freedom along which development evolution may proceed more easily than others, putting us in a position to understand not just the developmental basis of evolutionary changes but also the creative potential development possesses to facilitate diversity, and doing so not despite the conservation of its building blocks, but because of it. In our quest to understand why evolution unfolds the way it does and not some other way, a comparative developmental perspective thus makes a unique contribution, one no other framework can provide. I would like to close this section by highlighting two concrete case studies to fully illustrate the power and promise of this approach.

Creating a Landscape for Channeling Innovation and Diversification—Two Case Studies

(a)   The Evolution of *Drosophila* Wing Pigmentation Patterns

In an important series of studies, Prud'homme and colleagues (reviewed in Prud'homme et al. 2007; see also Wittkopp, Carroll, and Kopp 2003; Gompel et al. 2005; Prud'homme et al. 2006) explored how the developmental architecture that governs the patterning of the fly wing has channeled the subsequent evolution of wing pigmentation patterns. A large body of work has documented the patterning mechanisms that help establish different aspects of the *Drosophila* wing, such as compartmentalization into an anterior and posterior portion, the identity and location of veins, patterning of the distal margin, and so on. All of this is accomplished through transcription factors whose localized expression and action in conjunction with cofactors impart location-specific developmental fates. Collectively, this establishes what is often referred to as the transregulatory landscape of the wing, in existence at a time when most of the wing has grown, and in particular, when in a subset of cycclorrhaphan flies (to which *Drosophila* belongs) pigment synthesis begins and ultimately results in specific, highly diversified wing patterns.

Wing pigmentation patterns result from the conversion of precursor metabolites into visible pigments via the action of specific enzymes operating in a locally restricted manner along the two-dimensional surface of the wings, and as such constitute a developmentally and experimentally very tractable trait. Prud'homme and colleagues were able to show that even though pigment synthesis is a developmental process completely independent of that of wing formation and patterning, the evolution of *pigmentation patterns* on the wing nevertheless integrated both processes in a way that informs our understanding of how the nature of developmental processes both facilitates and biases evolutionary routes to novelty (Wittkopp et al. 2003; Gompel et al. 2005; Prud'homme et al. 2006). Specifically, this work showed that redeployment of preexisting transcription factors and their location-specific expression on the wing into the regulation of pigment-production facilitated the location-specific expression of pigmentation patterns. Recruitment of *different combinations of transcription factors* in different lineages then facilitated the corresponding lineage-specific divergence of pigmentation patterns, but one that was at least in part predictable by the transregulatory landscape already in existence on the wing prior to origins of the first pigment spot.

This work led Wagner and Lynch (2008) to postulate the *Christmas tree model of morphological evolution*, which equates the tree's branches as a reflection of the (preexisting) regulatory landscape of a trait or organism, characterized by the spatio-temporal distribution of transcription factor expression domains. Novel traits, or ornaments in the context of the model, can then be added most easily to existing branches. Or put into fly wing context: pigmentation pattern diversity is preconfigured by the combinatorial possibilities of this regulatory landscape (Wagner and Lynch 2008). This model explains not just the patterns that were able to evolve in different lineages but also the many examples of parallel pattern evolution, the ease with which some patterns can be lost and (re)gained, and so on. More generally it paints a picture of developmental evolution that converts a constraint into scaffolding for novel diversity, a notion we have by now encountered on several levels in this chapter. At the same time, it assumes a specific polarity: transcription factors create a regulatory landscape, and subsequent innovation and diversification are then shaped by this landscape. Or: branches specify locations for new ornaments, rather than the other way around. Work since suggests, however, that innovation itself, once successful, may well create novel regulatory landscapes, or in the language of the Christmas tree model, that the branches of the tree do not just provide opportunities for new ornaments but respond to new ornaments by growing into previously unoccupied space. Our last example seeks to illuminate this perspective.

(b) The Regulatory Landscape Is Not Static: Innovation on the Dorsal Head of Insects

From the stalks of stalk-eyed flies and the weevil rostrum to the cephalic horns of dung beetles, the dorsal head of adult insects has emerged as an evolutionary hotspot for innovation and diversification (Grimaldi and Engel 2005). At the same time, the insect head has been in existence for at least 420 million years, and the developmental genetic network that patterns head formation is even older, manifest in a remarkable degree of conservation across phyla. Recent work on the horned beetle genus *Onthophagus* has documented that the first position for head horn formation that evolved, and the one now most commonly observed in extant taxa, coincides with the boundary between the clypeolabral and ocular head segments (Busey, Zattara, and Moczek 2016). These segments are first specified during embryonic development, but the mechanisms that specify the boundary between them appear to have been repurposed much later in late larval and pupal development to provide positional information for where horns are to be integrated within the future adult head. Other horn positions also evolved, but did so much more recently and are thus found in far fewer extant species. Up to this point this narrative matches what we have seen thus far—a preexisting regulatory landscape channels novelty—in this case horns used as weapons in male combat— down specific evolutionary avenues. But subsequent studies paint a more complex picture.

Studies on embryonic head development in *Tribolium* beetles implicate the interplay between two transcription factors, *six3/optix* and *orthodenticle (otd)* in establishing the clypeolabral-ocular segment boundary (Posnien et al. 2010, 2011). Both genes play critical roles in embryonic head formation in all bilaterians studied, yet their roles in postembryonic development (e.g., during larval, pupal, and adult development) are much less well known. One major exception constitutes *otd*, which in *Drosophila* plays a critical role in late development through promoting the development of ocelli, three single-lens eyes positioned along the posterior midline of the dorsal adult head (Blanco et al. 2009). Following *otd*-inactivation, these ocelli no longer form. Most insect orders possess ocelli, though it is presently not known whether ocellar development is also under the control of *otd* in these other orders. What is known, however, is that almost all beetles have secondarily *lost* ocelli. Further, experimental down-regulation of *otd*, while lethal in *Tribolium* embryos, has no phenotypic consequences during the formation of the adult dorsal head of the same species (Zattara et al. 2016). Even though *otd* is expressed during adult head formation in *Tribolium*, this expression appears functionless. So far so good.

However, similar experiments in the horned scarab beetle genus *Onthophagus* yielded completely different outcomes. Here, *otd* emerged as absolutely critical for the proper patterning of the dorsal head, including the positioning of cephalic horns. Further, while its function in embryonic development is intimately tied to that of *six3*, that interdependence no longer exists at later developmental stages. It is tempting to speculate that *otd* may have been freed up to evolve this novel function because of the putative secondary loss of its role in regulating ocelli formation in the same part of the dorsal head, prior to the evolution of the first horns. Alternatively, *otd* expression in adult beetle heads may simply be an embryonic leftover, which *Onthophagus* capitalized upon and recruited into the context of horn formation. Regardless of how *otd* arrived at its novel role in *Onthophagus* development, the most important finding, however, is that things did not stop there. Instead, *otd*-specified horn-bearing head regions acquired all sorts of additional functions via the recruitment of a secondary set of pathways: for example, recruitment of the somatic sex-determination pathway enabled horns to be expressed solely in males and exaggerated under high nutrition, facilitating the evolution of both sexual dimorphisms and highly positive allometries in males only (Kijimoto, Moczek, and Andrews 2012). Further recruitment of the hedgehog-signaling pathway enabled the evolution of a complementary function—active suppression of horns under low nutrition only—enabling the evolution of alternative-horned and hornless male phenotypes cued by nutritional conditions experiences as larvae (Kijimoto and Moczek 2016). Additional pathways include signaling via the insulin and serotonin pathways, again pathways that as best as we know play no role in dorsal epidermal head development in insects, yet appear to have been recruited into this context once the opportunity to operate on a new module—horn-forming head regions—existed (Casasa et al. 2017). More generally, these observations suggest that the Christmas tree model of morphological evolution may need to be replaced perhaps by a Romanesco broccoli model, where each addition of a new ornament begets the fractal-like addition of a new branch, a new whirl, offering yet more opportunities for subsequent ornamentation.

What *Causes* Does *Evo Devo* Contribute?

This chapter was meant to explore the contributions made by *evo devo* to our understanding for why and how evolution unfolds the way it does. It is easy to get lost in the details and idiosyncrasies of the many case studies of developmental evolution, and thus worth to step back and ask: what *causes* in evolution does *evo devo* contribute that may not have been considered prior to the existence of the field? What here is truly new?

For starters, *evo devo* more than any other discipline offers a mechanistically concrete understanding of how traits come into being, and how the underlying processes have to be modified to yield novel trait variants. It is one thing to associate trait variation with allelic variation in a population or to map quantitative trait loci to disease phenotypes. It is a much deeper explanation to also understand how some genetic variants but not others result, for instance, in altered DNA binding of proteins, the circumstances under which this may facilitate the expression of old gene products in new locations or developmental time points, yield the corresponding induction of cellular differentiation events and respective organ formation, and so on. As such *evo devo* offers richer, perhaps more satisfying explanations of what mechanistic causes underlie evolutionary changes. But does it also offer qualitatively novel causes previously unconsidered?

I would posit that it does, on at least two levels. First, *evo devo* offers what has been called a lineage explanation (Calcott 2009) for biological diversity, including novelty. In other words, it offers the opportunity to understand developmental evolution and innovation as a sequence of events, where one event is needed to enable another to take place, leading eventually to a final outcome. Because the traits of most interest to *evo devo* practitioners are typically complex, lineage explanations that reside solely on the level of DNA sequences are insufficient to understand how developmental evolution transitioned from one state to the next: even though DNA sequence changes are of course an integral component of such an explanation it takes an understanding of how form comes into being— the *devo*—and how form-making is altered over generation,—the *evo*—to allow this level of causation in evolution to have explanatory power.

Second, *evo devo* contributes a novel type of causation that focuses on what we might call the degrees of freedom underlying developmental assembly, rather than simply the number and diversity of developmental component parts (Wagner 2000; Eble 2005). By recognizing that the nature of development is modular in developmental time, space, and regulation, *evo devo* is the first discipline to emphasize that evolutionary changes in aspects of this modular organization contribute critical degrees of freedom, or axes of variability, that enable and guide developmental evolution. Here the explanatory cause contributed by *evo devo* lies less with the discovery of the individual module: a module in it of itself contributes the same explanatory value as a gene might. Instead it comes with the discovery of the dimension within which a given modularity exists—space, time, cis-regulation—that contributes novel explanatory power, thereby enriching our understanding of what causes developmental evolution to unfold the way it does (Uller et al. 2018).

What Is Next? Current Conceptual Challenges to *Evo Devo*

Over the past 30 years, evolutionary developmental biology has provided context after context that establish development and evolution as both cause and effect of each other. Development is one of the many products of phenotypic evolution, which in turn is shaped by the nature of developmental processes. Viewed this way, *to build a phenotype requires development, while to evolve a change in phenotypes requires changes in the genetic basis of development.* At the same time *evo devo* is encountering its own challenges, both from within as well as from neighboring disciplines, which it must meet in order to remain relevant. I would like to close this chapter by highlighting the three challenges I consider most significant.

Too many genes, too little development Over the past 50 years, developmental biology has morphed into developmental genetics. Attending developmental biology meetings now one finds precious few talks or posters where the emphasis isn't on characterizing genes whose products and interactions contribute to the formation of some, preferentially experimentally very tractable, trait. *Evo devo* reflects the same trend and should perhaps be more aptly named *evolutionary developmental genetics. Evo devo* text books and courses are filled with examples of developmentally significant genes whose evolution has contributed in some way to important developmental changes in how traits are made. *But making a difference in a trait is not the same as making a trait*, especially not the complex traits *evo devo* is most concerned with (Keller 2010). Instead, traits are the products of developmental systems to which genes and their products contribute important interactants. Other products and interactions that are just as critical for trait formation emerge on other levels of biological organization as well, for example, through the communication among cells, or reciprocally inductive events among tissues, or the complex feedbacks commonplace among the component parts of organ system (Moczek 2015). While genes and genetic variation contribute to each of these as well, and therefore contribute to making such interactions reliably heritable, that relationship is not nearly as straightforward as that between a transcription factor and its binding site. Thus, while *evo devo* has managed to point us in productive directions as to how to better conceptualize and investigate the origins of diversity, novelty, and complexity, full realization of this goal will require a reorientation away from an understanding of traits and organisms as residing solely in genes and genomes, and toward an appreciation of traits as products of developmental systems. Viewed this way, *phenotypes emerge from developmental systems, whose evolution requires heritable changes in system functions.*

The contingent nature of development and developmental evolution: What, exactly, is environment? The proper functioning of all developmental processes ultimately depends on context. Context, in turn, is created by past developmental processes generating conditions for the next round of phenotype construction to take place. This constructive nature of development is so ubiquitous we tend to overlook that it is of profound significance in ensuring the proper progression of development. Only more recently has it come into the focus of developmental biologists and *evo devo* practitioners that this context—or *environment*-constructing ability of developmental processes—does not end with where we conventionally assume the organisms itself ends: instead we now recognize that organisms, through their behavior, metabolism, and choices, actively and non-randomly also modify their external environment in ways that in turn feed back to affect their own fitness. Such *niche construction* blurs our conventional understanding of where organisms end and their environment begins, and opens up additional routes to adaptation and inheritance: organisms may no longer adapt solely by modifying their traits to suit environmental conditions but modify environmental conditions to suit their traits. Similarly, organisms no longer endow their offspring just with a set of genes but pass on to them everything from methylation states to transcripts, antibodies to symbionts, and territories to positions within a social hierarchy (Laland et al. 2015). Put together, the constructive nature of organismal functioning thus transcends many dimensions both internal and external to the organism. Viewed this way, *to develop is to interact with (and often construct) internal and external environmental states. Developmental evolution then requires alteration of these interactions in a heritable manner.* Integrating the study of the mechanisms and consequences of these interactions into its portfolio of research programs will greatly enhance the explanatory power of *evo devo*.

Microecoevodevo *Evo devo* is correct in its assessment that our ability to understand, reconstruct, and predict the evolution of complex traits will be impossible without an explicit developmental, phenotype-*constructing* perspective. But population genetics is also correct in its assessment that all genetic evolution is subject to the rules and constraints imposed by population biology. And, as emphasized above, future models must better integrate the simultaneously environment-dependent and -constructing nature of development and developmental evolution. How best to achieve this is unclear, but the quantitative frameworks that already exist in population genetics and niche construction theory on one side, and the increased appreciation of developmental symbioses and phenotypic plasticity within *evo devo*, offer good starting points to continue and deepen the necessary conversations.

Conclusion

The novel ways of thinking advanced by evolutionary developmental biology are providing powerful, new approaches to expand and, in part, correct our thinking on cause and process in evolution, thereby putting us in a position to resolve long-standing questions across diverse biological disciplines, in particular in evolutionary biology. *Evo devo* itself has grown tremendously in the recent past (reviewed in Moczek et al. 2015), and I expect this transformation to further escalate as *evo devo* connects more thoroughly and thoughtfully with ecology, population genetics, and microbiology, at a time when an integrative and holistic understanding of evolutionary processes, and their causes and consequences, is more needed than ever.

Acknowledgments

I would like to thank the workshop organizers and the Konrad Lorenz Institute and its faculty and personnel for making this stimulating meeting possible. Research stemming from my own group and discussed in this chapter has been made possible with funding from the John Templeton Foundation and National Science Foundation grants IOS 1120209 and 1256689. This manuscript was written while I was supported by a Fulbright Distinguished Chair Scholarship. The opinions, interpretations, conclusions, and recommendations are those of the author and are not necessarily endorsed by the John Templeton Foundation, the National Science Foundation, or the Australian-American Fulbright Commission.

References

Abzhanov, A. M., W. P. Kuo, C. Hartmann, B. R. Grant, P. R. Grant, and C. J. Tabin. 2006. "The Calmodulin Pathway and Evolution of Elongated Beak Morphology in Darwin's Finches." *Nature* 442:563–567.

Abzhanov, A., M. B. R. Protas, P. R. Grant, and C. J. Tabin. 2004. "Bmp4 and Morphological Variation of Beaks in Darwin's Finches." *Science* 305 (5689): 1462–1465.

Blanco, J., M. Seimiya, T. Pauli, H. Reichert, and W. J. Gehring. 2009. "*Wingless* and *Hedgehog* Signaling Pathways Regulate Orthodenticle and Eyes Absent during Ocelli Development in *Drosophila*." *Developmental Biology* 329 (1): 104–115.

Brakefield, P. M., J. Gates, D. Keys, F. Kesbeke, P. J. Wijngaarden, A. Monteiro, V. French, and S. B. Carroll.1996. "Development, Plasticity and Evolution of Butterfly Eyespot Patterns." *Nature* 384 (6606): 236–242.

Busey, H. A., E. E. Zattara, and A. P. Moczek. 2016. "Conservation, Innovation, and Bias: Embryonic Segment Boundaries Position Posterior, but Not Anterior, Head Horns in Adult Beetles." *Journal of Experimental Zoology Part B: Molecular and Developmental Evolution* 326:271–279.

Calcott, B. 2009. "Lineage Explanations: Explaining How Biological Mechanisms Change." *British Journal for the Philosophy of Science* 60:51–78.

Carroll, S. B., J. K. Grenier, and S. D. Weatherbee. 2004. *From DNA to Diversity: Molecular Genetics and the Evolution of Animal Design*, 2nd ed. Malden, MA: Blackwell Scientific.

Casasa, S., D. B. Schwab, and A. P. Moczek. 2017. "Developmental Regulation and Evolution of Scaling: Novel Insights through the Study of Onthophagus Beetles." *Current Opinion in Insect Science* 19:52–60.

Darwin, C. 1859. *The Origin of Species.* London: John Murray.

Eble, G. 2005. "Morphological Modularity and Macroevolution: Conceptual and Empirical Aspects." In *Modularity: Understanding the Development and Evolution of Natural Complex Systems*, edited by W. Callebaut and D. Rasskin-Gutman, 221–239. Cambridge, MA: MIT Press.

Erclik, T., V. Hartenstein, H. D. Lipshitz, and R. R. McInnes. 2008. "Conserved Role of the Vsx Genes Supports a Monophyletic Origin for Bilaterian Visual Systems." *Current Biology* 18:1278–1287.

Gerhart, J. C., and M. W. Kirschner. 2010. "Facilitated Variation." In *Evolution: The Extended Synthesis*, edited by M. Pigliucci and B. G. Mueller, 253–280. Cambridge, MA: MIT Press.

Gilbert, S. F. 2013. *Developmental Biology*, 10th ed. Sunderland, MA: Sinauer.

Gilbert, S. F., and D. Epel. 2009. *Ecological Developmental Biology: Integrating Epigenetics, Medicine, and Evolution*. Sunderland, MA: Sinauer.

Gompel, N., B. Prud'homme, P. J. Wittkopp, V. A. Kassner, and S. B. Carroll. 2005. "Chance Caught on the Wing: Cis-regulatory Evolution and the Origin of Pigment Patterns in *Drosophila*." *Nature* 433:481–487.

Grimaldi, D., and M. S. Engel. 2005. *Evolution of the Insects*. Cambridge: Cambridge University Press.

Gullan, P. J., and P. S. Cranston. 2014. *The Insects: An Outline of Entomology*, 5th ed. Hoboken, NJ: Wiley.

Held, L. I. 2017. *Deep Homology? Uncanny Similarities of Humans and Flies Uncovered by Evo-Devo*. https://doi.org/10.1017/9781316550175.

Keller, E. F. 2010. *The Mirage of a Space between Nature and Nurture*. Durham, NC: Duke University Press.

Kijimoto, T., and A. P. Moczek. 2016. "Hedgehog Signaling Enables Nutrition-Responsive Inhibition of an Alternative Morph in a Polyphenic Beetle." *Proceedings of the National Academy of Sciences* 113:5982–5987.

Kijimoto, T., A. P. Moczek, and J. Andrews. 2012. "Diversification of *Doublesex* Function Regulates Morph-, Sex-, and Species-specific Expression of Beetle Horns." *Proceedings of the National Academy of Sciences*. 10.1073/pnas.1118589109.

Laland K. N., T. Uller, M. Feldman, K. Sterelny, G. B. Müller, A. P. Moczek, E. Jablonka, and J. Odling-Smee. 2015. "Darwin Review: The Extended Evolutionary Synthesis: Its Structure, Assumptions, and Predictions." *Proceedings of the Royal Society of London, Series B*, 282: 20151019.

Ledon-Rettig, C. C., E. E. Zattara, and A. P. Moczek. 2017. "Asymmetric Interactions between Doublesex and Sex- and Tissue-specific Target Genes Mediate Sexual Dimorphism in Beetles." *Nature Communications* 8:14593.

Mayr, E. 1960. "The Emergence of Evolutionary Novelties." In *Evolution after Darwin*, edited by S. Tax, 349–380. Chicago: University of Chicago Press.

Mayr, E. 1982. *The Growth of Biological Thought: Diversity, Evolution and Inheritance*. Cambridge, MA: Belknap Press.

Mercader, N., E. Leonardo, N. Azpiazu, A. Serrano, G. Morata, C. Martinez, and M. Torres. 1999. "Conserved Regulation of Proximodistal Limb Axis Development by Meis1/Hth." *Nature* 402:425–429.

Moczek, A. P. 2008. "On the Origin of Novelty in Development and Evolution." *Bioessays* 5:432–447.

Moczek, A. P. 2015. "Re-evaluating the Environment in Developmental Evolution." *Frontiers in Ecology and Evolution* 3:7.

Moczek, A. P., and T. Kijimoto. 2014. "Development and Evolution of Insect Polyphenisms: Novel Insights through the Study of Sex Determination Mechanisms." *Current Opinion in Insect Science* 1:52–58.

Moczek, A. P., and D. Rose. 2009. "Differential Recruitment of Limb Patterning Genes during Development and Diversification of Beetle Horns." *Proceedings of the National Academy of Sciences USA* 106:8992–8997.

Moczek, A. P., D. Rose, W. Sewell, and B. R. Kesselring. 2006. "Conservation, Innovation, and the Evolution of Horned Beetle Diversity." *Development Genes and Evolution* 216:655–665.

Moczek, A. P., K. E. Sears, A. Stollewerk, P. J. Wittkopp, P. Diggle, I. Dworkin, C. Ledon-Rettig, et al. 2015. "The Significance and Scope of Evolutionary Developmental Biology: A Vision for the 21st Century." *Evolution and Development*, 17:198–219.

Müller, G. B. 1990. "Developmental Mechanisms at the Origin of Morphological Novelty: A Side-Effect Hypothesis." In *Evolutionary Innovations*, edited by M. H. Nitecki, 99–130. Chicago: University of Chicago Press.

Müller, G. B., and G. P. Wagner. 1991. "Novelty in Evolution: Restructuring the Concept." *Annual Review of Ecology, Evolution, and Systematics* 22:229–256.

Oakley, T. H. 2003. "The Eye as a Replicating and Diverging, Modular Developmental Unit." *Trends in Ecology and Evolution* 18:623–627.

Panganiban, G., S. M. Irvine, C. Lowe, H. Roehl, L. S. Corley, B. Sherbon, J. K. Grenier, et al. 1997. "The Origin and Evolution of Animal Appendages." *Proceedings of the National Academy of Sciences USA* 13:5162–5166.

Posnien, N., N. D. B. Koniszewski, H. J. Hein, and G. Bucher. 2011. "Candidate Gene Screen in the Red Flour Beetle *Tribolium* Reveals Six3 as Ancient Regulator of Anterior Median Head and Central Complex Development." *PLoS Genetics* 7.

Posnien, N., J. B. Schinko, S. Kittelmann, and G. Bucher. 2010. "Genetics, Development and Composition of the Insect Head—A Beetle's View." *Arthropod Structure and Development* 39:399–410.

Prud'homme, B., N. Gompel, and S. B. Carroll. 2007. "Emerging Principles of Regulatory Evolution." *Proceedings of the National Academy of Sciences USA* 104:8605–8612.

Prud'homme, B., N. Gompel, A. Rokas, V. A. Kassner, T. M. Williams, S. D. Yeh, J. R. True, and S. B. Carroll. 2006. "Repeated Morphological Evolution through Cis-regulatory Changes in a Pleiotropic Gene." *Nature* 440:1050–1053.

Remane, A. 1952. *Die Grundlagen des natürlichen Systems der vergleichenden Anatomie and the Phylogenetik.* Leipzig, Germany: Geest und Portig.

Richardson, M. 1999. "Vertebrate Evolution: The Developmental Origins of Adult Variation." *BioEssays* 21:604–613.

Sears, K. E. 2004. "Constraints on the Evolution of Morphological Evolution of Marsupial Shoulder Girdles." *Evolution* 58:2353–2370.

Shapiro, M. D., J. Hanken, and N. Rosenthal. 2003. "Developmental Basis of Evolutionary Digit Loss in the Australian Lizard *Hemiergis.*" *Journal of Experimental Zoology* 279B:48–56.

Shubin, N., C. Tabin, and S. Carroll. 2009. "Deep Homology and the Origins of Evolutionary Novelty." *Nature* 457:818–823.

Smith, K. 2003. "Time's Arrow: Heterchrony and the Evolution of Development." *International Journal of Developmental Biology* 47:612–621.

True, J. R., and E. S. Haag. 2001. "Developmental System Drift and Flexibility in Evolutionary Trajectories." *Evolution and Development* 3:109–119.

Truman, J. W, and L. W. Riddiford. 1999. "The Origins of Insect Metamorphosis." *Nature* 401:447–452.

Uller, T., A. P. Moczek, R. A. Watson, P. M. Brakefield, and K. L. Laland. 2018. "Developmental Bias and Evolution: A Regulatory Network Perspective." *Genetics* 209:949–966.

Wagner, G. 2000. "Characters, Units and Natural Kinds: An Introduction." In *The Character Concept in Evolutionary Biology*, edited by G. Wagner, 1–10. San Diego, CA: Academic Press.

Wagner, G. P. 2014. *Homology Genes and Evolutionary Innovation.* Princeton, NJ: Princeton University Press.

Wagner, G. P., and V. J. Lynch. 2008. "The Gene Regulatory Logic of Transcription Factor Evolution." *Trends in Ecology and Evolution* 23 (7): 377–385.

Weiss, K. M., and S. M. Fullerton. 2000. "Phenogenetic Drift and the Evolution of Genotype-Phenotype Relationships." *Theoretical Population Biology* 57:187–195.

Wittkopp, P. J., S. B. Carroll, and A. Kopp. 2003. "Evolution in Black and White: Genetic Control of Pigment Patterns in *Drosophila.*" *Trends in Genetics* 19 (9): 495–504.

Wu, P., T. X. Jiang, J. Y. Shen, R. B. Widelitz, and C. M. Chuong. 2006. "Morphoregulation of Avian Beaks: Comparative Mapping of Growth Zone Activities and Morphological Evolution." *Developmental Dynamics* 235:1400–1412.

Wu, P., T. X. Jiang, S. Suksaweang, R. B. Widelitz, and C. M. Chuong. 2004. "Molecular Shaping of the Beak." *Science* 305:1465–1466.

Zattara, E. E., H. Busey, D. Linz, Y. Tomoyasu, and A. P. Moczek. 2016. "Neofunctionalization of Embryonic Head Patterning Genes Facilitates the Positioning of Novel Traits on the Dorsal Head of Adult Beetles." *Proceedings of the Royal Society of London,* Series B. 283:20160824.

5 Incorporating the Environmentally Sensitive Phenotype into Evolutionary Thinking

David I. Dayan, Melissa A. Graham, John A. Baker, and Susan A. Foster

Introduction

Understanding the origin of biological variation is one of the principal goals of biology. The modern synthesis provided an explanation for the evolution of biological diversity by reconciling Mendel's genetic mode of inheritance with Darwin's theory of natural selection (Mayr and Provine 1998). Crucial to the many successes of this synthesis was a focus on the gene as the sole means of inheritance but also as the primary determinant of phenotypes themselves. As a consequence, evolution is commonly studied in terms of genetic variation, yet what we hope to understand is the evolution of traits (Lewontin 1974; West-Eberhard 2003). The limits of an approach that links genes and traits in a one-to-one relationship through a linear and deterministic genotype-to-phenotype map are becoming clear in light of the wealth of investigations that can explain only a small portion of observed trait variation (Manolio et al. 2009; Zuk et al. 2012; Boyle, Li, and Pritchard 2017). At the same time, the study of phenotypic plasticity has re-emerged and offers novel avenues by which to study the causes of trait variation and evolution. The field has shifted its focus away from early debates on, for example, the relative merits of alternative means of characterizing phenotypic plasticity, and whether there exist genes for plasticity (for review see Via et al. 1995). Instead, current theory on phenotypic plasticity incorporates specific patterns of environmental heterogeneity (Snell-Rood et al. 2010; Van Dyken and Wade 2010), genetic characteristics of populations (Crispo 2008; Ledon-Rettig et al. 2014), trait genetic architectures (Snell-Rood et al. 2010; Anderson et al. 2013; Des Marais, Hernandez, and Juenger 2013), and historical and contemporary patterns of selection (Masel 2006; Ghalambor et al. 2007; Wund et al. 2008; Ghalambor et al. 2015).

We suggest that we are on the cusp of integrating these diverse factors into a framework that predicts how environmentally induced variation interacts with genetic variation to influence traits and evolutionary trajectories. Here, we discuss the potential of this new framework in shifting phenotypic plasticity from a distraction or nuisance to a paradigm in evolutionary biology. First, we examine the still prevailing view in which traits emerge from a genetic blueprint. We suggest that evidence from association and population genetics as well as evolutionary rates highlights the limits of such a view. We devote the majority of this chapter to the theoretical role of phenotypic plasticity in promoting adaptive evolution and shaping evolutionary mechanisms and trajectories. Then, we highlight a limited number of empirical cases that illustrate these theoretical mechanisms. Finally, we conclude by outlining avenues to address outstanding questions and providing an example of how better integrating phenotypic plasticity into our understanding of evolution provides new insights that strengthen evolutionary hypotheses and help resolve long-standing controversies.

Limits of the Genotype-to-Phenotype Map

One of the central organizing principles of the modern synthesis is the direct relationship between genotype and phenotype, in which variation in traits emerges as a direct consequence of information that is stored and inherited at the level of genes. This relationship between genes and traits, or a genetic blueprint, is often framed in terms of a simplifying heuristic referred to as the genotype-to-phenotype map. While contemporary biology has a more nuanced understanding of trait development, the concept of a genotype-to-phenotype map permeates biological thinking and is the core assumption underlying many fields of research (Orgogozo, Morizot, and Martin 2015). As a consequence, causal priority in evolutionary research is often given to natural selection among genetic variants that influence the distribution of traits in a population with less importance given to other factors that determine trait variation, such as phenotypic plasticity and developmental processes.

The utility of the genotype-to-phenotype map heuristic is evident in the number of population and quantitative genetic dissections of evolutionary-relevant traits such as coat coloration in mice (Linnen et al. 2013), lateral plate armor in stickleback (Cresko et al. 2004), wing patterning in *Heliconius* (Joron et al. 2011), flowering time in model plants such as *Mimulus guttatus* and *Arabidopsis thaliana* (Ausin, Alonso-Blanco, and Martinez-Zapater 2004), or the gap gene network in *Drosophila* (Jaeger 2011). Increasingly, however, the limits of this gene-centric view are becoming clear. While it is often a useful heuristic, the genotype-to-phenotype map is not a real attribute of biological organisms. Much of the phenotypic variation in nature is difficult to explain with an analytical framework that views genetic variation as the primary determinant of trait variation. We provide three perspectives drawn from largely independent fields to support this assertion.

Perhaps the most striking evidence comes from the quantitative trait nucleotide (QTN) or genome-wide association study (GWAS) program of research (Rockman 2012). The goal of this massive research effort is to identify the causal factors that underlie phenotypic variation by uncovering statistical associations among genetic variants and traits. While researchers have discovered thousands of such associations, we remain unable to explain the majority of variation even in extensively studied and highly heritable traits (Visscher et al. 2012; Boyle et al. 2017). For example, 60 to 80 percent of schizophrenia risk is heritable, which suggests that the genetic component of variation in the schizophrenia phenotype is very large, yet the causal variants discovered by GWAS can explain just 3 percent of this heritable variation (Schizophrenia Working Group of the Psychiatric Genomics 2014). Similarly, the genetic variants significantly associated with height, a complex trait that has been the subject of recent natural selection across many human populations (Stulp and Barrett 2016), explain only 20 percent of the heritable variation in this trait (Wood et al. 2014). Some explanations of this so-called missing heritability problem assume the genotype-to-phenotype map is a true aspect of organisms that can be uncovered, and that missing heritability stems from methodological shortcomings. Either we are simply querying the wrong type of genetic variation by focusing on common variant, ignoring structural variation, rare effect variants, microRNAs, and others (Manolio et al. 2009; Eichler et al. 2010), or the disconnect stems from a systematic over-estimation of trait heritability (Kumar et al. 2016).

However, an alternative set of explanations posit that the genotype-to-phenotype map is instead highly complex and dynamic; many genes contribute to traits and their effects vary across environments, genomic context, and ontogenetic and evolutionary time, and it is this complexity that creates the gap between explained and observed heritability. Under one such hypothesis, most traits have a nearly infinitesimal genetic architecture, that is, traits are determined by thousands, or tens of thousands of genetic loci, each with small individual impacts, making their detection statistically intractable (Rockman 2012; Wellenreuther and Hansson 2016; Shi, Kichaev, and Pasaniuc 2016). Under a second, not mutually exclusive hypothesis, non-additive, but heritable effects such as epistasis and interactions between genes and environments (Zuk et al. 2012; Bloom et al. 2013), as well as epigenetic inheritance (Bourrat, Lu, and Jablonka 2017) significantly contribute to trait variation, leading to a gap between narrow and broad sense heritability. These explanations call into question the utility of the genotype-to-phenotype map as a metaphor, because the map is either too complex or too dynamic to be interpretable and predictive.

Shifting focus from the production of phenotypes to their evolution, we observe a similar pattern: with the advent of next-generation sequencing, we have begun to identify the genetic variation that contributes to adaptive evolution in natural populations (Vitti, Grossman, and Sabeti 2013; Hoban et al. 2016), but with the exception of oligogenic traits, we have only limited understanding how this genetic variation is translated into adaptive variation in phenotypes. For example, in humans, natural selection has driven genetic differentiation among populations (Hancock, Witonsky et al. 2010; Wollstein and Stephan 2015), and this differentiation occurs within genomic regions associated with complex traits that may themselves be adaptively significant (Turchin et al. 2012). Yet, for most traits examined, differentiation at the complex trait-associated genomic regions does not appear to have been driven by selection on the trait itself (Zhang et al. 2013; Berg and Coop 2014). Therefore, even as new techniques in genomics have vastly increased the number of putatively adaptive genetic variants for evolutionary biologists to consider, it remains difficult to address the molecular basis of evolution in the complex traits that interest us most, because the links between selection at the level of the phenotype and its effects on genetic variation are often unclear.

Evolutionary rates observed in nature also highlight the challenges with a simple genotype-to-phenotype map. Population-genetic models often rely on an adaptive walk of successive fixations of new mutants (Orr 2005), resulting in slow evolutionary rates and evolutionary gradualism. Yet, this model cannot account for the observed rates of adaptive evolution of traits in new environments (Hendry 2016; Messer, Ellner, and Hairston 2016), nor does it fit with the observation that there is abundant heritable genetic variation for most traits (Hansen, Pélabon, and Houle 2011; Hendry 2013). Instead, the ubiquity of examples of rapid evolution, both in the lab and in examples of contemporary evolutionary responses to human impacts, strongly suggests highly polygenic trait architectures and selection on standing genetic variation.

As the networks of genes and their interactions that produce trait variation becomes more diffuse and evolution begins to rely more on selection on standing genetic variation, the genetic architecture of adaptation is expected to become more transient or dynamic across space and time. Under polygenic trait architecture, the effects of gene flow are predicted to make the impacts of selection on underlying genetic variation depend more on transient covariance of small effect alleles rather than the successive fixations of an adaptive walk (Nijhout and Paulsen 1997; Yeaman and Whitlock 2011; Yeaman 2015). Similarly, as rapid evolution becomes empowered by abundant targets on which natural selection can act, fluctuation in selection pressures mediated by short-term ecological variation, for example seasonality and diel temperature variation, becomes a more important factor in evolution (Bell 2010). These predictions are supported by genome scan studies that discover temporal adaptive shifts in allele frequency at genetic loci (Therkildsen et al. 2013; Bergland et al. 2014) and the prevalence of evidence for so-called soft sweeps and polygenic selection from the standing genetic variation over hard sweeps resulting from fixation of novel mutants (Hernandez et al. 2011; Hermisson, Pennings, and Kelley 2017).

Finally, the putatively adaptive genetic variants discovered in genome scans are enriched among regulatory rather than protein coding regions (Jones et al. 2012; Fraser 2013; Pickrell 2014). This finding, and others demonstrating the evolutionary significance of gene expression (Oleksiak, Churchill, and Crawford 2002; Fraser 2013; Dayan, Crawford, and Oleksiak 2015; Josephs et al. 2015; Kenkel and Matz 2016), point to regulatory interactions as the primary targets of natural selection. Rather than a simple one-to-one mapping with traits, regulatory variation is defined by extensive pleiotropy, epistasis, and gene-by-environment interaction (Phillips 2008), creating a complex relationship between genetic variation at regulatory sites and their ultimate adaptive significance at the level selection operates.

These observations, drawn from diverse fields (association genomics, quantitative genetics, and population genomics, respectively) cumulatively point to a relationship between genes and traits that is highly nonlinear, complex, and dynamic. Not only do thousands or more genetic loci commonly contribute to a trait but their impacts frequently depend on interactions with other loci and the environment. Therefore, while genetic variation may provide the primary basis of heritability, our current ability to empirically determine the genotype-to-phenotype map places limits on the utility of genetic variation as a predictor of trait variation or evolutionary trajectories. This is not to say that future efforts to uncover the relationship between genetic variation and evolutionarily significant trait variation are futile, nor do we wish to diminish the importance of the results of past investigation under the genotype-to-phenotype map framework. However, we suggest, as have others (e.g., Schlichting and Pigliucci 1998; Pigliucci 2001; West-Eberhard 2003; Sultan 2015), that previous efforts have largely ignored the central role of the environment in determining phenotypes, and that better incorporating the environment into our understanding of the production of traits has the potential to provide new insights.

Consider how incorporating the environment into our understanding of the production of traits has the potential to ameliorate or address each of the three limits outlined above.

(i) If the environment interacts with genetic variation in a non-additive fashion, then we would expect the causative variants discovered by GWAS to explain only a portion of heritable variation in phenotypes, because these gene-by-environment interactions lead to a disconnect between broad and narrow-sense heritability (Flint and Mackay 2009; Zuk et al. 2012). Understanding how such gene-by-environment interactions operate may provide an answer to the missing heritability problem (Eichler et al. 2010; Marigorta and Gibson 2014), yet the primacy of the gene in previous research efforts has meant that neither the frequency, nor the genetic architecture of gene-by-environment interactions in complex traits is clear (Des Marais, Hernandez, and Juenger 2013), despite a long recognition of their evolutionary potential (Waddington 1953; Bradshaw 1965).

(ii) Accumulating evidence suggests that novel or rare environments not only expose genetic variation that does not usually contribute to traits but that the amount of heritable genetic variation for traits increases in these environments (McGuigan et al. 2011; Rokholm et al. 2011; Kuttner et al. 2014; Parsons et al. 2016), for reviews see (Ledon-Rettig et al. 2014; Paaby and Rockman 2014). Future work that considers the role of the environment in generating and releasing this cryptic genetic variation may explain rapid evolution of traits that appear to have limited evolutionary potential under current environmental conditions (Paaby and Rockman 2014) or the origin of novelties and major evolutionary transitions (Moczek et al. 2011).

(iii) Gene-centric approaches to studying adaptation, such as genome scans for signatures of selection, have identified regulatory variation as a major target of natural selection. At the same time, research investigating the role of regulatory variation among species and populations within ecologically relevant environmental contexts has led to novel insights into trait production and evolution (e.g., Abouheif and Wray 2002; Wellmer and Riechmann 2010; Schneider, Meyer, and Gunter 2014). Findings from these and related fields (e.g., eco-evo devo) are only beginning to be incorporated into broader evolutionary thought (Abouheif et al. 2014; Gilbert and Epel 2015), but they have the potential to make the connections among genetic variation and traits that would prove impossible to decipher without explicit consideration of the environment.

Evolutionary Implications of Phenotypic Plasticity

The brief discussion above highlights some of the limitations of a gene-centric view of the production and evolution of traits and introduces how increased consideration of the environment may offer novel insights. In this section, we summarize the body of theory that hypothesizes how environmental sensitivity, or phenotypic plasticity, influences evolutionary trajectories.

Genetic variation in phenotypic plasticity, usually viewed as the statistical notion of a gene-by-environment interaction (GxE), is commonly observed at the trait level (Schlichting and Pigliucci 1998). The consequences of GxE are twofold. First, by providing the variation for plasticity on which selection can act, GxE permits the evolution of plasticity itself (Pigliucci 2001). Perhaps more importantly, the existence of GxE means that genetic variation is expressed differently across environmental contexts (Bradshaw 1965). Traditional perspectives hold that phenotypic plasticity functions in evolution only to shield genotypes from selection, thus reducing the effectiveness of directional selection and slowing adaptation (S. Wright 1931; Jong 1995; Orr 1999). Yet, there has been a revival of theoretical and empirical research, based on early models (Baldwin 1896; Morgan 1896; Schmalhausen 1949; Waddington 1953), indicating that phenotypic plasticity can play a major role in shaping evolutionary trajectories because it mediates the relationship between selection and genetic variation. This effect is only part of the broader body of evolutionary theory concerning plasticity, but we choose to focus here.

The mechanisms by which phenotypic plasticity can potentially influence evolution fall into three broad categories depending on whether phenotypic plasticity: (i) reduces the cost of selection in novel environments and allows populations to persist; (ii) determines the phenotypes on which selection can act, guiding the pattern of evolution through genetic accommodation *sensu* (West-Eberhard 2003); or (iii) alters the genetic variation available to selection. These mechanisms have been hypothesized to play a role in many evolutionary phenomena with varying levels of empirical support, including adaptation to new environments (Ghalambor et al. 2007), the evolutionary origin of complex traits and novel innovations (Moczek 2008; Moczek et al. 2011; Schlichting and Wund 2014), and the divergence of specialized ecotypes, speciation and adaptive radiations (Pfennig et al. 2010;

Thibert-Plante and Hendry 2011; Fitzpatrick 2012; Schneider and Meyer 2017). The mechanisms are briefly reviewed below.

Much of the theory concerning the role of phenotypic plasticity in evolution is focused on whether adaptive plasticity allows populations to persist in novel environments or after an environmental shift. As a consequence, phenotypic plasticity is often viewed as a phenomenon promoting adaptive evolution because it reduces the cost of selection (*sensu* Haldane 1957) after an environmental shift and allows populations to persist and subsequently adapt through either new mutations or selection on the standing genetic variation (Baldwin 1896; Robinson and Dukas 1999; Price, Qvarnström, and Irwin 2003; Chevin and Lande 2011). For example, in the model of Chevin and Lande (2010), a population with a partially adaptive plastic response can maintain higher population size relative to a population without plasticity, reducing the probability of extinction and therefore increasing the rate of evolutionary adaptation. Extending these predictions to scenarios of a heterogeneous environment with gene flow produces similar results. Gene flow from populations in the ancestral habitat reduces mean fitness and imposes a selection cost in the population experiencing the new conditions, but this cost is offset by adaptive plasticity (Chevin and Lande 2011; Thibert-Plante and Hendry 2011). From this perspective, phenotypic plasticity flattens the adaptive landscape *sensu* (Simpson 1944) and provides a valley crossing mechanism, allowing populations to explore adaptive peaks other than the local optimum (Price, Qvarnström, and Irwin 2003; Frank 2011). This role of plasticity in promoting adaptive evolution may be especially important for populations with high rates of migration and gene flow from ancestral habitats (Sultan and Spencer 2002; Crispo 2008) and for traits with moderately beneficial plasticity (Price, Qvarnström, and Irwin 2003).

After populations are exposed to new environmental conditions, phenotypic plasticity can influence evolutionary trajectories through genetic accommodation (West-Eberhard 2003). This mechanism views "phenotypes as leaders and genes as followers in evolution" (West-Eberhard 2003); novel phenotypes arise in a population, either because of new mutations or exposure to different environmental conditions, and are subsequently refined by selection through quantitative genetic changes at many loci. If the phenotypic variants arise because environmental changes, genetic accommodation can have a major influence on the rate of adaptive evolution because the emergence of the phenotype occurs within a single generation at high frequency in the population and across diverse genetic backgrounds (West-Eberhard 2003). Thus, populations do not need to wait for the emergence of a single adaptive mutation to arise, which would be vulnerable to stochastic loss (Orr 2005).

Instead, because the environment plays a role in both the production of and selection on a trait, adaptation can occur through quantitative genetic changes from the standing genetic variation that adaptively refines the regulatory architecture of the trait's expression (Pfennig et al. 2010; Moczek et al. 2011; Wund 2012; Ehrenreich and Pfennig 2015). This refinement (accommodation) is possible because plastic phenotypes frequently have complex genetic architectures that not only provide many genetic targets for selection but are also likely to exhibit substantial genetic variation owing to their conditional expression (Windig, De Kovel, and Jong 2004; Aubin-Horth and Renn 2009; Hodgins-Davis and Townsend 2009; Snell-Rood et al. 2010; Van Dyken and Wade 2010) and because plasticity may lead to a bias among genetic effects along the axis of adaptive phenotypic variation (Draghi and Whitlock 2012; Lind et al. 2015). Furthermore, phenotypic plasticity promotes increased genetic variation in populations experiencing novel conditions because adaptive plasticity can increase gene flow from other populations (Crispo 2008; Colautti and Barrett 2011).

Genetic accommodation of environmentally sensitive phenotypes can have several outcomes depending on how the regulation of the phenotype is altered. A loss of plasticity or decreased threshold of induction for the phenotype such that it is constitutively expressed in the new environment may occur if the plastic response is costly to produce or alternate environments are rare, that is, genetic assimilation (Waddington 1953; Pigliucci, Murren, and Schlichting 2006). Alternatively, selection may favor enhanced phenotypic plasticity in the direction of the trait optimum. In this case, accommodation reduces the costs and limits of phenotypic plasticity (Suzuki and Nijhout 2006; Murren et al. 2015). Finally, not all phenotypic plasticity is adaptive. In fact, maladaptive phenotypic plasticity that arises as a consequence of passive responses to the environment by biological molecules may be the primary form of plasticity (Van Kleunen and Fischer 2005; Schulte, Healy, and Fangue 2011). Thus, genetic accommodation may function to reduce the effects of maladaptive phenotypic plasticity, such that local adaptation of populations through accommodation leads to the stabilization of phenotypes across heterogeneous environments (Ho and Zhang 2018). This pattern is frequently observed in nature and is termed counter-gradient variation because the directions of plastic and genetic effects on traits are in opposite directions with respect to the environment (Conover and Schultz 1995). Genetic accommodation of this form is distinguished from others under the term genetic compensation (Grether 2005). Genetic compensation may be particularly relevant to the role of plasticity in promoting divergence, because it establishes reproductive barriers to gene flow between locally adapted populations (Fitzpatrick 2012).

In the final mechanism, phenotypic plasticity is predicted to impact evolutionary adaptation because it promotes the accumulation and release of genetic variation (Gibson and Dworkin 2004; Hermisson and Wagner 2004; Le Rouzic and Carlborg 2008; McGuigan and Sgro 2009; Paaby and Rockman 2014). Phenotypic plasticity contributes to an increase in genetic variation in two ways. First, homeostatic mechanisms exerted through phenotypic plasticity might buffer the effects of new mutations, reducing genetic constraints. The best studied instance of this phenomenon is the heat shock protein Hsp90's function as a "genetic capacitor" (Rutherford and Lindquist 1998). Thermal induction of Hsp90 canalizes the outcome of protein folding during gene expression and stabilizes protein populations in the cell, thereby simultaneously reducing the effects of environmental and genetic perturbation and promoting the accumulation of genetic variation by reducing the efficacy of purifying and/or stabilizing selection (Queitsch, Sangster, and Lindquist 2002). Such genetic capacitors may be common (Sangster et al. 2008; Chen et al. 2013). In a second, more pervasive effect, phenotypic plasticity may lead to the conditional expression of genetic variation across space and time such that some genetic variation has an effect in only a subset of individuals or populations (Snell-Rood et al. 2011; Des Marais, Hernandez, and Juenger 2013). This conditional expression can result in relaxed selection in the non-inducing environment and an increase in polymorphism at these loci because purifying selection can only remove deleterious alleles in the subset of individuals experiencing the inducing conditions (Kawecki 1994; Lahti et al. 2009; Snell-Rood et al. 2010; Van Dyken and Wade 2010). Conditional expression of genetic variation may also lead to the evolution of developmental systems that bias the effects of both the standing genetic variation and novel mutants along the axis of plasticity, and ultimately, adaptive phenotypic variation (Draghi and Whitlock 2012).

The genetic variation that accumulates by these mechanisms is studied under the phenomenon of cryptic genetic variation (CGV) (Gibson and Dworkin 2004; Hermisson and Wagner 2004; Paaby and Rockman 2014; Paaby and Gibson 2016). While such variation is simply part of the standing genetic variation, it is considered cryptic when the inducing conditions are uncommon in the evolutionary history of the species. The release of CGV is likely to have a major impact on the rate of evolution because it may be accompanied by an increase in heritable variation of phenotypes under rarely encountered or novel conditions where selection may be strongest (i.e., genetic variance; Waddington 1953; Hoffmann and Merila 1999; Rokholm et al. 2011).

In many ways the concepts of evolution via cryptic genetic variation and genetic accommodation are closely linked. While discussion of CGV in the literature is typically framed in terms that give causal priority to the role of genes in trait production and evolution, the core assumption underlying CGV is that the environment determines how and which genetic variants contribute to a trait. In this way, CGV can be thought of as a population-genetic and quantitative genetic account of how evolution might be thought to proceed in a scenario where phenotypes are leaders, not followers during evolutionary transitions. There are, of course, aspects of genetic accommodation, and plasticity-led evolution more generally, that are not adequately encompassed by cryptic genetic variation. However, in this chapter we focus on how cryptic genetic variation can be leveraged as a bridge for exploring the impacts of genetic accommodation using the empirical tools of population genomics and quantitative genetics. For example, an important aspect of genetic accommodation of an environmentally induced trait is the simultaneous induction of the trait across many members of a population. This prediction is paralleled by the phenomenon that alleles that form the basis of cryptic genetic variation can be present at intermediate frequencies in a population because they are conditionally neutral before the onset of selection (Fry, Heinsohn, and Trudy 1996). Furthermore, the increased genetic variance provided by the release of CGV in a population experiencing novel or rare environments provides the basis for quantitative genetic changes in the refinement of environmentally induced traits through genetic accommodation. Also, because alleles are free to recombine across multiple genetic backgrounds and remain neutral until the inducing conditions are encountered, CGV might underlie the evolution of complex traits that require multiple simultaneous genetic changes (Paaby and Rockman 2014). Under the adaptive landscape metaphor, this latter process provides a mechanism for demonstrating how plasticity allows populations to escape local optima and explore broader phenotypic spaces.

Demonstrating the Role of Plasticity in Evolution: The Challenge

The extent to which these theoretical frameworks are predictive in natural populations depends upon the relative importance of phenotypic plasticity and gene-by-environment interaction in explaining trait variation and its evolution. That plasticity is variable in magnitude, virtually ubiquitous, and can evolve is nowhere better demonstrated than in West-Eberhard's (2003) volume on the role of plasticity in evolution. Other reviews offer similar evidence of the pervasive nature of plasticity and its evolution (Schlichting and Pigliucci 1998; Whitman and Agrawal 2009; Sultan 2015). However, demonstrating that plasticity plays an important role in facilitating and directing evolution by (i) permitting population persistence, (ii) determining the array of phenotypes available to selection, or (iii) increasing the genetic variance and determining which genetic variants are available to selection is challenging (Schlichting and Pigliucci 1998; West-Eberhard 2003; Sultan 2015), and despite extensive empirical research each of these remains controversial (Wund 2012; Laland et al. 2014; Laland et al. 2015; Kovaka 2017). In this section we offer examples of points (i) and (ii). Point (iii) is discussed in detail in an ensuing section.

Many studies on a variety of taxa provide evidence compatible with a relationship between plasticity and population persistence. Orizaola and Laurila (2016), for example, demonstrated that pool frogs at their northern range margin in Europe possessed the greatest plasticity in developmental rate with respect to temperature, an attribute that may have permitted them to take advantage of the rare warm periods in that marginal environment. Similarly, a study of the response of three arid-zone floodplain plant species (Nicol et al. 2017) suggests a role for plasticity in persistence in this unpredictable environment. Further, the traits for which plasticity is expressed differed across the three species, illustrating multiple plastic pathways to persistence in a single environment. Finally, plasticity with respect to host plant species in seed beetles appears to mediate the fitness consequences of egg laying on suboptimal host plant seeds (smaller seeds than the native host plant), and in doing so, may shield populations from selection in the novel environment. This again may facilitate persistence and potentially allow time for adaption to the new environment to occur (Amarillo-Suárez and Fox 2006). This example is perhaps most convincing, as plasticity is "ancestral" (was present prior to colonization of novel host plants) and was expressed as clearly adaptive phenotypes.

A number of studies offer evidence that plasticity can alter the array of phenotypes available to selection, and hence have potentially altered the direction of evolution through genetic accommodation (point [ii] above). We elaborate two examples we find compelling that are largely compatible with the hypothesis. The first involves phenotypic plasticity of morphology in threespine stickleback—research that also highlights aspects of evolution via genetic accommodation (Wund et al. 2008, 2012). Laboratory-reared oceanic ("ancestral") stickleback exhibit modest but appropriate patterns of plasticity (figure 5.1) when raised in environments simulating those at the extremes of a benthic-limnetic foraging axis (Wund et al. 2008, 2012). This ancestral plasticity parallels the evolutionary transitions made by benthic and limnetic freshwater ecotypes, transitions that have evolved repeatedly and independently in derived freshwater populations. This pattern suggests that an ancestral "flexible stem" (*sensu* West-Eberhard 2003) might have influenced the direction of evolutionary change in the adaptive radiation in freshwater, promoting the environmental match. Further, the patterns of induced phenotypic plasticity in freshwater populations (figure 5.1) differ from those in oceanic populations under the alternative laboratory rearing conditions, demonstrating genetic accommodation (Wund et al. 2008; Wund 2012). Collectively, these studies suggest both that the direction of evolutionary change could have been mediated by plasticity and that genetic accommodation has occurred.

A particularly intriguing example of the way in which plasticity could have influenced subsequent evolution involves the major evolutionary transition in vertebrates from aquatic to terrestrial locomotion. *Polypterus senegalus*, an extant analogue of the fish likely to have given rise to tetrapods, exhibit different limb development when raised in an aquatic environment versus a moist terrestrial environment. In a terrestrial environment, these fish use their fins differently than in the aquatic environment, and they develop limb anatomical characteristics that are similar to those observed in stem tetrapod lineages during the Devonian Period (Standen, Du, and Larsson 2014). Thus, plasticity could well have influenced the ultimate shape and function of tetrapod limbs by altering the range of phenotypes, and underlying patterns of gene expression exposed to selection in the novel, terrestrial environment.

Using Population Genomics to Address the Plasticity Hypothesis

There is substantial evidence that adaptive phenotypic plasticity readily evolves (Schlichting and Pigliucci 1998; Pigliucci 2001; DeWitt and Scheiner 2004) and is often a component of adaptive divergence (Torres-Dowdall et al. 2012; Dayan et al. 2015; Kenkel and Matz 2016). Additionally, a great wealth of studies demonstrate how environmental variation

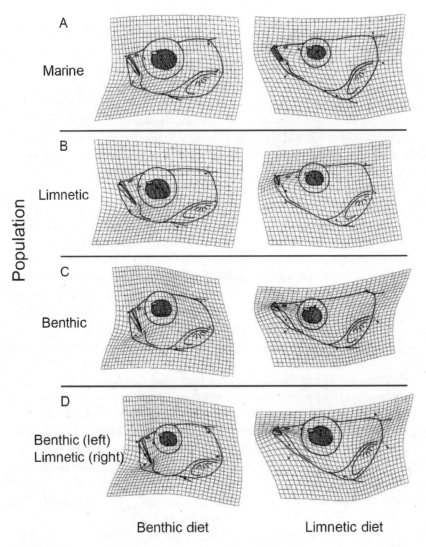

Figure 5.1
Head shape plasticity with respect to food type (benthic items vs. limnetic plankton) in oceanic (ancestral surrogate), limnetic (freshwater plankton feeding ecotype), and benthic (freshwater benthic feeding ecotype) threespine stickleback raised in the laboratory. Deformation grids are based on consensus landmark positions for each experimental group and are exaggerated 4x to make differences more apparent. Benthic diet left column, limnetic diet right column. From Wund et al. (2008).

integrates with developmental process to produce variation in evolutionarily relevant phenotypes (reviewed in Abouheif et al. 2014). However, despite the rich theoretical framework surrounding phenotypic plasticity, empirical evidence pointing to the role of plasticity in directing evolution varies among the models outlined above (see Schlichting and Wund 2014; Levis and Pfennig 2016 for recent reviews). There is suggestive evidence that (i) phenotypic plasticity can promote population persistence in novel or marginal environments (Réale et al. 2003; Yeh and Price 2004; Amarillo-Suárez and Fox 2006; Geng et al. 2006; Orizaola and Laurila 2016), but see (Davidson, Jennions, and Nicotra 2011), that (ii) ancestral plasticity mirrors adaptive divergence (Losos et al. 2000; Gomez-Mestre and Buchholz 2006; Rajakumar et al. 2012) and (iii) that genetic variance increases when current environmental conditions are rare in a population's history (Hoffmann and Merila 1999; Ledón-Rettig, Pfennig, and Crespi 2010; McGuigan et al. 2011; Takahashi 2015; Rowinski and Rogell 2017). Yet many other questions remained unanswered. Below, we discuss how contemporary and emerging techniques in quantitative genetics and population genomics can be leveraged to address some of these outstanding questions.

First, *to what extent does the environment determine genetic architecture of adaptive traits?* In quantitative genetic terms, the extent to which traits are determined by constitutively versus conditionally expressed genetic variation has profound implications for the origin and maintenance of biological variation (Colautti, Lee, and Mitchell-Olds 2012; Paaby and Rockman 2014). Where trait variation is mediated by the same genes across environments (i.e., antagonistic pleiotropy or allelic sensitivity), phenotypic plasticity can slow the rate of adaptation because of pleiotropic constraints (Scarcelli et al. 2007) or drive divergence by establishing reproductive barriers to gene flow (Kawecki and Ebert 2004). On the other hand, conditionally-expressed variation leads to relaxed selection and polymorphism accumulation at conditionally-expressed loci because genetic variation will only be subject to selection in a subset of individuals (Snell-Rood et al. 2010; Van Dyken and Wade 2010). Understanding the extent to which the environment determines which genes matter for evolution provides much needed inference into the relative importance of genetic variation and the environment as causal forces in evolution. At one extreme, the genetic architecture of a trait is static among the environments experienced by a population across time or space. In this case, it is clear how genetic variation can be said to cause the trait variation that is subject to selection and conceptual frameworks such as the genotype-to-phenotype map will be sufficient for evolutionary inference. At the other extreme, the genetic variants that contribute to a trait may completely be determined by the environment. In this case, a clear causal relationship between genes and traits cannot be drawn without invoking the environment, because it is the environment that determines both the distribution of phenotypes and the genetic variants that selection acts on.

Theoretical research into this topic is thorough, but the extent to which it is predictive has yet to be addressed across diverse study systems and at a genome-wide perspective. Recombinant mapping crosses of highly divergent ecotypes or crop varieties suggest that different genes influence traits across environments (Anderson et al. 2013; Des Marais et al. 2013; Parsons et al. 2016), but aspects of these study populations might predispose them toward conditional expression (Kassen 2002; Hall, Lowry, and Willis 2010; Colautti et al. 2012), leading to ascertainment bias. Recent results in quantitative genetics (Pavličev and Cheverud 2015; Wood and Brodie 2015) point to a greater role for conditional neutrality than antagonistic pleiotropy, and some experimental evolution results suggest that selection acts on conditionally expressed variation (Sikkink et al. 2015), but relatively little is known about the influence of the environment on genetic architecture in highly outbred, large natural populations that are key to our understanding of ecologically relevant mechanisms of evolution. Application of emerging techniques in population genomics and quantitative genetics across environments in such organisms provide promising avenues to address the role of the environment in shaping genetic architecture of ecologically relevant traits. These include pedigree-free estimation of trait heritability (Stanton-Geddes et al. 2013) and polygene-trait association techniques such as regional heritability analysis (Berenos et al. 2015), chromosome partitioning (Santure et al. 2015), and machine learning–based classification and regression algorithms (Brieuc et al. 2018).

Second, *how common is genetic accommodation in nature* (Schlichting and Wund 2014; Levis and Pfennig 2016)? Does plasticity determine evolutionary trajectories? There are ample examples of parallelism between adaptive divergence and patterns of adaptive phenotypic plasticity within a lineage (Losos et al. 2000; Gomez-Mestre and Buchholz 2006; Wund et al. 2012). This suggests that divergence is promoted by high levels of phenotypic plasticity (Pfennig et al. 2010), but these studies do not establish a causal relationship between divergence and plasticity (Kovaka 2017). In order to avoid this pitfall, studies investigating genetic accommodation often rely on comparisons of phenotypic plasticity between derived and extant ancestral populations (Yeh and Price 2004; Wund et al. 2008; Badyaev 2009; McCairns and Bernatchez 2010; Aubret 2015), or cases in which the ancestral state of plasticity can be inferred phylogenetically (Rajakumar et al. 2012). These studies suggest indirectly that adaptation through genetic accommodation may be common, because phenotypic divergence in derived groups mirrors ancestral plasticity. However, the distinguishing feature of genetic accommodation is the initiation of phenotypic diversity followed by refinement through selection (West-Eberhard 2003). Phenotypes are the leaders, not the followers in evolution by genetic accommodation. Therefore, comparisons between populations or species after the onset of selection cannot definitively demonstrate genetic accommodation, because they do not reveal how the phenotype in question arose (Wund 2012; Kovaka 2017). For example, evolution of plasticity in derived lineages may be the result of relaxed selection at genetic loci underlying plastic traits, and not vice versa (Hunt et al. 2011; Leichty et al. 2012).

More direct evidence of genetic accommodation in nature can be inferred from studies that comprehensively assess the evidence of genetic accommodation in natural populations by simultaneously characterizing the genetic and molecular mechanisms underlying plastic traits while demonstrating adaptive refinement of the traits by selection. For example, work in the horned beetles of the genus *Onthophagus* has revealed in detail how the modularity of developmental plasticity has permitted the reutilization of conserved developmental processes to produce an evolutionarily novel trait and how the emergence of this trait has contributed to diversification within the lineage (Moczek 2009; Snell-Rood et al. 2011).

While such detailed analyses of adaptive traits provide strong direct evidence that plasticity-led evolution can shape evolutionary trajectories for a given trait in a particular lineage, they cannot address the broader significance of genetic accommodation in nature because the intensive nature of this study limits the number of experimental systems in which the question can be addressed. Similarly, experimental evolution studies also provide a strong test of genetic accommodation when demonstrable genetic accommodation of introduced experimental populations mimics the evolved differences in plasticity between ancestral and derived natural populations and the molecular mechanisms underlying the plastic phenotype in both the experimental and natural populations are shared. Such analyses were recently completed by Ghalambor et al. (2015) and Ho and Zhang (2018), but similarly appropriate natural study systems may be too rare to provide easily generalizable examples (Ehrenreich and Pfennig 2015).

We propose that alternative approaches that leverage the power of contemporary population genomic tools provide complementary evidence that can discriminate traditional gene-centered models of evolution and evolution by genetic accommodation. These approaches are readily extendable to non-model systems where little is known. The cryptic genetic variation framework, in particular, provides a link between empirically available evidence, in the form of gene-trait associations and genomic signatures of selection, and the higher order theory of genetic accommodation (Levis and Pfennig 2016). In the case of ancestral-derived comparisons, for example, future studies should strive to address whether cryptic genetic variation is not only released under environmental conditions experienced by the derived populations, but that this cryptic genetic variation is under selection and contributes to adaptive variation in plastic traits in derived populations. If phenotypes are leaders rather than followers during adaptation, genetic variants that conditionally contribute to the focal trait will be enriched for signatures of selection over variants that contribute to the trait in the ancestral environment. In this way, the signature of selection left on genomic variation is a distinctive empirical trace, *sensu* Kovaka (2017), which can be used to discriminate between adaptive traits that emerge via genetic mechanisms or via environmental induction. Until recently, researchers pursuing this research question are faced with the limitation of working with traits for which the mechanisms and underlying genetic architecture are either well-known or experimentally tractable through QTL mapping or association analysis. Such integrated approaches have already suggested that genetic accommodation may shape variation at single loci underlying known adaptive traits (Parsons et al. 2016; Pespeni, Ladner, and Moczek 2017). However, the ever-decreasing cost of acquiring genomic information makes such study at genome-wide scales a possibility for research going forward.

Plasticity Can Inform Population Genetics

Just as population genetics may inform our understanding of the environment in evolution, better incorporation of the environment into evolutionary thinking may inform long-standing questions in population genetics. Traditional understanding of the adaptive evolutionary process is that of a single new adaptive mutation that arises after the onset of selection in a new environment and is subsequently driven to high frequency (Orr 1998). Yet, as outlined above, evidence of such selective sweeps is rare in natural populations (Pritchard, Pickrell, and Coop 2010; Hernandez et al. 2011; Druet et al. 2013; Garud et al. 2015; but see Fagny et al. 2014; Jensen 2014) and complex traits under selection commonly have a polygenic basis (Rockman 2012; Shi et al. 2016). Moreover, the long lag time associated with the initial phase of adaptation by this process is at odds with the prevalence of rapid adaptation by natural selection (Lee 2002; Ellner, Geber, and Hairston 2011; Colautti and Barrett 2013; Messer, Ellner, and Hairston 2016) and rapid changes in quantitative traits under artificial selection (Falconer and Mackay 1996; Wright et al. 2005).

Adaptation from standing genetic variation can account for rapid adaption to novel conditions provided there is sufficient variation present in populations (Barrett and Schluter 2008), but is likely to have a population genomic signature that is not as clearly detected in genome scans (Hermisson and Pennings 2005; Przeworski, Coop, and Wall 2005; Teshima, Coop, and Przeworski 2006; Messer and Petrov 2013). Additionally, many subtle allele-frequency changes from standing genetic variation can drive evolution and may be sufficient to explain the adaptive evolution of phenotypes (Hancock, Alkorta-Aranburu et al. 2010; Hancock, Witonsky et al. 2010; Pritchard and Di Rienzo 2010; Berg and Coop 2014; Stephan 2016).

The evolutionary consequences of phenotypic plasticity discussed above may predispose populations toward evolution from standing genetic variation rather than selective sweeps from a single new mutation of large effect. Yet, there is a conspicuous lack of theoretical research that explicitly links these phenomena (Paaby and Gibson 2016). The parameters determining the probability of adaptation from standing variation are the population size (N), the mutation rate toward an adaptive allele at the relevant locus (μ), and the relative selective advantage of the mutation after the environmental shift R_α (Hermisson and Pennings 2005). Specifically, where $2N\mu = \Theta > 1/\log(1+ R_\alpha)$, we expect adaptation from the standing genetic variation to be frequent (Hermisson and Pennings 2005; Messer and Petrov 2013).

Phenotypic plasticity influences all three of these parameters in a direction that promotes evolution from standing genetic variation. Phenotypic plasticity increases population size

(N) in new environments by reducing the cost of selection when plastic responses are adaptive under the new conditions (Chevin and Lande 2010). Phenotypic plasticity also increases the mutational target size, thereby increasing μ, because traits with phenotypic plasticity are likely to have more complex genetic architectures than more constitutively expressed traits (Sultan and Stearns 2005; Moczek 2008). The most profoundly affected of these parameters, however, is likely to be R_α. As discussed above, phenotypic plasticity may be due to conditional expression of genetic variation (Des Marais et al. 2013), leading to accumulation of polymorphism due to conditional neutrality and relaxed selection in the ancestral environment (Snell-Rood et al. 2010; Van Dyken and Wade 2010). Depending on the frequency with which the population has been historically exposed to the new environmental conditions, the fitness effects of genetic variation accumulated under conditional trait expression may not be symmetrically distributed around zero. If environmental conditions are sufficiently rare or entirely novel in the species' evolutionary history, this body of variation may be enriched for deleterious variants, including fixed lethal variants (Kawecki 1994), thereby reducing R_α. On the other hand, previous bouts of purifying selection on this variation during prior exposures can lead to an enrichment of adaptive alleles (Masel 2006). Taken together, it is possible that phenotypic plasticity increases R_α because it reduces deleterious effects of polymorphism in ancestral environments and increases adaptive effects in the new environment. Similarly, evolution of plastic developmental systems can lead to a bias toward adaptive effects of genetic variation (Draghi and Whitlock 2012).

We believe this discussion encapsulates why it is crucial to deepen the integration between the environment and the genotype in our understanding of the production and evolution of traits. From a traditional perspective, the empirical evidence of the role of standing genetic variation in evolutionary adaptation is contentious because it calls into question the utility of many of our best-understood models. Yet, when one appreciates that the environment may be central in determining the genetic architecture of traits and the patterns of genetic variation among and within populations, selection on the standing genetic variation becomes an intuitive component of adaptation.

Conclusion

It is increasingly clear that a gene-centric view of trait development is limited in its power to describe the diversity of phenotypic variation and therefore evolution. Better incorporating phenotypic plasticity into evolutionary thinking offers a potential solution, not only because the environment plays a critical role in determining traits, but because the theoretical development surrounding phenotypic plasticity has provided novel directions for inquiry. Empirical research investigating these questions is accumulating, but much about plasticity-led evolution remains unknown. In this chapter, we have attempted to highlight how the concept of cryptic genetic variation provides a connection between readily available empirical evidence and theoretical predictions about the role of plasticity in evolution that complements and extends a more developmental, trait-focused approach. We suggest that other investigators should similarly focus on diversifying the set of empirically testable hypotheses that address extent to which plasticity-led evolution is a widespread phenomenon in nature.

Ackowledgments

We would like to acknowledge Marjorie Oleksiak for her contribution to an earlier version of this chapter. We also acknowledge an anonymous reviewer and the editors, whose comments substantially improved this chapter. We thank Matthew Wund, whose work and many valuable discussions have shaped our thinking about phenotypic plasticity.

This work was completed with support from the John Templeton Foundation (grant no. 60501).

References

Abouheif, E., M.-J. Favé, A. Sofía Ibarrarán-Viniegra, M. P. Lesoway, A. Matteen Rafiqi, and R. Rajakumar. 2014. "Eco-evo-devo: The Time Has Come." *Advances in Experimental Medicine and Biology* 781:107–125.

Abouheif, E., and G. A. Wray. 2002. "Evolution of the Gene Network Underlying Wing Polyphenism in Ants." *Science* 297 (5579): 249–252. doi: 10.1126/science.1071468.

Amarillo-Suárez, A. R., and C. W. Fox. 2006. "Population Differences in Host Use by a Seed-Beetle: Local Adaptation, Phenotypic Plasticity and Maternal Effects." *Oecologia* 150 (2): 247–258. doi: 10.1007/s00442-006-0516-y.

Anderson, J. T., C. R. Lee, C. A. Rushworth, R. I. Colautti, and T. Mitchell-Olds. 2013. "Genetic Trade-Offs and Conditional Neutrality Contribute to Local Adaptation." *Molecular Ecology* 22 (3): 699–708. doi: 10.1111/j.1365–294X.2012.05522.x.

Aubin-Horth, N., and S. C. Renn. 2009. "Genomic Reaction Norms: Using Integrative Biology to Understand Molecular Mechanisms of Phenotypic Plasticity." *Molecular Ecology* 18 (18): 3763–3780. doi: 10.1111/j.1365–294X.2009.04313.x.

Aubret, F. 2015. "Island Colonisation and the Evolutionary Rates of Body Size in Insular Neonate Snakes." *Heredity* 115 (4): 349–356. doi: 10.1038/hdy.2014.65.

Ausin, I., C. Alonso-Blanco, and J.-M. Martinez-Zapater. 2004. "Environmental Regulation of Flowering." *International Journal of Developmental Biology* 49 (5–6): 689–705.

Badyaev, A. V. 2009. "Evolutionary Significance of Phenotypic Accommodation in Novel Environments: An Empirical Test of the Baldwin Effect." *Philosophical Transactions of the Royal Society of London B: Biological Sciences* 364 (1520): 1125–1141. doi: 10.1098/rstb.2008.0285.

Baldwin, J. M. 1896. "A New Factor in Evolution." *The American Naturalist* 30 (354): 441–451. doi: doi:10.1086/276408.

Barrett, R. D., and D. Schluter. 2008. "Adaptation from Standing Genetic Variation." *Trends in Ecology and Evolution* 23 (1): 38–44. doi: 10.1016/j.tree.2007.09.008.

Bell, G. 2010. "Fluctuating Selection: The Perpetual Renewal of Adaptation in Variable Environments." *Philosophical Transactions of the Royal Society of London B: Biological Sciences* 365 (1537): 87–97. doi: 10.1098/rstb.2009.0150.

Berenos, C., P. A. Ellis, J. G. Pilkington, S. H. Lee, J. Gratten, and J. M. Pemberton. 2015. "Heterogeneity of Genetic Architecture of Body Size Traits in a Free-living Population." *Molecular Ecology* 24 (8): 1810–1830. doi: 10.1111/mec.13146.

Berg, J. J., and G. Coop. 2014. "A Population Genetic Signal of Polygenic Adaptation." *PLoS Genetics* 10 (8): e1004412. doi: 10.1371/journal.pgen.1004412.

Bergland, A. O., E. L. Behrman, K. R. O'Brien, P. S. Schmidt, and D. A. Petrov. 2014. "Genomic Evidence of Rapid and Stable Adaptive Oscillations over Seasonal Time Scales in Drosophila." *PLoS Genetics* 10 (11): e1004775. doi: 10.1371/journal.pgen.1004775.

Bloom, J. S., I. M. Ehrenreich, W. T. Loo, T. L. Lite, and L. Kruglyak. 2013. "Finding the Sources of Missing Heritability in a Yeast Cross." *Nature* 494 (7436): 234–237. doi: 10.1038/nature11867.

Bourrat, P., Q. Lu, and E. Jablonka. 2017. "Why the Missing Heritability Might Not Be in the DNA." *Bioessays* 39 (7). doi: 10.1002/bies.201700067.

Boyle, E. A., Y. I. Li, and J. K. Pritchard. 2017. "An Expanded View of Complex Traits: From Polygenic to Omnigenic." *Cell* 169 (7): 1177–1186. doi: 10.1016/j.cell.2017.05.038.

Bradshaw, A. D. 1965. "Evolutionary Significance of Phenotypic Plasticity in Plants." *Advances in Genetics* 13 (1): 115–155.

Brieuc, M. S. O., C. D. Waters, D. P. Drinan, and K. A. Naish. 2018. "A Practical Introduction to Random Forest for Genetic Association Studies in Ecology and Evolution." *Molecular Ecology Resources*. doi: 10.1111/1755-0998.12773.

Chen, Y. S., J. D. Racca, P. W. Sequeira, N. B. Phillips, and M. A. Weiss. 2013. "Microsatellite-encoded Domain in Rodent Sry Functions as a Genetic Capacitor to Enable the Rapid Evolution of Biological Novelty." *Proceedings of the National Academy of Sciences USA* 110 (33): E3061–E3070. doi: 10.1073/pnas.1300860110.

Chevin, L. M., and R. Lande. 2010. "When Do Adaptive Plasticity and Genetic Evolution Prevent Extinction of a Density-regulated Population?" *Evolution* 64 (4): 1143–1150. doi: 10.1111/j.1558-5646.2009.00875.x.

Chevin, L. M., and R. Lande. 2011. "Adaptation to Marginal Habitats by Evolution of Increased Phenotypic Plasticity." *Journal of Evolutionary Biology* 24 (7): 1462–1476. doi: 10.1111/j.1420-9101.2011.02279.x.

Colautti, R. I., and S. C. H. Barrett. 2011. "Population Divergence along Lines of Genetic Variance and Covariance in the Invasive Plant *Lythrum salicaria* in Eastern North America." *Evolution* 65 (9): 2514–2529.

Colautti, R. I., and S. C. H. Barrett. 2013. "Rapid Adaptation to Climate Facilitates Range Expansion of an Invasive Plant." *Science* 342 (6156): 364–366. doi: 10.1126/science.1242121.

Colautti, R. I., C.-R. Lee, and T. Mitchell-Olds. 2012. "Origin, Fate, and Architecture of Ecologically Relevant Genetic Variation." *Current Opinion in Plant Biology* 15 (2): 199–204. doi: http://dx.doi.org/10.1016/j.pbi.2012.01.016.

Conover, D. O., and E. T. Schultz. 1995. "Phenotypic Similarity and the Evolutionary Significance of Countergradient Variation." *Trends in Ecology and Evolution* 10 (6): 248–252.

Cresko, W. A., A. Amores, C. Wilson, J. Murphy, M. Currey, P. Phillips, M. A. Bell, C. B. Kimmel, and J. H. Postlethwait. 2004. "Parallel Genetic Basis for Repeated Evolution of Armor Loss in Alaskan Threespine Stickleback Population Divergence Along Lines of Genetic Variance and Covariance in the Invasive Plant Lythrum salicaria in Eastern North Americaopulations." *Proceedings of the National Academy of Sciences USA* 101 (16): 6050–6055.

Crispo, E. 2008. "Modifying Effects of Phenotypic Plasticity on Interactions among Natural Selection, Adaptation and Gene Flow." *Journal of Evolutionary Biology* 21 (6): 1460–1469. doi: 10.1111/j.1420-9101.2008.01592.x.

Davidson, A. M., M. Jennions, and A. B. Nicotra. 2011. "Do Invasive Species Show Higher Phenotypic Plasticity than Native Species and, If So, Is It Adaptive? A Meta-analysis." *Ecology Letters* 14 (4): 419–431. doi: 10.1111/j.1461-0248.2011.01596.x.

Dayan, D. I., D. L. Crawford, and M. F. Oleksiak. 2015. "Phenotypic Plasticity in Gene Expression Contributes to Divergence of Locally Adapted Populations of *Fundulus heteroclitus*." *Molecular Ecology* 24 (13): 3345–3359. doi: 10.1111/mec.13188.

Des Marais, D. L., K. M. Hernandez, and T. E. Juenger. 2013. "Genotype-by-Environment Interaction and Plasticity: Exploring Genomic Responses of Plants to the Abiotic Environment." *Annual Review of Ecology, Evolution, and Systematics* 44:5–29.

DeWitt, T. J., and S. M. Scheiner. 2004. *Phenotypic Plasticity: Functional and Conceptual Approaches*. New York: Oxford University Press.

Draghi, J. A., and M. C. Whitlock. 2012. "Phenotypic Plasticity Facilitates Mutational Variance, Genetic Variance, and Evolvability along the Major Axis of Environmental Variation." *Evolution* 66 (9): 2891–2902. doi: 10.1111/j.1558-5646.2012.01649.x.

Druet, T., L. Pérez-Pardal, C. Charlier, and M. Gautier. 2013. "Identification of Large Selective Sweeps Associated with Major Genes in Cattle." *Animal Genetics* 44 (6): 758–762. doi: 10.1111/age.12073.

Ehrenreich, I. M., and D. W. Pfennig. 2015. "Genetic Assimilation: A Review of Its Potential Proximate Causes and Evolutionary Consequences." *Annals of Botany*. doi: 10.1093/aob/mcv130.

Eichler, E. E., J. Flint, G. Gibson, A. Kong, S. M. Leal, J. H. Moore, and J. H. Nadeau. 2010. "Missing Heritability and Strategies for Finding the Underlying Causes of Complex Disease." *Nature Reviews Genetics* 11 (6): 446–450. doi: 10.1038/nrg2809.

Ellner, S. P., M. A. Geber, and N. G. Hairston. 2011. "Does Rapid Evolution Matter? Measuring the Rate of Contemporary Evolution and Its Impacts on Ecological Dynamics." *Ecology Letters* 14 (6): 603–614. doi: 10.1111/j.1461-0248.2011.01616.x.

Fagny, M., E. Patin, D. Enard, L. B. Barreiro, L. Quintana-Murci, and G. Laval. 2014. "Exploring the Occurrence of Classic Selective Sweeps in Humans Using Whole-Genome Sequencing Data Sets." *Molecular Biology and Evolution* 31 (7): 1850–1868. doi: 10.1093/molbev/msu118.

Falconer, D. S., and T. F. C. Mackay. 1996. *Introduction to Quantitative Genetics*. London: Longman.

Fitzpatrick, B. M. 2012. "Underappreciated Consequences of Phenotypic Plasticity for Ecological Speciation." *International Journal of Ecology* 2012:12. doi: 10.1155/2012/256017.

Flint, J., and T. F. Mackay. 2009. "Genetic Architecture of Quantitative Traits in Mice, Flies, and Humans." *Genome Research* 19 (5): 723–733. doi: 10.1101/gr.086660.108.

Frank, S. A. 2011. "Natural Selection. ii. Developmental Variability and Evolutionary Rate." *Journal of Evolutionary Biology* 24:2310–2320.

Fraser, H. B. 2013. "Gene Expression Drives Local Adaptation in Humans." *Genome Research* 23 (7): 1089–1096.

Fry, J. D., S. L. Heinsohn, and F. C. Mackay Trudy. 1996. "The Contribution of New Mutations to Genotype-Environment Interaction for Fitness in *Drosophila melanogaster*." *Evolution* 50 (6): 2316–2327. doi: 10.2307/2410700.

Garud, N. R., P. W. Messer, E. O. Buzbas, and D. A. Petrov. 2015. "Recent Selective Sweeps in North American *Drosophila melanogaster* Show Signatures of Soft Sweeps." *PLoS Genetics* 11 (2) :e1005004. doi: 10.1371/journal.pgen.1005004.

Geng, Y.-P., Xiao-Yun Pan, C.-Y. Xu, W.-J. Zhang, B. Li, J.-K. Chen, B.-R. Lu, and Z.-P. Song. 2006. "Phenotypic Plasticity Rather Than Locally Adapted Ecotypes Allows the Invasive Alligator Weed to Colonize a Wide Range of Habitats." *Biological Invasions* 9 (3): 245–256. doi: 10.1007/s10530-006-9029-1.

Ghalambor, C. K., K. L. Hoke, E. W. Ruell, E. K. Fischer, D. N. Reznick, and K. A. Hughes. 2015. "Non-adaptive Plasticity Potentiates Rapid Adaptive Evolution of Gene Expression in Nature." *Nature* 525 (7569): 372–375. doi: 10.1038/nature15256.

Ghalambor, C. K., J. K. McKay, S. P. Carroll, and D. N. Reznick. 2007. "Adaptive versus Non-adaptive Phenotypic Plasticity and the Potential for Contemporary Adaptation in New Environments." *Functional Ecology* 21 (3): 394–407. doi: 10.1111/j.1365-2435.2007.01283.x.

Gibson, G., and I. Dworkin. 2004. "Uncovering Cryptic Genetic Variation." *Nature Reviews Genetics* 5 (9): 681–690. doi: 10.1038/nrg1426.

Gilbert, S. F., and D. Epel. 2015. *Ecological Developmental Biology: The Environmental Regulation of Development, Health, and Evolution*. Sunderland: Sinauer Associates Inc.

Gomez-Mestre, I., and D. R. Buchholz. 2006. "Developmental Plasticity Mirrors Differences among Taxa in Spadefoot Toads Linking Plasticity and Diversity." *Proceedings of the National Academy of Sciences USA* 103 (50): 19021–19026.

Grether, G. F. 2005. "Environmental Change, Phenotypic Plasticity, and Genetic Compensation." *The American Naturalist* 166 (4): E115-E123. doi: 10.1086/432023.

Haldane, J. B. S. 1957. "The Cost of Natural Selection." *Journal of Genetics* 55 (3): 511–524.

Hall, M. C., D. B. Lowry, and J. H. Willis. 2010. "Is Local Adaptation in *Mimulus guttatus* Caused by Trade-Offs at Individual Loci?" *Molecular Ecology* 19 (13): 2739–2753. doi: 10.1111/j.1365–294X.2010.04680.x.

Hancock, A. M., G. Alkorta-Aranburu, D. B. Witonsky, and A. Di Rienzo. 2010. "Adaptations to New Environments in Humans: The Role of Subtle Allele Frequency Shifts." *Philosophical Transactions of the Royal Society of London B: Biological Sciences* 365 (1552): 2459–2468.

Hancock, A. M., D. B. Witonsky, E. Ehler, G. Alkorta-Aranburu, C. Beall, A. Gebremedhin, R. Sukernik, G. Utermann, J. Pritchard, and G. Coop. 2010. "Human Adaptations to Diet, Subsistence, and Ecoregion Are Due to Subtle Shifts in Allele Frequency." *Proceedings of the National Academy of Sciences USA* 107 (Supplement 2): 8924–8930.

Hansen, T. F., C. Pélabon, and D. Houle. 2011. "Heritability Is Not Evolvability." *Evolutionary Biology* 38 (3): 258.

Hendry, A. P. 2013. "Key Questions in the Genetics and Genomics of Eco-Evolutionary Dynamics." *Heredity* 111 (6): 456–466. doi: 10.1038/hdy.2013.75.

Hendry, A. P. 2016. *Eco-evolutionary Dynamics*. Princeton, NJ: Princeton University Press.

Hermisson, J., and P. S. Pennings. 2005. "Soft Sweeps Molecular Population Genetics of Adaptation from Standing Genetic Variation." *Genetics* 169 (4): 2335–2352.

Hermisson, J., P. S. Pennings, and J. Kelley. 2017. "Soft Sweeps and Beyond: Understanding the Patterns and Probabilities of Selection Footprints under Rapid Adaptation." *Methods in Ecology and Evolution* 8 (6): 700–716. doi: 10.1111/2041–210x.12808.

Hermisson, J., and G. P. Wagner. 2004. "The Population Genetic Theory of Hidden Variation and Genetic Robustness." *Genetics* 168 (4): 2271–2284.

Hernandez, R. D., J. L. Kelley, E. Elyashiv, S. C. Melton, A. Auton, G. McVean, G. Sella, and M. Przeworski. 2011. "Classic Selective Sweeps Were Rare in Recent Human Evolution." *Science* 331 (6019): 920–924.

Ho, W. C., and J. Zhang. 2018. "Evolutionary Adaptations to New Environments Generally Reverse Plastic Phenotypic Changes." *Nature Communications* 9 (1): 350. doi: 10.1038/s41467-017-02724-5.

Hoban, S., J. L. Kelley, K. E. Lotterhos, M. F. Antolin, G. Bradburd, D. B. Lowry, M. L. Poss, L. K. Reed, A. Storfer, and M. C. Whitlock. 2016. "Finding the Genomic Basis of Local Adaptation: Pitfalls, Practical Solutions, and Future Directions." *The American Naturalist* 188 (4): 379–397. doi: 10.1086/688018.

Hodgins-Davis, A., and J. P. Townsend. 2009. "Evolving Gene Expression: From G to E to GxE." *Trends in Ecology and Evolution* 24 (12): 649–658. doi: 10.1016/j.tree.2009.06.011.

Hoffmann, A. A., and J. Merila. 1999. "Heritable Variation and Evolution under Favourable and Unfavourable Conditions." *Trends in Ecology and Evolution* 14 (3): 96–101.

Hunt, B. G., L. Ometto, Y. Wurm, D. Shoemaker, S. V. Yi, L. Keller, and M. A. Goodisman. 2011. "Relaxed Selection Is a Precursor to the Evolution of Phenotypic Plasticity." *Proceedings of the National Academy of Sciences USA* 108 (38): 15936–15941. doi: 10.1073/pnas.1104825108.

Jaeger, J. 2011. "The Gap Gene Network." *Cellular and Molecular Life Sciences* 68 (2): 243–274. doi: 10.1007/s00018-010-0536-y.

Jensen, J. D. 2014. "On the Unfounded Enthusiasm for Soft Selective Sweeps." *Nature Communications* 5:5281.

Jones, F. C., M. G. Grabherr, Y. F. Chan, P. Russell, E. Mauceli, J. Johnson, R. Swofford, et al. 2012. "The Genomic Basis of Adaptive Evolution in Threespine Sticklebacks." *Nature* 484 (7392): 55–61. doi: 10.1038/nature10944.

Jong, G. de. 1995. "Phenotypic Plasticity as a Product of Selection in a Variable Environment." *The American Naturalist* 145 (4): 493–512. doi: 10.2307/2462965.

Joron, M., L. Frezal, R. T. Jones, N. L. Chamberlain, S. F. Lee, C. R. Haag, A. Whibley, et al. 2011. "Chromosomal Rearrangements Maintain a Polymorphic Supergene Controlling Butterfly Mimicry." *Nature* 477:203–206.

Josephs, E. B., Y. W. Lee, J. R. Stinchcombe, and S. I. Wright. 2015. "Association Mapping Reveals the Role of Purifying Selection in the Maintenance of Genomic Variation in Gene Expression." *Proceedings of the National Academy of Sciences USA* 112 (50): 15390–15395. doi: 10.1073/pnas.1503027112.

Kassen, R. 2002. "The Experimental Evolution of Specialists, Generalists, and the Maintenance of Diversity." *Journal of Evolutionary Biology* 15 (2): 173–190. doi: 10.1046/j.1420–9101.2002.00377.x.

Kawecki, T. J. 1994. "Accumulation of Deleterious Mutations and the Evolutionary Cost of Being a Generalist." *The American Naturalist* 144 (5): 833–838.

Kawecki, T. J., and D. Ebert. 2004. "Conceptual Issues in Local Adaptation." *Ecology Letters* 7 (12): 1225–1241. doi: 10.1111/j.1461–0248.2004.00684.x.

Kenkel, C. D., and M. V. Matz. 2016. "Gene Expression Plasticity as a Mechanism of Coral Adaptation to a Variable Environment." *Nature Ecology and Evolution* 1:0014.

Kovaka, K. 2017. "Underdetermination and Evidence in the Developmental Plasticity Debate." *The British Journal for the Philosophy of Science*: axx038. doi: 10.1093/bjps/axx038.

Kumar, S. K., M. W. Feldman, D. H. Rehkopf, and S. Tuljapurkar. 2016. "Correction for Krishna Kumar et al., Limitations of GCTA as a Solution to the Missing Heritability Problem." *Proceedings of the National Academy of Sciences USA* 113 (6): E813. doi: 10.1073/pnas.1600634113.

Kuttner, E., K. J. Parsons, A. A. Easton, S. Skulason, R. G. Danzmann, and M. M. Ferguson. 2014. "Hidden Genetic Variation Evolves with Ecological Specialization: The Genetic Basis of Phenotypic Plasticity in Arctic Charr Ecomorphs." *Evolution and Development* 16 (4): 247–257. doi: 10.1111/ede.12087.

Lahti, D. C., N. A. Johnson, B. C. Ajie, S. P. Otto, A. P. Hendry, D. T. Blumstein, R. G. Coss, K. Donohue, and S. A. Foster. 2009. "Relaxed Selection in the Wild." *Trends in Ecology and Evolution* 24 (9): 487–496. doi: 10.1016/j.tree.2009.03.010.

Laland, K, N., T. Uller, M. W. Feldman, K. Sterelny, G. B. Muller, A. Moczek, E. Jablonka, and J. Odling-Smee. 2015. "The Extended Evolutionary Synthesis: Its Structure, Assumptions and Predictions." *Proceedings of the Royal Society B: Biological Sciences* 282 (1813): 20151019. doi: 10.1098/rspb.2015.1019.

Laland, K., T. Uller, M. Feldman, K. Sterelny, G. B. Müller, A. Moczek, E. Jablonka, J. Odling-Smee, G. A. Wray, and H. E. Hoekstra. 2014. "Does Evolutionary Theory Need a Rethink?" *Nature* 514 (7521): 161.

Ledon-Rettig, C. C., D. W. Pfennig, A. J. Chunco, and I. Dworkin. 2014. "Cryptic Genetic Variation in Natural Populations: A Predictive Framework." *Integrative and Comparitive Biology* 54 (5): 783–793. doi: 10.1093/icb/icu077.

Ledón-Rettig, C. C., D. W. Pfennig, and E. J. Crespi. 2010. "Diet and Hormonal Manipulation Reveal Cryptic Genetic Variation: Implications for the Evolution of Novel Feeding Strategies." *Proceedings of the Royal Society of London B: Biological Sciences* 277 (1700): 3569–3578. doi: 10.1098/rspb.2010.0877.

Lee, C. E. 2002. "Evolutionary Genetics of Invasive Species." *Trends in Ecology and Evolution* 17 (8): 386–391. doi: 10.1016/s0169–5347(02)02554–5.

Leichty, A. R., D. W. Pfennig, C. D. Jones, and K. S. Pfennig. 2012. "Relaxed Genetic Constraint Is Ancestral to the Evolution of Phenotypic Plasticity." *Integrative and Comparative Biology* 52 (1): 16–30. doi: 10.1093/icb/ics049.

Le Rouzic, A., and O. Carlborg. 2008. "Evolutionary Potential of Hidden Genetic Variation." *Trends in Ecology and Evolution* 23 (1): 33–37. doi: 10.1016/j.tree.2007.09.014.

Levis, N. A., and D. W. Pfennig. 2016. "Evaluating 'Plasticity-First' Evolution in Nature: Key Criteria and Empirical Approaches." *Trends in Ecology and Evolution* 31 (7): 563–574. doi: 10.1016/j.tree.2016.03.012.

Lewontin, R. C. 1974. *The Genetic Basis of Evolutionary Change*. Vol. 560: New York: Columbia University Press.

Lind, M. I., K. Yarlett, J. Reger, M. J. Carter, and A. P. Beckerman. 2015. "The Alignment between Phenotypic Plasticity, the Major Axis of Genetic Variation and the Response to Selection." *Proceedings of the Royal Society B: Biological Sciences* 282 (1816): 20151651. doi: 10.1098/rspb.2015.1651.

Linnen, C. R., Y. P. Poh, B. K. Peterson, R. D. Barrett, J. G. Larson, J. D. Jensen, and H. E. Hoekstra. 2013. "Adaptive Evolution of Multiple Traits through Multiple Mutations at a Single Gene." *Science* 339 (6125): 1312–1316. doi: 10.1126/science.1233213.

Losos, J. B., D. A. Creer, D. Glossip, R. Goellner, A. Hampton, G. Roberts, N. Haskell, P. Taylor, and J. Ettling. 2000. "Evolutionary Implications of Phenotypic Plasticity in the Hindlimb of the Lizard *Anolis sagrei*." *Evolution* 54 (1): 301–305.

Manolio, T. A., F. S. Collins, N. J. Cox, D. B. Goldstein, L. A. Hindorff, D. J. Hunter, M. I. McCarthy, E. M. Ramos, Lon R. Cardon, and A. Chakravarti. 2009. "Finding the Missing Heritability of Complex Diseases." *Nature* 461 (7265): 747–753.

Marigorta, U. M., and G. Gibson. 2014. "A Simulation Study of Gene-by-Environment Interactions in GWAS Implies Ample Hidden Effects." *Frontiers in Genetics* 5:225. doi: 10.3389/fgene.2014.00225.

Masel, J. 2006. "Cryptic Genetic Variation Is Enriched for Potential Adaptations." *Genetics* 172 (3): 1985–1991. doi: 10.1534/genetics.105.051649.

Mayr, E., and W. B. Provine. 1998. *The Evolutionary Synthesis: Perspectives on the Unification of Biology.* Cambridge, MA: Harvard University Press.

McCairns, R. J., and L. Bernatchez. 2010. "Adaptive Divergence between Freshwater and Marine Sticklebacks: Insights into the Role of Phenotypic Plasticity from an Integrated Analysis of Candidate Gene Expression." *Evolution* 64 (4): 1029–1047. doi: 10.1111/j.1558–5646.2009.00886.x.

McGuigan, K., N. Nishimura, M. Currey, D. Hurwit, and W. A. Cresko. 2011. "Cryptic Genetic Variation and Body Size Evolution in Threespine Stickleback." *Evolution* 65 (4): 1203–1211. doi: 10.1111/j.1558–5646.2010.01195.x.

McGuigan, K., and C. M. Sgro. 2009. "Evolutionary Consequences of Cryptic Genetic Variation." *Trends in Ecology and Evolution* 24 (6): 305–311.

Messer, P. W., S. P. Ellner, and N. G. Hairston, Jr. 2016. "Can Population Genetics Adapt to Rapid Evolution?" *Trends in Genetics* 32 (7): 408–418. doi: 10.1016/j.tig.2016.04.005.

Messer, P. W., and D. A. Petrov. 2013. "Population Genomics of Rapid Adaptation by Soft Selective Sweeps." *Trends in Ecology and Evolution* 28 (11): 659–669.

Moczek, A. P. 2008. "On the Origins of Novelty in Development and Evolution." *BioEssays* 30 (5): 432–447.

Moczek, A. P. 2009. "Phenotypic Plasticity and the Origins of Diversity: A Case Study on Horned Beetles." In *Phenotypic Plasticity in Insects: Mechanisms and Consequences*, edited by Douglas W. Whitman and T. N. Ananthakrishnan, 81–134. Enfield, NH: CRC Press.

Moczek, A. P., S. Sultan, S. Foster, C. Ledón-Rettig, I. Dworkin, H. F. Nijhout, E. Abouheif, and D. W. Pfennig. 2011. "The Role of Developmental Plasticity in Evolutionary Innovation." *Proceedings of the Royal Society B: Biological Sciences* 278 (1719): 2705–2713. doi: 10.1098/rspb.2011.0971.

Morgan, C. L. 1896. "On Modification and Variation." *Science*, 733–740.

Murren, C. J., J. R. Auld, H. Callahan, C. K. Ghalambor, C. A. Handelsman, M. A. Heskel, J. G. Kingsolver, et al. 2015. "Constraints on the Evolution of Phenotypic Plasticity: Limits and Costs of Phenotype and Plasticity." *Heredity* 115 (4): 293–301. doi: 10.1038/hdy.2015.8.

Nicol, J. M., G. G. Ganf, K. F. Walker, and B. Gawne. 2017. "Response of Three Arid Zone Floodplain Plant Species to Inundation." *Plant Ecology* 219 (1): 57–67. doi: 10.1007/s11258-017-0777-z.

Nijhout, H. F., and S. M. Paulsen. 1997. "Developmental Models and Polygenic Characters." *The American Naturalist* 149 (2): 394–405. doi: 10.1086/285996.

Oleksiak, M. F., G. A. Churchill, and D. L. Crawford. 2002. "Variation in Gene Expression within and among Natural Populations." *Nature Genetics* 32 (2): 261–266. doi: 10.1038/ng983.

Orgogozo, V., B. Morizot, and A. Martin. 2015. "The Differential View of Genotype-Phenotype Relationships." *Front Genet* 6:179.

Orizaola, G., and A. Laurila. 2016. "Developmental Plasticity Increases at the Northern Range Margin in a Warm-Dependent Amphibian." *Evolutionary Applications* 9 (3): 471–478. doi: 10.1111/eva.12349.

Orr, H. A. 1998. "The Population Genetics of Adaptation: The Distribution of Factors Fixed during Adaptive Evolution." *Evolution* 52:935–949.

Orr, H. A. 1999. "An Evolutionary Dead End?" *Science* 285 (5426): 343–344. doi: 10.1126/science.285.5426.343.

Orr, H. A. 2005. "The Genetic Theory of Adaptation: A Brief History." *Nature Reviews Genetics* 6 (2): 119–127. doi: 10.1038/nrg1523.

Paaby, A. B., and G. Gibson. 2016. "Cryptic Genetic Variation in Evolutionary Developmental Genetics." *Biology (Basel)* 5 (2). doi: 10.3390/biology5020028.

Paaby, A. B., and M. V. Rockman. 2014. "Cryptic Genetic Variation: Evolution's Hidden Substrate." *Nature Reviews Genetics* 15 (4): 247–258. doi: 10.1038/nrg3688.

Parsons, K. J., M. Concannon, D. Navon, J. Wang, I. Ea, K. Groveas, C. Campbell, and R. C. Albertson. 2016. "Foraging Environment Determines the Genetic Architecture and Evolutionary Potential of Trophic Morphology in Cichlid fishes." *Molecular Ecology.* doi: 10.1111/mec.13801.

Pavličev, M., and J. M. Cheverud. 2015. "Constraints Evolve: Context Dependency of Gene Effects Allows Evolution of Pleiotropy." *Annual Review of Ecology, Evolution, and Systematics* 46 (1): 413–434. doi: 10.1146/annurev-ecolsys-120213-091721.

Pespeni, M. H., J. T. Ladner, and A. P. Moczek. 2017. "Signals of Selection in Conditionally Expressed Genes in the Diversification of Three Horned Beetle Species." *Journal of Evolutionary Biology* 30 (9): 1644–1657. doi: doi:10.1111/jeb.13079.

Pfennig, D. W., M. A. Wund, E. C. Snell-Rood, T. Cruickshank, C. D. Schlichting, and A. P. Moczek. 2010. "Phenotypic Plasticity's Impacts on Diversification and Speciation." *Trends in Ecology and Evolution* 25 (8): 459–467. doi: 10.1016/j.tree.2010.05.006.

Phillips, P. C. 2008. "Epistasis—The Essential Role of Gene Interactions in the Structure and Evolution of Genetic Systems." *Nature Reviews Genetics* 9 (11): 855–867. doi: 10.1038/nrg2452.

Pickrell, J. K. 2014. "Joint Analysis of Functional Genomic Data and Genome-wide Association Studies of 18 Human Traits." *American Journal of Human Genetics* 94 (4): 559–573. doi: 10.1016/j.ajhg.2014.03.004.

Pigliucci, M. 2001. *Phenotypic Plasticity: Beyond Nature and Nurture.* Baltimore, MD: Johns Hopkins University Press.

Pigliucci, M., C. J. Murren, and C. D. Schlichting. 2006. "Phenotypic Plasticity and Evolution by Genetic Assimilation." *Journal of Experimental Biology* 209 (Pt 12): 2362–2367. doi: 10.1242/jeb.02070.

Price, T. D., A. Qvarnström, and D. E. Irwin. 2003. "The Role of Phenotypic Plasticity in Driving Genetic Evolution." *Proceedings of the Royal Society of London B: Biological Sciences* 270 (1523): 1433–1440.

Pritchard, J. K., and A. Di Rienzo. 2010. "Adaptation–Not by Sweeps Alone." *Nature Reviews Genetics* 11 (10):665–667.

Pritchard, J. K., J. K. Pickrell, and G. Coop. 2010. "The Genetics of Human Adaptation: Hard Sweeps, Soft Sweeps, and Polygenic Adaptation." *Current Biology* 20 (4): R208–R215.

Przeworski, M., G. Coop, and J. D. Wall. 2005. "The Signature of Positive Selection on Standing Genetic Variation." *Evolution* 59 (11): 2312–2323.

Queitsch, C., T. A. Sangster, and S. Lindquist. 2002. "Hsp90 as a Capacitor of Phenotypic Variation." *Nature* 417 (6889): 618–624. doi: http://www.nature.com/nature/journal/v417/n6889/suppinfo/nature749_S1.html.

Rajakumar, R., D. San Mauro, M. B. Dijkstra, M. H. Huang, D. E. Wheeler, F. Hiou-Tim, A. Khila, M. Cournoyea, and E. Abouheif. 2012. "Ancestral Developmental Potential Facilitates Parallel Evolution in Ants." *Science* 335 (6064): 79–82.

Réale, D., A. G. McAdam, S. Boutin, and D. Berteaux. 2003. "Genetic and Plastic Responses of a Northern Mammal to Climate Change." *Proceedings of the Royal Society of London B: Biological Sciences* 270 (1515): 591–596. doi: 10.1098/rspb.2002.2224.

Robinson, B. W., and R. Dukas. 1999. "The Influence of Phenotypic Modifications on Evolution: The Baldwin Effect and Modern Perspectives." *Oikos* 85 (3): 582–589. doi: 10.2307/3546709.

Rockman, M. V. 2012. "The QTN Program and the Alleles That Matter for Evolution: All That's Gold Does Not Glitter." *Evolution* 66 (1): 1–17.

Rokholm, B., K. Silventoinen, L. Ängquist, A. Skytthe, K. Ohm Kyvik, and T. I. A. Sørensen. 2011. "Increased Genetic Variance of BMI with a Higher Prevalence of Obesity." *PLoS ONE* 6 (6): e20816.

Rowinski, P. K., and B. Rogell. 2017. "Environmental Stress Correlates with Increases in Both Genetic and Residual Variances: A Meta-analysis of Animal Studies." *Evolution* 71 (5): 1339–1351. doi: 10.1111/evo.13201.

Rutherford, S. L., and S. Lindquist. 1998. "Hsp90 as a Capacitor for Morphological Evolution." *Nature* 396 (6709): 336–342. doi: http://www.nature.com/nature/journal/v396/n6709/suppinfo/396336a0_S1.html.

Sangster, T. A., N. Salathia, H. N. Lee, E. Watanabe, K. Schellenberg, K. Morneau, H. Wang, S. Undurraga, C. Queitsch, and S. Lindquist. 2008. "HSP90-buffered Genetic Variation Is Common in *Arabidopsis thaliana*." *Proceedings of the National Academy of Sciences USA* 105 (8): 2969–2974.

Santure, A. W., J. Poissant, I. De Cauwer, K. van Oers, M. R. Robinson, J. L. Quinn, M. A. Groenen, M. E. Visser, B. C. Sheldon, and J. Slate. 2015. "Replicated Analysis of the Genetic Architecture of Quantitative Traits in Two Wild Great Tit Populations." *Molecular Ecology* 24 (24): 6148–6162. doi: 10.1111/mec.13452.

Scarcelli, N., J. M. Cheverud, B. A. Schaal, and P. X. Kover. 2007. "Antagonistic Pleiotropic Effects Reduce the Potential Adaptive Value of the FRIGIDA Locus." *Proceedings of the National Academy of Sciences USA* 104 (43): 16986–16991. doi: 10.1073/pnas.0708209104.

Schizophrenia Working Group of the Psychiatric Genomics, Consortium. 2014. "Biological Insights from 108 Schizophrenia-Associated Genetic Loci." *Nature* 511 (7510): 421–427. doi: 10.1038/nature13595.

Schlichting, C. D., and M. Pigliucci. 1998. *Phenotypic Evolution: A Reaction Norm Perspective.* Sunderland, MA: Sinauer Associates. ·

Schlichting, C. D., and M. A. Wund. 2014. "Phenotypic Plasticity and Epigenetic Marking: An Assessment of Evidence for Genetic Accommodation." *Evolution* 68 (3): 656–672. doi: 10.1111/evo.12348.

Schmalhausen, I. 1949. *Factors of Evolution: The Theory of Stabilizing Selection.* Philadelphia: Blakiston.

Schneider, R. F., Y. Li, A. Meyer, and H. M. Gunter. 2014. "Regulatory Gene Networks That Shape the Development of Adaptive Phenotypic Plasticity in a Cichlid Fish." *Molecular Ecology* 23 (18): 4511–4526. doi: 10.1111/mec.12851.

Schneider, R. F., and A. Meyer. 2017. "How Plasticity, Genetic Assimilation and Cryptic Genetic Variation May Contribute to Adaptive Radiations." *Molecular Ecology* 26 (1): 330–350. doi: 10.1111/mec.13880.

Schulte, Patricia M., Timothy M. Healy, and Nann A. Fangue. 2011. "Thermal Performance Curves, Phenotypic Plasticity, and the Time Scales of Temperature Exposure." *Integrative and Comparative Biology* 51 (5): 691–702.

Shi, H., G. Kichaev, and B. Pasaniuc. 2016. "Contrasting the Genetic Architecture of 30 Complex Traits from Summary Association Data." *American Journal of Human Genetics* 99 (1): 139–153. doi: 10.1016/j.ajhg.2016.05.013.

Sikkink, Kristin L., Rose M. Reynolds, William A. Cresko, and Patrick C. Phillips. 2015. "Environmentally Induced Changes in Correlated Responses to Selection Reveal Variable Pleiotropy across a Complex Genetic Network." *Evolution* 69 (5): 1128–1142. doi: 10.1111/evo.12651.

Simpson, G. G. 1944. *Tempo and Mode in Evolution.* New York: Columbia University Press.

Snell-Rood, E. C., A. Cash, M. V. Han, T. Kijimoto, J. Andrews, and A. P. Moczek. 2011. "Developmental Decoupling of Alternative Phenotypes: Insights from the Transcriptomes of Horn-Polyphenic Beetles." *Evolution* 65 (1): 231–245. doi: 10.1111/j.1558–5646.2010.01106.x.

Snell-Rood, E. C., J. D. Van Dyken, T. Cruickshank, M. J. Wade, and A. P. Moczek. 2010. "Toward a Population Genetic Framework of Developmental Evolution: The Costs, Limits, and Consequences of Phenotypic Plasticity." *Bioessays* 32 (1): 71–81. doi: 10.1002/bies.200900132.

Standen, E. M., T. Y. Du, and H. C. E. Larsson. 2014. "Developmental Plasticity and the Origin of Tetrapods." *Nature* 513 (7516): 54–58.

Stanton-Geddes, J., J. B. Yoder, R. Briskine, N. D. Young, and P. Tiffin. 2013. "Estimating Heritability using Genomic Data." *Methods in Ecology and Evolution* 4 (12): 1151–1158.

Stephan, W. 2016. "Signatures of Positive Selection: From Selective Sweeps at Individual Loci to Subtle Allele Frequency Changes in Polygenic Adaptation." *Molecular Ecology* 25 (1): 79–88. doi: 10.1111/mec.13288.

Stulp, G., and L. Barrett. 2016. "Evolutionary Perspectives on Human Height Variation." *Biological Reviews* 91 (1): 206–234. doi: 10.1111/brv.12165.

Sultan, S. E. 2015. *Organism and Environment: Ecological Development, Niche Construction, and Adaptation.* New York: Oxford University Press.

Sultan, S. E., and H. G. Spencer. 2002. "Metapopulation Structure Favors Plasticity over Local Adaptation." *The American Naturalist* 160 (2): 271–283. doi: 10.1086/341015.

Sultan, S. E., and S. C. Stearns. 2005. "Environmentally Contingent Variation: Phenotypic Plasticity and Norms of Reaction." In *Variation: A Central Concept in Biology,* edited by B. Hallgrimsson and B. K. Hall, 303–332. Burlington, MA: Elsevier Academic Press.

Suzuki, Y., and H. F. Nijhout. 2006. "Evolution of a Polyphenism by Genetic Accommodation." *Science* 311 (5761): 650–652. doi: 10.1126/science.1118888.

Takahashi, K. H. 2015. "Novel Genetic Capacitors and Potentiators for the Natural Genetic Variation of Sensory Bristles and Their Trait Specificity in *Drosophila melanogaster*." *Molecular Ecology* 24 (22): 5561–5572. doi: 10.1111/mec.13407.

Teshima, K. M., G. Coop, and M.Przeworski. 2006. "How Reliable Are Empirical Genomic Scans for Selective Sweeps?" *Genome Research* 16 (6): 702–712.

Therkildsen, N. O., J. Hemmer-Hansen, T. D. Als, D. P. Swain, M. J. Morgan, E. A. Trippel, S. R. Palumbi, D. Meldrup, and E. E. Nielsen. 2013. "Microevolution in Time and Space: SNP Analysis of Historical DNA Reveals Dynamic Signatures of Selection in Atlantic Cod." *Molecular Ecology* 22 (9): 2424–2440. doi: 10.1111/mec.12260.

Thibert-Plante, X., and A. P. Hendry. 2011. "The Consequences of Phenotypic Plasticity for Ecological Speciation." *Journal of Evolutionary Biology* 24 (2): 326–342. doi: 10.1111/j.1420–9101.2010.02169.x.

Torres-Dowdall, J., C. A. Handelsman, D. N. Reznick, and C. K. Ghalambor. 2012. "Local Adaptation and the Evolution of Phenotypic Plasticity in Trinidadian Guppies (*Poecilia reticulata*)." *Evolution* 66 (11): 3432–3443. doi: 10.1111/j.1558–5646.2012.01694.x.

Turchin, M. C., C. W. K. Chiang, C. D. Palmer, S. Sankararaman, D. Reich, J. N. Hirschhorn, and Genetic Investigation of Anthropometric Traits Consortium. 2012. "Evidence of Widespread Selection on Standing Variation in Europe at Height-Associated SNPs." *Nature Genetics* 44 (9): 1015–1019.

Van Dyken, J. D., and M. J. Wade. 2010. "The Genetic Signature of Conditional Expression." *Genetics* 184 (2): 557–570. doi: 10.1534/genetics.109.110163.

Van Kleunen, M., and M. Fischer. 2005. "Constraints on the Evolution of Adaptive Phenotypic Plasticity in Plants." *New Phytologist* 166 (1): 49–60. doi: 10.1111/j.1469–8137.2004.01296.x.

Via, S., R. Gomulkiewicz, G. De Jong, S. M. Scheiner, C. D. Schlichting, and P. H. van Tienderen. 1995. "Adaptive Phenotypic Plasticity: Consensus and Controversy." *Trends in Ecology and Evolution* 10:212–217.

Visscher, P. M., M. A. Brown, M. I. McCarthy, and J. Yang. 2012. "Five Years of GWAS Discovery." *American Journal of Human Genetics* 90 (1): 7–24. doi: 10.1016/j.ajhg.2011.11.029.

Vitti, J. J., S. R. Grossman, and P. C. Sabeti. 2013. "Detecting Natural Selection in Genomic Data." *Annual Review of Genetics* 47:97–120. doi: 10.1146/annurev-genet-111212–133526.

Waddington, C. H. 1953. "Genetic Assimilation of an Acquired Character." *Evolution* 7:118–126.

Wellenreuther, M., and B. Hansson. 2016. "Detecting Polygenic Evolution: Problems, Pitfalls, and Promises." *Trends in Genetics* 32 (3): 155–164. doi: 10.1016/j.tig.2015.12.004.

Wellmer, F., and J. L. Riechmann. 2010. "Gene Networks Controlling the Initiation of Flower Development." *Trends in Genetics* 26 (12): 519–527. doi: 10.1016/j.tig.2010.09.001.

West-Eberhard, M. J. 2003. *Developmental Plasticity and Evolution.* New York: Oxford University Press.

Whitman, D. W., and A. A. Agrawal. 2009. "What Is Phenotypic Plasticity and Why Is It Important." In *Phenotypic Plasticity of Insects: Mechanisms and Consequences*, edited by Douglas W. Whitman and T. N. Ananthakrishnan, 1–63. Enfield, NH: CRC Press.

Windig, J. J., C. G. F. De Kovel, and G. de Jong. 2004. "Genetics and Mechanics of Plasticity." In *Phenotypic Plasticity: Functional and Conceptual Approaches*, edited by T. J. DeWitt and S. M. Scheiner, 31–49. New York: Oxford University Press.

Wollstein, A., and W. Stephan. 2015. "Inferring Positive Selection in Humans from Genomic Data." *Investigative Genetics* 6:5. doi: 10.1186/s13323-015-0023-1.

Wood, A. R., T. Esko, J. Yang, S. Vedantam, T. H. Pers, S. Gustafsson, A. Y. Chu, K. Estrada, J. Luan, Z. Kutalik, et al. 2014. "Defining the Role of Common Variation in the Genomic and Biological Architecture of Adult Human Height." *Nature Genetics* 46 (11): 1173–1186. doi: 10.1038/ng.3097.

Wood, C. W., and E. D. Brodie. 2015. "Environmental Effects on the Structure of the G-matrix." *Evolution* 69 (11): 2927–2940. doi: 10.1111/evo.12795.

Wright, S. 1931. "Evolution in Mendelian Populations." *Genetics* 16 (2): 97.

Wright, S. I., I. Vroh Bi, S. G. Schroeder, M. Yamasaki, J. F. Doebley, M. D. McMullen, and B. S. Gaut. 2005. "The Effects of Artificial Selection on the Maize Genome." *Science* 308 (5726): 1310–1314. doi: 10.1126/science.1107891.

Wund, M. A. 2012. "Assessing the Impacts of Phenotypic Plasticity on Evolution." *Integrative and Comparative Biology* 52 (1): 5–15. doi: 10.1093/icb/ics050.

Wund, M. A., J. A. Baker, B. Clancy, J. L. Golub, and S. A. Foster. 2008. "A Test of the "Flexible Stem" Model of Evolution: Ancestral Plasticity, Genetic Accommodation, and Morphological Divergence in the Threespine Stickleback Radiation." *The American Naturalist* 172 (4): 449–462.

Wund, M. A., S. Valena, S. Wood, and J. A. Baker. 2012. "Ancestral Plasticity and Allometry in Threespine Stickleback Reveal Phenotypes Associated with Derived, Freshwater Ecotypes." *Biological Journal of the Linnean Society* 105 (3): 573–583. doi: 10.1111/j.1095–8312.2011.01815.x.

Yeaman, S. 2015. "Local Adaptation by Alleles of Small Effect." *The American Naturalist* 186 Suppl 1:S74–89. doi: 10.1086/682405.

Yeaman, S., and M. C. Whitlock. 2011. "The Genetic Architecture of Adaptation under Migration-Selection Balance." *Evolution* 65 (7): 1897–1911. doi: 10.1111/j.1558–5646.2011.01269.x.

Yeh, P. J., and T. D. Price. 2004. "Adaptive Phenotypic Plasticity and the Successful Colonization of a Novel Environment." *The American Naturalist* 164 (4): 531–542.

Zhang, G., L. J. Muglia, R. Chakraborty, J. M. Akey, and S. M. Williams. 2013. "Signatures of Natural Selection on Genetic Variants Affecting Complex Human Traits." *Applied and Translational Genomics* 2:78–94. doi: http://dx.doi.org/10.1016/j.atg.2013.10.002.

Zuk, O., E. Hechter, S. R. Sunyaev, and E. S. Lander. 2012. "The Mystery of Missing Heritability: Genetic Interactions Create Phantom Heritability." *Proceedings of the National Academy of Sciences USA* 109 (4): 1193–1198.

6 Genotype-Environment Interaction and the Unscripted Reaction Norm

Sonia E. Sultan

Introduction: Development as Genotype-Environment Interaction

It is a biological truism that an organism's environment affects its development. This general observation is now understood in molecular terms, as environmental conditions are known to influence the intricate pathways of gene expression in eukaryotes, both directly and via physiologically mediated effects on internal and intracellular states (Lewontin 2000; Schlichting and Smith 2002; Nijhout 2003; Gottlieb 2004; Carroll, Grenier, and Weatherbee 2005; Sultan and Stearns 2005; Lemos et al. 2008; Gilbert 2012). Consequently, a given genotype will produce somewhat different phenotypes in different environments, to an extent that depends on the organism, environmental factor, and trait in question. Over the past few decades, this *phenotypic plasticity* has been documented in all types of organism in response to environmental factors ranging from temperature, pH, light and atmospheric composition to food type and availability, population density, predator presence, and social interactions (reviewed by Gilbert and Epel 2015; Sultan 2015).

Environmentally contingent phenotypic outcomes can be empirically characterized as an organism's pattern of responses to alternative conditions or *norm of reaction* (Woltereck 1909). As a result of DNA sequence differences that affect pathways of environmental perception and transduction, genotypes express distinct and generally nonparallel norms of reaction for specific phenotypic traits (Haldane 1946; Barton and Turelli 1989; Kruuk, Slate, and Wilson 2008; Moczek et al. 2011; Des Marais, Hernandez, and Juenger 2013) and for patterns of trait covariation (Plaistow and Collin 2014). Norm of reaction studies carried out in diverse systems and conditions have provided two key insights. First, genotype and environment co-determine individual phenotypes and hence, at the population level, the adaptive variation on which natural selection acts. Second, nonparallel reaction norms cause the size and rank order of genotypic differences to vary among environments, making the genetic consequences of selection environment dependent. In statistical terms, genotypic differences in patterns of environmental response are termed *genotype-environment interaction* (*gxe*). More fundamentally, phenotypic variation among individual organisms reflects the actual, causal interaction between genotype and environment that comprises development.

This recognition would seem to pose a considerable challenge to the mid-twentieth-century notion of the genotype as a self-contained developmental "program"—a set of instructions written in the linear sequence of DNA base pairs, and passed as such from generation to generation (Keller 2000; Griffiths 2006; Sarkar 2006). Yet the concept of a genetic script for development remains dominant in evolutionary biology: ascribing to genes a unique role as the "biological carriers of [developmental] information" across generations (Müller-Wille and Rheinberger 2012, 215) underpins the Modern Synthesis model of adaptive evolution as change over time in the proportions of genetic alleles (Sultan 2017 and references therein). In this chapter, I (i) describe how the environmental context-dependency of phenotypic expression has heretofore been fitted into a "genetic program" model; (ii) discuss new insights that reveal genotype-environment interaction as a complex entanglement that cannot be encompassed by this model; and (iii) briefly raise some practical and conceptual implications.

The Norm of Reaction as a Genetic Program

Evolutionary biologists have reconciled their awareness of genotype-environment interaction with a genetically determinist model for development by defining the norm of reaction itself as a self-contained "property of the genotype" (Nager, Keller, and Van Noordwijk 2000)—an "environmental response program in the genes" (de Jong 1999). This view allows for the environmental context-dependency of phenotypic outcomes while still ascribing control to genes alone: "The answer to the question, 'Is variation plastic or genetic?' is simple—it's genetic" (i.e., "plastic responses are underlain by genes") (DeWitt and Scheiner 2004, 4–5). Clearly, this "localization" of control within the genotype "obscures the context-dependency of causation" (Oyama, Griffiths, and Gray 2001, 5). How did the joint causation of phenotypes revealed by reaction norms come to be imputed to the genotype alone as a developmental script contained within the DNA molecule?

This "asymmetric" emphasis on genes as the source of developmental instructions (Griffiths 2006; Nathalie Feiner, personal communication) reflects a shift away from the unified early twentieth-century concept of heredity and development, to a midcentury focus on genes as heritable biochemical units of information. For instance, at the 1907–1908 meeting of the American Association for the Advancement of Science, chair of Zoology E. G. Conklin characterized heredity as a similarity of development in parents and progeny that is guided by an "intracellular dialogue" among cytoplasmic, environmental, and nuclear factors rather than by nuclear "genes" alone (Conklin 1908), and in 1909 Woltereck clarified that the new concept of "genotype" should be understood not as fixed information but as a heritable *Reaktionsnorm*—a contingent developmental repertoire

(Sarkar 2004, 2006). Following the "molecularization" of biology in the era of Watson and Crick (Müller-Wille and Rheinberger 2012), when evolutionary geneticist A. D. Bradshaw returned to environmental response as a source of phenotypic variation he did so in gene-based terms, proposing that plasticity was "under its own specific genetic control" (Bradshaw 1965,119, discussed by Sarkar 2004). Bradshaw's view accorded with the foundational *operon* model that had been recently propounded by Jacob and Monod (1961). Their model brought the potentially destabilizing discovery of differential gene activity into line with a deterministic, "master molecule" framework by making the regulatory elements themselves genetic (Keller 2000; Sarkar 2006).

In quantitative genetics, D. S. Falconer (1952) developed an ingenious way to reconcile the environmental context-dependency of phenotypes with an internally contained "genetic program" model. The issue at hand was a practical problem for livestock and crop breeding caused by genotypic differences in what Falconer termed "environmental sensitivity": because nonparallel reaction norms caused the performance advantages of specific strains to vary among alternative environments, it was not clear which environment would be the most effective in which to carry out selective breeding. His solution was to express such genotype-environment interaction as an imperfect (< 1.0) "genetic correlation" between the alternative trait states expressed by a given genotype in two different environments, and then compare the relative impact of direct and indirect (correlated) selection in each environment. This solution maintained an implicit focus on genotype as the agent of developmental control: "When so formulated the genetic aspect of the situation becomes clear" (Falconer 1952, 293). It worked because, analytically, as long as most pairs of genotypic norms of reaction in a sample do not actually cross, statistical genetic correlations remain high (Falconer 1990). The genetic correlation approach could be generalized to numerous environments by using a genetic covariance matrix (Gomulkiewicz and Kirkpatrick 1992).

In terms of biological mechanism, this correlation approach required stretching the notion of the genotype as a deterministic developmental "program." Falconer and Via (who later extended Falconer's approach to examine evolution of plasticity; Via and Lande 1985) explained that the norm of reaction resulted from a different "genetic program" in each alternative environment; depending on how many of the same genes were expressed in both environments, the programs would be more or less "genetically correlated" (Via 1994). In this view, a genotype's environmental response repertoire is a sort of compendium of partially overlapping genetic "programs." Via also noted that a given genetic allele could be expressed differently in alternative environments (Via 1987; 1994), an observation that agrees with current molecular insights (see below). Other researchers at the time argued against the possibility of such allelic sensitivity as a mechanism for context-dependent outcomes, instead speculating that a trait's plasticity was governed by a distinct form of genetic control—"plasticity genes" governing a trait's responsiveness that were separate from the loci that dictated the trait's mean value (reviewed by Sarkar 2004;

Nicoglou 2015). In both mechanistic schemes, the reaction norm was considered to be under purely genetic control.

Defining the norm of reaction as a genotypic property has allowed plasticity to be "successfully integrated into the Modern Synthesis" (Berrigan and Scheiner 2004). Instead of natural selection on alleles that affect a trait's fixed value, selection could be understood to act on the alleles underpinning the norm of reaction (either indirectly as mediated by their correlation across environments, or via direct effects on alternative "plasticity genes"; Van Tienderen and Koelwijn 1994). The adaptive evolution of reaction norms as genetically determined traits could be modeled by including aspects of trait expression considered unique to plasticity such as reliable environmental cues and their accurate perception, response lag times, presumed special "costs," and statistical gxe variance, together with patterns of environmental heterogeneity and/or gene flow among environments. Arguments about genetic load and effects of population structure could be transferred to the reaction norm (e.g., Scheiner 1998; de Jong 1999).

Although the "plasticity genes" concept persists (e.g., Abouheif et al. 2014; Sikkink et al. 2014), the 1990s dispute about the genetic architecture of reaction norms has been resolved by the molecular revelation that gene expression is inherently context dependent. Both intracellular and external environmental factors contribute importantly to this exceedingly complex regulatory context. Environmental sensitivity of transcription factors and other regulatory elements, epigenetic modifications such as acetylation of histone proteins and DNA methylation, and post-transcriptional processing of mRNA (including alternative splicing and editing as well as regulatory interactions with noncoding small RNAs) all modulate the expression of a given genotype in ways that reflect environmental inputs (references in Sultan 2015). Recognizing that environmentally sensitive gene expression is a general feature of organisms, some researchers are investigating the evolutionary genetics of functional and fitness traits by treating the reaction norm as a continuous function of a specified environmental gradient such as temperature or salinity (Stinchcombe, Function-Valued Trait Working Group, and Kirkpatrick 2012; e.g., Carter et al. 2017). These statistically sophisticated approaches to norm of reaction parameters, along with careful critiques of alternative statistics for the shape and variation of response patterns (e.g., Morrissey and Liefting 2016), continue to presuppose that the reaction norm is a fixed genotypic property—that is, "the expected phenotype of a given genotype as a function of the environment" (Chevin, Lande, and Mace 2010).

That even these technically impressive, cutting-edge approaches locate the cause of gxe interaction strictly within the genotype reflects a continuing allegiance to the twentieth-century view of genes as self-contained instructions for development. When the norm of reaction is viewed as a "property of the genotype," the organism's developmental plasticity becomes simply an "extended phenotype" emanating from the genes, and the evolution of these response patterns can be understood in standard population-genetic terms. However, a closer look at the complexity of genotype-environment interaction reveals that a view of the reaction norm as genetically scripted is not supported empirically, making it far more problematic to reconcile context-dependent phenotypic variation with the Modern Synthesis evolutionary model.

The Unscripted Norm of Reaction: Higher Order *GxE* Interaction Effects

Recent studies of environmental effects and epigenetic modifications both across and within generations have provided stunning insights to the complexities of developmental causation. These insights make clear why the norm of reaction cannot be viewed as a genetically determined set of rules for development in specific environments. Instead, as explained below, a genotype's realized norm of reaction takes shape actively, modulated by several interacting layers of environmental and epigenetic effects.

One influence on an organism's phenotypic expression in various conditions is the particular environmental state encountered by its parents. In humans, for instance, a recent cohort study found that infant size at birth (i.e., the interaction of fetal genotype with uterine environment) was significantly influenced by the nutrient environment experienced by the mothers when *they* were in utero, while infant growth (24-month height) was influenced by the fathers' prenatal nutrient environment (Eriksen et al. 2017). Effects of parental environment on progeny development may be inherited via changes to cytoplasmic factors transmitted via gametes (ova or sperm), egg, or seed tissues, including nutrient reserves, signaling compounds such as hormones and proteins, and regulatory RNAs; substantial transgenerational effects of environment have been documented in hundreds of taxa across 32 biological orders (recently reviewed by Badyaev and Uller 2009; Maestripieri and Mateo 2009; Herman and Sultan 2011; Salinas et al. 2013; Blake and Watson 2016; Auge et al. 2017). These effects may persist throughout progeny development (Walter et al. 2016) and, via neural and behavioral changes, into maturity (e.g., Champagne 2008).

When parental effects on specific genotypes are studied in alternative progeny environments, it becomes clear that they affect progeny norms of reaction (e.g., Plaistow et al. 2015; Sultan 2017). Such effects can also be inferred from datasets that pool genotypes. For example, the mean growth rate of juvenile anemonefish dipped sharply in elevated versus standard CO_2 when parents had been raised in standard conditions, but when parents had been exposed to high concentrations of dissolved CO_2, their progeny developed normally in both rearing treatments, resulting in flatter norms of reaction across CO_2 treatments (Miller et al. 2012). In general, the impact of parental environment on development varies in alternative progeny conditions, revealing an interactive effect of parental and progeny environments (e.g., Groot et al. 2016; Baker, Berg, and Sultan 2018). Depending on the system, these interacting environmental influences may be cumulative, antagonistic, or compensatory (Auge et al. 2017; e.g., Sultan, Barton, and Wilczek 2009; Ezard et al. 2014).

Moreover, inherited effects of previous environments can persist across several generations, resulting in grandparental, great-grandparental, and potentially even more remote ancestral effects on an individual's developmental response to its own environment. Such multigeneration effects may include changes to phenotypic expression patterns for growth and morphology (e.g., Herman et al. 2012; Akkerman et al. 2016) as well as aspects of brain activity and behavioral response. In one mammalian study (Crews et al. 2012), rats were either briefly exposed to a commonly used pesticide or kept in control conditions, and their descendants monitored for three generations. Great-grandchildren (F_3-generation descendants) of these individuals showed dramatically different behavioral and physiological reactions to chronic stress imposed during their adolescence, responding either by increased anxiety and social avoidance or decreased anxiety, respectively. At present, little is known about the possible duration of such inherited effects on reaction norms, but in some cases it may be surprisingly long. In one of the few available longer-term studies, an environmentally induced increase in migratory response to a specific olfactory cue was found to persist for more than 40 generations in the "model system" nematode *C. elegans* (Remy 2010; see Jablonka and Raz 2009 for further examples).

Multigeneration effects on phenotypic response can be transmitted via heritable epigenetic modifications such as DNA methylation, histone acetylation, and micro-RNAs (Cortijo et al. 2014; Soubry et al. 2014; Kawakatsu et al. 2016). Such modifications may be induced by specific environmental stimuli (e.g., Akkerman et al. 2016) or arise spontaneously (Becker et al. 2011; Schmitz et al. 2011); it is clear that animal and plant populations harbor very high levels of epigenetic variation (Richards, Verhoeven, and Bossdorf 2012; e.g., Cortijo et al. 2014). Although little is known about the induction/reversal dynamics of epigenetic "marks" in natural systems, the degree to which they are stably transmitted appears to be quite variable (Becker and Weigel 2012; Turck and Coupland 2013; see Klironomos, Berg, and Collins 2013; Herman et al. 2014). Epigenetic modifications may profoundly influence phenotypic responses by "orchestrat[ing] genome accessibility, functionality, and three-dimensional structure" (Kawakatsu et al. 2016, 492). As these modifications are set and reset across the genome, they may cumulatively alter gene-expression pathways that affect norms of reaction in ways that reflect specific sequences of environmental states and stochastic events over several past generations.

Along with inherited influences, factors that arise within an organism's lifetime may influence its response to current environmental conditions. Because they alter the activity of transcription factors and other genes and gene interactions, immediate stochastic or induced epigenetic effects may comprise one such factor (references in Herman et al. 2014). More generally, the environment an organism encounters early in life may shape its physiological, developmental, and behavioral responses to conditions later in the life-cycle (e.g., Weinig and Delph 2001; Gluckman et al. 2007; Bateson, Gluckman, and Hanson 2014). In humans, for example, a nutrient-poor versus nutrient-rich environment while *in utero* shapes a child's later protein, amino acid, and lipid metabolism, resulting in profoundly different chances of survival in children who experience famine (Forrester et al. 2012).

In light of these insights, it is clearly inaccurate to characterize the norm of reaction as a preexisting program for phenotypic response that is scripted in the DNA. However, the developmental effects of early, parental, and ancestral environments—like those of an organism's current environment—are mediated by the genotype rather than independent of it, resulting in *genotype x parental environment* and *genotype x ancestral environment* interactions (e.g., Schmitt, Niles, and Wulff 1992; Sultan 1996; Stjernman and Little 2011; Plaistow et al. 2015; Vu et al. 2015; Herman and Sultan 2016). Inherited environmental effects vary by genotype because a parent individual's DNA sequence influences (a) its production of cytoplasmic factors in response to environmental conditions, and (b) available sites for methylation and other epigenetic modifications it may transmit to progeny (Gutierrez-Arcelus et al. 2013; Ladd-Acosta and Fallin 2016). Genotypes may differ not only in the inducibility and persistence of specific epigenetic modifications but also in the effects of these epigenetic factors in combination with each other and with cytoplasmic elements (Herman et al. 2014), resulting in higher order interactive effects of genotype, parental environment, and one to several ancestral environments (e.g., Herman et al. 2012). Although few studies exist, the interacting phenotypic effects of successive environments within a generation may also vary by genotype (e.g., genotype x $env_{[TIME\ 1]}$ x $env_{[TIME\ 2]}$ interaction effects on nematode development; Sikkink et al. 2014), potentially adding temporal feedbacks to this causal complexity.

Genotype-Environment "Entanglement"

Environmental information inherited from parental and ancestral generations, immediate epigenetic modifications to the genome, and conditions encountered previously within the lifecycle all combine to shape phenotypic expression in any particular environment. As a result, an organism's norm of reaction—its repertoire of responses to alternative conditions—is not a genotypic property but arises contextually from a particular layering of these environmental and epigenetic factors. Yet DNA sequence affects the developmental impact of these layered factors, resulting in a complex causal interplay that may be best described as a genotype-environment *entanglement* (Keller 2010; Sultan 2017). The difficulty of disentangling these sources of phenotypic variation as distinct threads raises fundamental questions about both empirical approaches and evolutionary causation.

If parental, ancestral, and early-life effects on development were trivial or very subtle, biologists could, in practice, continue to treat norms of reaction as genotypic properties that were just somewhat noisy in expression. However, this solution will not suffice, because the interacting effects of previous environments on realized norms of reaction may be substantial. In one well-studied plant system, for instance, different combinations and sequences of parental and grandparental environment (in genetically uniform individuals) resulted in significantly different survival probabilities in stressful conditions (Herman et al. 2012). Although studies testing prior environments across multiple generations are scarce, existing data (see previous section) show that even great-grandparental or more remote ancestral environments may strongly affect critical developmental, behavioral, and physiological traits—including in humans—suggesting that this unlooked-for causal complexity cannot be ignored.

A troubling implication of this recognition is that, in the vast majority of research contexts, the entities we consider "genotypes" are actually entanglements between genotypes and their environmental and epigenetic histories. This will be true even if these experimental entities are bred for a generation in uniform conditions with the goal of standardizing parental effects (e.g., Sultan et al. 1998; Sultan 2001), because their norms of reaction will still be influenced—and differently influenced—by that particular parental environment, as well as by uncontrolled, potentially different grandparental and ancestral conditions. In light of these interacting, variably persistent legacies, it is unclear how to study the phenotypic expression of genotypes as such, either in response to single environments or as norms of reaction. Is it possible to "disentangle" this nexus of influences to reveal a purely genotypic signal on phenotypic expression?

A straightforward solution might be to propagate experimental material over several generations of uniform conditions, in order to standardize any inherited environmental effects. Yet this approach raises further questions. Would it be practical for most labs to carry out multiple generations of rearing in uniform conditions? How many generations would be long enough to wipe clean induced epigenetic marks—three? Five? Twenty? Even if this number could be determined for each study species, what exactly would be the "neutral" or "control" epigenetic setting, or the "neutral" environmental state? More fundamentally, the existence of inherited, nongenetic developmental information, however meticulously standardized over previous generations, raises a critical interpretive question: would such "disentangled" genotypes provide meaningful insight to the causes of adaptive variation? If phenotypic expression arises from the various interactions of genotypes with their previous and current environments, why should researchers work so hard to experimentally isolate that partial, genotypic signal from other developmental information? And whatever laboratory approaches are taken, in real populations biotic and abiotic factors inevitably vary such that individuals exist as confounded, entangled entities. Because it is these genotype-environment entanglements that give rise to functional and fitness differences, it is these multigenerational entanglements that cause selective change in the genetic composition of populations, shaping the genotypic component of subsequent generations of entanglement. In other words, to understand the causes of variation and selection, it may be necessary to study genotype-environment entanglements as such.

A Case for Distributed Causation

Empirically based insights regarding the formative role of complex genotype-environment interaction suggests a re-evaluation of genetic control as a causal principle. Developmental outcomes are shaped by multiple types of information and not by DNA sequence alone (further discussion and references in Laland et al. 2015; Karola Stotz, this volume): phenotypes emerge from the real-time regulatory interactions of the evolved genotype with transient environmental and/or epigenetic influences that occur at timescales from within a generation to several or many generations. These factors cannot be pulled apart as individual causes, because they contribute interactively to phenotypic expression and hence to norms of reaction (Lewontin 2000; Griffiths 2006). Consequently, phenotypic outcomes can be seen to arise from "distributed control" in each generation (Oyama, Griffiths, and Gray 2001, 5; Sarkar 2006; see also Stotz 2006).

Replacing the "genetic program" model with distributed causation will undoubtedly complicate research design and interpretation. Practically speaking, documenting how environments may participate in gene expression patterns and phenotypic outcomes entails testing genotypes of interest not in a single "control" environment, but in at least two alternative environmental states. Because these environmental states are recognized to be sources of specific developmental inputs rather than neutral backdrops, this choice is in itself a significant aspect of experimental design, and one that requires some knowledge of the organism's biology. This kind of expanded design can be implemented within one or more generations in an experimental lineage, depending on which developmental interactions are the primary focus. For instance, to test for inherited (e.g., epigenetic) effects of specific parental conditions on offspring development (i.e., genotype by parent environment effects), the researcher might expose parent individuals of each genotype to different environments, and then compare their progeny within a single environment; to illuminate the higher order interaction of genotype, parent environment, and immediate environment (i.e., parental-environment effects on norms of reaction), developmental outcomes could be compared within several different progeny environments. This approach can be taken to test the inherited effects of specific conditions during any one previous generation (e.g., great-grandparental toxin exposure, Crews et al. 2012), or the combined effects of two or more prior environments (e.g., genotype by grandparental by parental environment interaction; Herman et al. 2012). (Note that, even for designs that control parental and grandparental conditions, effects of uncontrolled ancestral environments may be persistently confounded with genotype; the potential importance of this point will be unknown until more is known about epigenetic persistence and impact.)

Clearly, interrogating distributed causation—even in the necessarily simplified format of controlled experimentation—will require larger studies with more complex designs. Is this shift warranted? The limited findings to date from a diversity of biological systems suggest that it is. When the relevant influences on an intricate process such as development cannot all be controlled (nor even identified), it is reasonable to attribute causation to individual elements such as genotypes if those elements have predictable effects on the outcome regardless of the states of other variables—that is, when interactions with other factors have comparatively weak effects, and can be ignored as background noise (Wagner 1999). However, such weakness is clearly not the case for the interactions of genotypes with inherited environmental and epigenetic factors, which can profoundly alter phenotypic expression. In general, strong interaction effects preclude this kind of causal separability. Instead, "in a distributed process, the role or behavior of an individual or component is determined by its interactions with others rather than its inherent attributes" (Gordon 2016, 516). This is clearly the case for the genetic and environmental factors whose complex interplay mediates realized norms of reaction. Interestingly, distributed causation is associated with flexibility and resilience to disruption (Gordon 2016), two essential features of biological development.

What is fascinating about such causal structures is that it is the interactions themselves that are most formative, rather than individual elements such as genotype or environmental state. For example, Zhu, Ingelmo, and Rand (2014) found that the complex interaction between nuclear genotype, mitochondrial genotype, and diet resulted in 2-way and 3-way interaction effects on longevity (i.e., survival) in *Drosophila* that were much greater than the "main effects" of either the two genetic components or the environment individually; in this system "gene-by-gene and gene-by-environment interactions are not simply modifiers of key factors affecting longevity, but these interactions themselves are the very factors that underlie important variation" (p. 1). Experimental designs grounded in distributed causal hypotheses may illuminate the bases of human health outcomes as well; following decades of intense focus on genetic factors alone, biomedical researchers are increasingly testing parental or grandparental environment, present environment, and/or genotype as interactive causes of disease phenotypes such as cancer or obesity (e.g., Schoeps et al. 2014; Teh et al. 2014; Ding et al. 2017). Perhaps ironically, it is molecular biologists who have most clearly come to recognize "the dense entanglement of developmental processes and, in that recognition, to appreciate the limits of centralized control" by DNA (Keller 2000, 95; e.g., Cropley et al. 2006; Lemos et al. 2008; Mattick 2010; Jones 2012; Nadeau 2015). This awareness is equally central to progress in evolutionary biology.

Genotype-Environment Interaction and the Genotype Signal in Evolution

If phenotypic outcomes arise from genotype-environment interaction rather than genetically scripted "programs," questions about variation and selective evolution must be framed to reflect this understanding. Evolutionary biologists study the fitness variation that leads to adaptive change and differentiation in population and species composition (Sober 1984).

Such longer-term evolutionary trajectories can be rigorously tracked in terms of change in DNA sequence, which comprises a uniquely stable, continuous "information stream" across many generations (Bonduriansky 2012). Although this is the most constant stream of information, and therefore provides the best record of evolutionary history, it is not the only causally relevant one; along with genotype, environmental and epigenetic factors co-determine the phenotypic differences expressed within each generation, and consequently the strength and direction of natural selection (Denis Walsh, this volume). For example, the presence of heritable epialleles can alter selection on genetic alleles with which they interact developmentally (Day and Bonduriansky 2011; Geoghegan and Spencer 2012). In general, epigenetic and environmental influences that decouple phenotypic outcomes from specific allelic differences alter the impact of selection on such differences within populations (Klironomos et al. 2013; Herman et al. 2014; McNamara et al. 2016). Such decoupling can also affect divergence at the population level: adaptive phenotypic differences between environmentally distinct populations can result from individual plastic responses rather than genetic divergence (McIntyre and Strauss 2014; Lind et al. 2015) or from epigenetic rather than genetic variation (Kawakatsu et al. 2016; Baldanzi et al. 2017).

As long as phenotypic cause was located in the genotype (for norms of reaction as well as constitutive traits), DNA sequence variation was understood to be both the cause and the consequence of selective evolution, and revealing this variation was the central goal of evolutionary research (see Badyaev 2011). A distributed view of developmental causation alters this problematic. Now the key challenge is to understand how the stable, continuous DNA information stream that most robustly tracks evolutionary lineage and diversification is altered by selective evolution, when that DNA stream comprises only a *partial, context-dependent component* of selective causation—that is, when fitness variation reflects genotype-environment entanglement rather than genetic differences alone. In other words, the investigation of interactive selective *cause* must now detach from the tracking of evolved genetic *consequences* (Sultan 2015). Not surprisingly, perhaps, this detachment has been attended by a certain amount of conceptual uncertainty. Even researchers who explicitly recognize that "it is on the product of genetic and environmental influences that natural selection acts" maintain a need to isolate or "disentangle" the "relevant genetic effects" in order to predict selective change (Nager, Keller, and Van Noordwijk 2000, 95–96). Yet to precisely study evolution as change at the genetic level, it is essential to consider the actual, interactive causes of that change, not only "the relevant *genetic* effects" [italics added].

This may be particularly demanding in the case of norms of reaction. Because genotypes evolve due to drift, gene flow, and natural selection, the phenotypes they express in various multigenerational contexts will inevitably reflect those historical processes, resulting in realized norm of reaction differences among genotypes, populations, and species. Unmistakably, many aspects of both immediate and transgenerational genotype-environment interaction (including induced epigenetic changes) contribute to adaptive phenotypes, reflecting the past impact of selection. This need not mean that plasticity "must have a genetic basis" (i.e., be controlled by genotype; Nager, Keller, and Van Noordwijk 2000), but only that it has a genetic *component*, which is obviously the case. Isolating this DNA sequence component from its causal interactors, either experimentally or theoretically, will not reveal the variational basis of selective evolution as intended. Rather, it is necessary to investigate how selection will act on the actual entanglements that shape phenotypic expression.

The impact of selection on the genetic component of reaction norms will be modulated by parental effects and by the interactions between these effects and both previous and immediate environments. These complex interactions may include transient or persistent epigenetic variants, which may strongly interfere with a direct genetic signal on fitness. Indirect genetic signals will include the effects of sequence variation on the impact and dynamics of environmental and epigenetic factors. Just as variable expression of fitness-related traits in multiple environments can constrain selective evolution (Gomulkiewicz and Kirkpatrick 1992), these higher order interaction effects of genotype, previous environments, and current environment(s) are likely to increase the dimensionality of phenotypic variation so as to constrain selective change in genetic components. Multiple layers of environmental and epigenetic information may lead to complex patterns of conditionally expressed variation, adding to the amount of existing genetic variation that is contextually "cryptic" to selection (Van Dyken and Wade 2010; Ledón-Rettig et al. 2014; Paaby and Rockman 2014; further references in Sultan 2015). Indeed, natural populations may contain unexpectedly high levels of genetic variation because these multiple interactive influences on phenotypes may obscure potential fitness effects of allelic variants *per se* and hence deflect selection from winnowing out such variants.

Although the dimensionality of genotype-environment interactions may be greater than previously recognized, this dimensionality need not be intractable to research. Empirical studies can be designed to vary genetic, environmental, and epigenetic factors across several generations (see examples in previous sections), and efforts have begun to model the selective consequences of these complex interactions by using an expanded causal framework. For instance, Leimar and McNamara (2015) modeled the genotype as a "genetic cue" that the developing organism integrates with immediate and inherited environmental cues (see also Shea, Pen, and Uller 2011; Uller, English, and Pen 2015). Extending their model to temporally variable environments (McNamara et al. 2016), they showed that the most effective strategy for utilizing these various cues as adaptive developmental information depended on several factors. In one scenario, for example, when maternal effects accurately reflected the maternal environment (i.e., were specifically induced by it) and conditions were strongly autocorrelated across generations, the developmental combination of maternal effects and immediate environmental cues led to the highest progeny fitness. These models explicitly recognize that genotypes hold partial developmental information, which takes phenotypic form only when "combined and integrated" with environmental information across generations (Leimar and McNamara 2015, 1267).

Conclusion

A "genetic program" model has, explicitly or implicitly, guided research on norms of reaction and their adaptive evolution. However, as a result of new insights to inherited nongenetic factors, it is no longer possible to view the norm of reaction as a deterministic property of the genotype. Instead, the developmental responses of organisms to their environments are unscripted—they arise actively, through real-time regulatory events. Researchers have only recently begun to examine the causal interplay between genetic, environmental and epigenetic factors within and across generations to understand the distributed basis of phenotypic expression. An important next step will be to determine the prevalence and impact of such complex effects across biological systems, through experiments that "disentangle" genotypes from environmental and epigenetic influences and then "re-tangle" them in known, controlled contexts. Recent insights to epigenetic and environmental influences on developmental pathways have opened the door to a richer, post-Mendelian understanding of biological inheritance and of genes as developmental factors. To illuminate more fully the causes and dynamics of adaptive evolution, the norm of reaction must be released from its presumed genotypic script, to allow for the actual complexity of regulatory interactions.

Acknowledgments

I thank the organizers of the 2017 KLI Workshop *Cause and Process in Evolution,* Kevin Laland and Tobias Uller, for the invitation to participate, and the John F. Templeton Foundation for my workshop and travel funding. I also thank Karola Stotz and Denis Walsh for kindly sharing with me their chapter manuscripts in progress.

References

Abouheif, E., M. J. Favé, A. S. Ibarrarán-Viniegra, M. P. Lesoway, A. M. Rafiqi, and R. Rajakumar. 2014. "Eco-evo-devo: The Time Has Come." In *Ecological Genomics: Ecology and the Evolution of Genes and Genomes.* Advances in Experimental Medicine & Biology, no. 781, edited by C. R. Landry and N. Aubin-Horth, 107–125. Netherlands: Springer.

Akkerman, K. C., A. Sattarin, J. K. Kelly, and A. G. Scoville. 2016. "Transgenerational Plasticity Is Sex-dependent and Persistent in Yellow Monkeyflower (*Mimulus guttatus*)." *Environmental Epigenetics* 2:dvw003.

Auge, G. A., L. D. Leverett, B. R. Edwards, and K. Donohue. 2017. "Adjusting Phenotypes via Within- and Across-Generational Plasticity." *New Phytologist* 216:343–349.

Badyaev, A. V. 2011. "On the Origin of the Fittest: Link between Emergent Variation and Evolutionary Change as a Critical Question in Evolutionary Biology." *Proceedings of the Royal Society of London B* 278:1921–1929.

Badyaev, A. V., and T. Uller. 2009. "Parental Effects in Ecology and Evolution: Mechanisms, Processes and Implications." *Philosophical Transactions of the Royal Society (B)* 364:1169–1177.

Baker, B. H., L. J. Berg, and S. E. Sultan. 2018. "Context-Dependent Developmental Effects of Parental Shade versus Sun Are Mediated by DNA Methylation." *Frontiers in Plant Science* 9 (1251). doi: 10.3389/fpls.2018.01251.

Baldanzi, S., R. Watson, C. D. McQuaid, G. Gouws, and F. Porri. 2017. "Epigenetic Variation among Natural Populations of the South African Sandhopper *Talorchestia capensis*." *Evolutionary Ecology* 31:77–91.

Barton, N. H., and M. Turelli. 1989. "Evolutionary Quantitative Genetics: How Little Do We Know?" *Annual Review of Genetics* 23:337–370.

Bateson, P., P. Gluckman, and M. Hanson. 2014. "The Biology of Developmental Plasticity and the Predictive Adaptive Response Hypothesis." *Journal of Physiology* 592:2357C–2368C.

Becker, C., J. Hagmann, J. Müller, D. Koenig, O. Stegle, K. Borgwardt, and D. Weigel. 2011. "Spontaneous Epigenetic Variation in the *Arabidopsis thaliana* Methylome." *Nature* 480:245–249.

Becker, C., and D. Weigel. 2012. "Epigenetic Variation: Origin and Transgenerational Inheritance." *Current Opinion in Plant Biology* 15:562–567.

Berrigan, D., and S. M. Scheiner. 2004. "Modeling the Evolution of Phenotypic Plasticity." In *Phenotypic Plasticity: Functional and Conceptual Approaches*, edited by T. M. DeWitt and S. M. Scheiner, 82–97. New York: Oxford University Press.

Blake, G. E., and E. D. Watson. 2016. "Unravelling the Complex Mechanisms of Trans-generational Epigenetic Inheritance." *Current Opinion in Chemical Biology* 33:101–107.

Bondouriansky, R. 2012. "Rethinking Heredity, Again." *Trends in Ecology and Evolution* 27:330–336.

Bradshaw, A. D. 1965. "Evolutionary Significance of Phenotypic Plasticity in Plants." *Advances in Genetics* 13:115–155.

Carroll, S. B, J. K. Grenier, and S. D. Weatherbee. 2005. *From DNA to Diversity: Molecular Genetics and the Evolution of Animal Design,* 4th ed. Malden, MA: Blackwell.

Carter, M. J., M. I. Lind, S. R. Dennis, W. Hentley, and A. P. Beckerman. 2017. "Evolution of a Predator-induced, Nonlinear Reaction Norm." *Proceedings of the Royal Society B* 284:20170859.

Champagne, F. A. 2008. "Epigenetic Mechanisms and the Transgenerational Effects of Maternal Care." *Frontiers in Neuroendocrinology* 29:386–397.

Chevin, L.-M., R. Lande, and G. M. Mace. 2010. "Adaptation, Plasticity, and Extinction in a Changing Environment: Towards a Predictive Theory." *Public Library of Science Biology* 8:e1000357.

Conklin, E. G. 1908. "The Mechanism of Heredity." *Science* 27:89–99.

Cortijo, S., R. Wardenaar, M. Colomé-Tatché, A. Gilly, M. Etcheverry, K. Labadie, E. Caillieux, et al. 2014. "Mapping the Epigenetic Basis of Complex Traits." *Science* 343:1145–1148.

Crews, D., R. Gillette, S. V. Scarpino, M. Manikkam, M. I. Savenkova, and M. K. Skinner. 2012. "Epigenetic Transgenerational Inheritance of Altered Stress Responses." *Proceedings of the National Academy of Sciences (USA)* 109:9143–9148.

Cropley, J. E., C. M. Suter, K. B. Beckam, and D. I. Martin. 2006. "Germ-line Epigenetic Modification of the Murine *Avy* Allele by Nutritional Supplementation." *Proceedings of the National Academy of Sciences (USA)* 103:17308–17312.

Day, T., and R. Bondouriansky. 2011. "A Unified Approach to the Evolutionary Consequences of Genetic and Non-Genetic Inheritance." *The American Naturalist* 178:E18–E36.

De Jong, G. 1999. "Unpredictable Selection in a Structured Population Leads to Local Genetic Differentiation in Evolved Reaction Norms." *Journal of Evolutionary Biology* 12:839–851.

Des Marais, D. L., K. M. Hernandez, and T. E. Juenger. 2013. "Genotype-by-Environment Interaction and Plasticity: Exploring Genomic Responses of Plants to the Abiotic Environment." *Annual Review of Ecology, Evolution and Systematics* 44:5–29.

DeWitt, T. J., and S. M. Scheiner. 2004. "Phenotypic Variation from Single Genotypes, a Primer." In *Phenotypic Plasticity: Functional and Conceptual Approaches*, edited by T. J. DeWitt and S. M. Scheiner, 1–9. New York: Oxford University Press.

Ding, M., C. Yuan, A. J. Gaskins, A. E. Field, S. A. Missmer, K. B. Michels, F. Hu, C. Zhang, M. W. Gillman, and J. Chavarro. 2017. "Smoking during Pregnancy in Relation to Grandchild Birth Weight and BMI Trajectories." *PLoS ONE* 12:e0179368.

Eriksen, K. G., E. J. Radford, M. J. Silver, A. J. C. Fulford, R. Wegmüller, and A. M. Prentice. 2017. "Influence of Intergenerational *in utero* Parental Energy and Nutrient Restriction on Offspring Growth in Rural Gambia." *The FASEB Journal* 31:4928–4934.

Ezard, T. H. G., R. Prizak, R. B. Hoyle, and D. Marshall. 2014. "The Fitness Costs of Adaptation via Phenotypic Plasticity and Maternal Effects." *Functional Ecology* 28:693–701.

Falconer, D. S. 1952. "The Problem of Environment and Selection." *The American Naturalist* 86:293–298.

Falconer, D. S. 1990. "Selection in Different Environments: Effects on Environmental Sensitivity (Reaction Norm) and on Mean Performance." *Genetical Research, Cambridge* 56:57–70.

Forrester, T. E., A. V. Badaloo, M. S. Boyne, C. Osmond, D. Thompson, C. Green, C. Taylor-Bryan, et al. 2012. "Prenatal Factors Contribute to the Emergence of Kwashiorkor or Marasmus in Severe Undernutrition: Evidence for the Predictive Adaptation Model." *PLoS ONE* 7:e0035907.

Geoghegan, J. L., and H. G. Spencer. 2012. "Population-Epigenetic Models of Selection." *Theoretical Population Biology* 81:232–242.

Gilbert, S. F. 2012. "Ecological Developmental Biology: Environmental Signals for Normal Animal Development." *Evolution and Development* 14:20–28.

Gilbert, S. F., and D. Epel. 2015. *Ecological Developmental Biology: Integrating Epigenetics, Medicine, and Evolution*, 2nd ed. Sunderland, MA: Sinauer Associates.

Gluckman, P. D., K. A. Lillycrop, M. H. Vickers, A. B. Pleasants, E. S. Phillips, A.S. Beedle, G. C. Burdge, and M. A. Hanson. 2007. "Metabolic Plasticity during Mammalian Development Is Directionally Dependent on Early Nutritional Status." *Proceedings of the National Academy of Sciences (USA)*104:12796–12800.

Gomulkiewicz, R., and M. Kirkpatrick. 1992. "Quantitative Genetics and the Evolution of Reaction Norms." *Evolution* 46:390–411.

Gordon, D. M. 2016. "The Evolution of the Algorithms for Collective Behavior." *Cell Systems* 3:514–520.

Gottlieb, G. 2004. "Normally Occurring Environmental and Behavioral Influences on Gene Activity: From Central Dogma to Probabilistic Epigenesis." In *Nature and Nurture: The Complex Interplay of Genetic and Environmental Influences on Human Behavior and Development*, edited by C. G. Coll, E. L. Bearer, and R. M. Lerner, 85–106. London: Lawrence Erlbaum Associates.

Griffiths, P. E. 2006. "Philip Kitcher, Genetic Determinism, and the Informational Gene." In *Genes in Development: Re-reading the Molecular Paradigm*, edited by E. M. Neumann-Held and C. Rehmann-Sutter, 175–198. Durham, NC: Duke University Press.

Groot, M. P., R. Kooke, N. Knoben, P. Vergeer, J. J. Keurentjes, N. J. Ouborg, and K. J. Verhoeven. 2016. "Effects of Multi-generational Stress Exposure on the Expression and Persistence of Transgenerational Effects in *Arabidopsis thaliana*." *PLoS ONE* 11:e0151566.

Gutierrez-Arcelus, M., T. Lappalainen, S. B. Montgomery, A. Buil, H. Ongen, A. Yurovsky, J. Bryois, et al. 2013. "Passive and Active DNA Methylation and the Interplay with Genetic Variation in Gene Regulation." *Elife* 2:e00523.

Haldane, J. B. S. 1946. "The Interaction of Nature and Nurture." *Annals of Eugenics* 13:197–205.

Herman, J. J., H. G. Spencer, K. Donohue, and S. E. Sultan. 2014. "How Stable 'Should' Epigenetic Modifications Be? Insights from Adaptive Plasticity and Bet Hedging." *Evolution* 68:632–643.

Herman, J. J., and S. E. Sultan. 2011. "Adaptive Transgenerational Plasticity: Case Studies, Mechanisms, and Implications for Natural Populations." *Frontiers in Plant Genetics and Genomics* 2:e102.

Herman, J. J., and S. E. Sultan. 2016. "DNA Methylation Mediates Genetic Variation for Adaptive Transgenerational Plasticity." *Proceedings of the Royal Society B* 283:20160988.

Herman, J. J., S. E. Sultan, T. Horgan-Kobelski, and C. E. Riggs. 2012. "Adaptive Transgenerational Plasticity in an Annual Plant: Grandparental and Parental Drought Stress Enhance Performance of Seedlings in Dry Soil." *Integrative and Comparative Biology* 52:1–12.

Jablonka, E., and G. Raz. 2009. "Transgenerational Epigenetic Inheritance: Prevalence, Mechanisms, and Implications for the Study of Heredity and Evolution." *Quarterly Review of Biology* 84:1331–1176.

Jacob, F., and J. Monod. 1961. "Genetic Regulatory Mechanisms in the Synthesis of Proteins." *Journal of Molecular Biology* 3:318–356.

Jones, P. A. 2012. "Functions of DNA Methylation: Islands, Start Sittes, Gene Bodies and Beyond." *Nature Reviews Genetics* 13:484–492.

Kawakatsu, T., S. S. Huang, F. Jupe, E. Sasaki, R. J. Schmitz, M. A. Urich, R. Castanon, et al. 2016. "Epigenomic Diversity in a Global Collection of *Arabidopsis thaliana* Accessions." *Cell* 166:492–505.

Keller, E. F. 2000. *The Century of the Gene*. Cambridge, MA: Harvard University Press.

Keller, E. F. 2010. *The Mirage of a Space between Nature and Nurture*. Durham, NC: Duke University Press.

Klironomos, F. D., J. Berg, and S. Collins. 2013. "How Epigenetic Mutations Can Affect Genetic Evolution: Model and Mechanism." *Bioessays* 35:571–578.

Kruuk, L. E. B., J. Slate, and A. J. Wilson. 2008. "New Answers for Old Questions: The Evolutionary Quantitative Genetics of Wild Animal Populations." *Annual Review of Ecology, Evolution and Systematics* 39:525–548.

Ladd-Acosta, C., and M. D. Fallin. 2016. "The Role of Epigenetics in Genetic and Environmental Epidemiology." *Epigenomics* 8:271–283.

Laland, K. N., T. Uller, M. W. Feldman, K. Sterelny, G. B. Müller, A. Moczek, E. Jablonka, and J. Odling-Smee. 2015. "The Extended Evolutionary Synthesis: Its Structure, Assumptions and Predictions." *Proceedings of the Royal Society (B)* 282:20151019.

Ledón-Rettig, C. C., D. W. Pfennig, A. J. Chunco, and I. Dworkin. 2014. "Cryptic Genetic Variation in Natural Populations: A Predictive Framework." *Integrative and Comparative Biology* 54:783–793.

Leimar, O., and J. M. McNamara. 2015. "The Evolution of Transgenerational Integration of Information in Heterogeneous Environments." *The American Naturalist* 185:E55–E69.

Lemos, B., C. R. Landry, P. Fontanillas, S. C. P. Renn, R. Kulathinal, K. M. Brown, and D. Hartl. 2008. "Evolution of Genomic Expression." In *Evolutionary Genomics and Proteomics,* edited by M. Pagel and A. Pomiankowski, 81–118. Sunderland, MA: Sinauer Associates.

Lewontin, R. C. 2000. *The Triple Helix: Gene, Organism and Environment*. Cambridge, MA: Harvard University Press.

Lind, M. I., K. Yarlett, J. Reger, M. J. Carter, and A. P. Beckerman. 2015. "The Alignment between Phenotypic Plasticity, the Major Axis of Genetic Variation and the Response to Selection." *Proceedings of the Royal Society (B)* 282:20151651.

Maestripieri, D., and J. M. Mateo (editors). 2009. *Maternal Effects in Mammals*. Chicago: University of Chicago Press.

Mattick, J. S. 2010. "RNA as the Substrate for Epigenome-Environment Interactions." *Bioessays* 32:548–552.

McIntyre, P. J., and S. Y. Strauss. 2014. "Phenotypic and Transgenerational Plasticity Promote Local Adaptation to Sun and Shade Environments." *Evolutionary Ecology* 28:229–246.

McNamara, J. M., S. R. X. Dall, P. Hammerstein, and O. Leimar. 2016. "Detection vs. Selection: Integration of Genetic, Epigenetic and Environmental Cues in Fluctuating Environments." *Ecology Letters* 19:1267–1276.

Miller, G. M., S.-A. Watson, J. M. Donelson, M. I. McCormick, and P. L. Munday. 2012. "Parental Environment Mediates Impacts of Increased Carbon Dioxide on a Coral Reef Fish." *Nature Climate Change* 2:858–861.

Moczek, A. P., S. E. Sultan, S. Foster, C. Ledón-Rettig, I. Dworkin, H. F. Nijhout, E. Abouheif, and D. W. Pfennig. 2011. "The Role of Developmental Plasticity in Evolutionary Innovation." *Proceedings of the Royal Society (B)* 278:2705–2713.

Morrissey, M. B., and M. Liefting. 2016. "Variation in Reaction Norms: Statistical Considerations and Biological Interpretation." *Evolution* 70:1944–1959.

Müller-Wille, S., and H. J. Rheinberger. 2012. *A Cultural History of Heredity*. Chicago: University of Chicago Press.

Nadeau, J. H. 2015. "The Nature of Evidence for and Against Epigenetic Inheritance." *Genome Biology* 16:137.

Nager, R. G., L. F. Keller, and A. J. Van Noordwijk. 2000. "Understanding Natural Selection on Traits That Are Influenced by Environmental Conditions." In *Adaptive Genetic Variation in the Wild*, edited by T. Mousseau, B. Sinervo, and J. A. Endler, 5–115. New York: Oxford University Press.

Nicoglou, A. 2015. "The Evolution of Phenotypic Plasticity: Genealogy of a Debate in Genetics." *Studies in History and Philosophy of Biological and Biomedical Sciences* 50:67–76.

Nijhout, H. F. 2003. "Development and Evolution of Adaptive Polyphenisms." *Evolution and Development* 5:9–18.

Oyama, S., P. E. Griffiths, and R. D. Gray. 2001. "Introduction: What Is Developmental Systems Theory?" In *Cycles of Contingency: Developmental Systems and Evolution*, edited by S. Oyama, P. E. Griffiths, and R. D. Gray, 1–11. Cambridge, MA: MIT Press.

Paaby, A. B., and M. V. Rockman. 2014. "Cryptic Genetic Variation: Evolution's Hidden Substrate." *Nature Reviews Genetics* 15:247–258

Plaistow, S. J,. and H. Collin. 2014. "Phenotypic Integration Plasticity in *Daphnia magna*: An Integral Facet of GxE Interactions." *Journal of Evolutionary Biology* 27:1913–1920.

Plaistow, S. J., C. Shirley, H. Collin, S. J. Cornell, and E. D. Harney. 2015. "Offspring Provisioning Explains Clone-Specific Maternal Age Effects on Life History and Life Span in the Water Flea, *Daphnia pulex*." *The American Naturalist* 186:376–389.

Remy, J.-J. 2010. "Stable Inheritance of an Acquired Behavior in *Caenorhabditis elegans*." *Current Biology* 20:R877–R878.

Richards, C. L., K. J. F. Verhoeven, and O. Bossdorf. 2012. "Evolutionary Significance of Epigenetic Variation." In *Plant Genome Diversity Volume 1*, 257–274. Vienna: Springer-Verlag.

Salinas, S., S. C. Brown, M. Mangel, and S. B. Munch. 2013. "Non-genetic Inheritance and Changing Environments." *Non-Genetic Inheritance* 1:38–50.

Sarkar, S. 2004. "From the Reaktionsnorm to the Evolution of Adaptive Plasticity: A Historical Sketch, 1909–1999." In *Phenotypic Plasticity: Functional and Conceptual Approaches*, edited by T. J. DeWitt and S. M. Scheiner, 10–30. Oxford: Oxford University Press.

Sarkar, S. 2006. "From Genes as Determinants to DNA as Resource." In *Genes in Development: Re-reading the Molecular Paradigm*, edited by E. M. Neumann-Held and C. Rehmann-Sutter, 77–95. Durham, NC: Duke University Press.

Scheiner, S. M. 1998. "The Genetics of Phenotypic Plasticity. VII. Evolution in a Spatially-Structured Environment." *Journal of Evolutionary Biology* 11:303–320.

Schlichting, C. D., and H. Smith. 2002. "Phenotypic Plasticity: Linking Molecular Mechanisms with Evolutionary Outcomes." *Evolutionary Ecology* 16:189–211.

Schmitt, J., J. Niles, and R. Wulff. 1992. "Norms of Reaction of Seed Traits to Maternal Environments in *Plantago lanceolata*." *The American Naturalist* 139:451–466.

Schmitz, R. J., M. D. Schultz, M. G. Lewsey, R. C. O'Malley, M. A. Urich, O. Libiger, N. J. Schork, and J. R. Ecker. 2011. "Transgenerational Epigenetic Instability Is a Source of Novel Methylation Variants." *Science* 334:369–373.

Schoeps, A., A. Rudolph, P. Seibold, A. M. Dunning, R. L. Milne, S. E. Bojesen, A. Swerdlow, et al. 2014. "Identification of New Genetic Susceptibility Loci for Breast Cancer through Consideration of Gene-Environment Interactions." *Genetic Epidemiology* 38:84–93.

Shea, N., I. Pen, and T. Uller. 2011. "Three Epigenetic Information Channels and Their Different Roles in Evolution." *Journal of Evolutionary Biology* 24:1178–1187.

Sikkink, K. L., R. M. Reynolds, C. M. Ituarte, W. M. Cresko, and P. C. Phillips. 2014. "Rapid Evolution of Phenotypic Plasticity and Shifting Thresholds of Genetic Assimilation in the Nematode *Caenorhabditis remainei*." *Genes/Genomes/Genetics* 4:1103–1112.

Sober, E. 1984. *The Nature of Natural Selection*. Cambridge, MA: MIT Press.

Soubry, A., C. Hoyo, R. L. Jirtle, and S. K. Murphy. 2014. "A Paternal Environmental Legacy: Evidence for Epigenetic Inheritance through the Male Germ Line." *Bioessays* 36:359–371.

Stinchcombe, J. R., Function-Valued Trait Working Group, and M. Kirkpatrick. 2012. "Genetics and Evolution of Function-Valued Traits: Understanding Environmentally Responsive Phenotypes." *Trends in Ecology and Evolution* 27:637–647.

Stjernman, M., and T. J. Little. 2011. "Genetic Variation for Maternal Effects on Parasite Susceptibility." *Journal of Evolutionary Biology* 24:2357–2363.

Stotz, K. 2006. "Molecular Epigenesis: Distributed Specificity as a Break in the Central Dogma." *History and Philosophy of the Life Sciences* 26:527–544.

Sultan, S. E. 1996. "Phenotypic Plasticity for Offspring Traits in *Polygonum persicaria.*" *Ecology* 77:1791–1807.

Sultan, S. E. 2001. "Phenotypic Plasticity for Fitness Components in *Polygonum* Species of Contrasting Ecological Breadth." *Ecology* 82:328–343.

Sultan, S. E. 2015. *Organism and Environment: Ecological Development, Niche Construction and Adaptation.* London: Oxford University Press.

Sultan, S. E. 2017. "Developmental Plasticity: Re-conceiving the Genotype." *Interface Focus* 7:20170009.

Sultan, S. E., K. Barton, and A. M. Wilczek. 2009. "Contrasting Patterns of Transgenerational Plasticity in Ecologically Distinct Congeners." *Ecology* 90:1831–1839.

Sultan, S. E., and S. C. Stearns. 2005. "Environmentally Contingent Variation: Phenotypic Plasticity and Norms of Reaction." In *Variation: A Hierarchical Examination of a Central Concept in Biology*, edited by B. Hall and B. Hallgrimsson, 303–332. New York: Elsevier Academic Press.

Sultan, S. E., A. M. Wilczek, D. L. Bell, and G. Hand. 1998. "Physiological Response to Complex Environments in Annual *Polygonum* Species of Contrasting Ecological Breadth." *Oecologia* 115:564–578.

Teh, A. L., H. Pan, L. Chen, M. Ong, S. Dogra, J. Wong, J. L. MacIsaac, et al. 2014. "The Effect of Genotype and *in utero* Environment on Interindividual Variation in Neonate DNA Methylomes." *Genome Research* 24:1064–1074.

Turck, F., and G. Coupland. 2013. "Natural Variation in Epigenetic Gene Regulation and Its Effects on Plant Developmental Traits." *Evolution* 68:620–631.

Uller, T., S. English, and I. Pen. 2015. "When Is Incomplete Epigenetic Resetting in Germ Cells Favoured by Natural Selection?" *Proceedings of the Royal Society B* 282:20150682.

Van Dyken, J. D., and M. J. Wade. 2010. "The Genetic Signature of Conditional Expression." *Genetics* 184:557–570.

Van Tienderen, P. H., and H. P. Koelwijn. 1994. "Selection on Reaction Norms, Genetic Correlations and Constraints." *Genetical Research, Cambridge* 64:115–125.

Via, S. 1987. "Genetic Constraints on the Evolution of Phenotypic Plasticity." In *Genetic Constraints on Adaptive Evolution*, edited by V. Loeschke, 47–71. Berlin: Springer-Verlag.

Via, S. 1994. "The Evolution of Phenotypic Plasticity: What Do We Really Know?" In *Ecological Genetics*, edited by L. Real, 35–57. Princeton, NJ: Princeton University Press.

Via, S., and R. Lande. 1985. "Genotype-Environment Interaction and the Evolution of Phenotypic Plasticity." *Evolution* 39:505–522.

Vu, W. T., P. L. Chang, K. S. Moriuchi, and M. L. Friesen. 2015. "Genetic Variation of Transgenerational Plasticity of Offspring Germination in Response to Salinity Stress and the Seed Transcriptome of *Medicago truncatula.*" *BMC Evolutionary Biology* 15:59.

Wagner, A. S. 1999. "Causality in Complex Systems." *Biology and Philosophy* 14:83–101.

Walter, J., D. E. V. Harter, C. Beierkuhnlein, A. Jentsch, and H. de Kroon. 2016. "Transgenerational Effects of Extreme Weather: Perennial Plant Offspring Show Modified Germination, Growth and Stoichiometry." *Journal of Ecology* 104:1032–1040.

Weinig, C., and L. F. Delph. 2001. "Phenotypic Plasticity Early in Life Constrains Developmental Responses Later." *Evolution* 55:930–936.

Woltereck, R. 1909. "Weitere experimentelle Untersuchungen über Arrveranderung, speziell über das Wesen quntitativer Artuntershiede bei Daphniden." *Verhandlungen der Deutschen Zoologischen Gesellschaft* 19:110–172.

Zhu, C.-T., P. Ingelmo, and D. M. Rand. 2014. "G x G x E for Lifespan in *Drosophila*: Mitochondrial, Nuclear, and Dietary Interactions That Modify Longevity." *PLoS Genetics* 10:e1004354.

7 Understanding Niche Construction as an Evolutionary Process

Kevin N. Laland, John Odling-Smee, and Marcus W. Feldman

Few scientists today would contest that organisms modify their environments and those of other organisms, that this has important ecological and evolutionary consequences, and that the concomitant interactions, including feedbacks, feed-forwards, and general nonlinearities are worthy of scientific investigation. However, this has not always been the case. This article provides historical context for the emergence of Niche Construction Theory (henceforth NCT[1]), the rationale behind the formulation of niche construction as an evolutionary process, and discusses the various responses to this theory from different subsections of the scientific community. We discuss what we regard as three major causes of these different responses, which range from enthusiastic endorsement to disdainful resistance: (i) acceptance or rejection of the notion of organismic agency; (ii) acceptance or rejection of an "explanatory gap" between determinate gene effects and the realized impacts of organisms on their environments (and hence concomitant views of development); and (iii) different assumptions concerning the independence or interdependence of the causes of phenotypic variation, differential fitness, and inheritance. These points are essential to understand what is new and important about NCT, and why a Modern Synthesis framework would need to change, and not simply expand, to embrace NCT fully.

At the heart of much of the contention lies the vexed issue of organismal agency. Traditional analyses of evolution usually ignore the agency of organisms, which for some, but not all, researchers create an explanatory gap in evolutionary accounts. We suggest that whether or not this explanatory gap is recognized may depend, at least in part, on how researchers understand the processes of development. NCT seeks to fill this gap by treating niche construction as a process within evolutionary biology. NCT offers a broader interpretation of evolutionary causation than conventional neo-Darwinism, an interpretation that recognizes a role for organisms as imposing a bias on selection through systematically shaping the properties of selective environments (Odling-Smee 1988; Laland et al. 2015; Laland, Odling-Smee, and Endler 2017). This bias results in part from the co-dependence of the processes underlying phenotypic variation, differential fitness, and inheritance, to all of which organismal activities contribute. However, the claim that niche construction operates as an evolutionary bias has generated contention among scholars of evolutionary biology, and making sense of this contention sheds light on wider debates about the structure of evolutionary theory, including those surrounding developmental constraints/bias, developmental plasticity, and expanded inheritance.

We go on to suggest means by which to resolve these disagreements, clarify the roles that niche construction plays in evolution, and evaluate the extent to which niche construction should be regarded as bias. We describe empirical methods that may consolidate the evidence that niche construction operates as an evolutionary bias and suggest that theoretical approaches to this question can also aid this endeavor.

1. Historical and Philosophical Context to the Debates over Niche Construction

When Richard Lewontin wrote his first influential essays on niche construction (1982, 1983), the study of evolutionary feedbacks was still highly restricted. At the time, there were only a handful of mathematical models of sexual selection (O'Donald 1980; Lande 1980; Kirkpatrick 1982), frequency-dependent selection (Slatkin 1979; Wright 1984), coevolution (Futuyma and Slatkin 1983), and gene-culture coevolution (Feldman and Cavalli-Sforza 1976). Evolutionary game theory was in its infancy (Maynard Smith 1982), and adaptive dynamics had barely been conceived.

That is not to suggest that there had been no scientific investigation of organisms' impact on their environments. The scientific study of niche construction dates back to the early writings on ecological succession by William King in the seventeenth century. There were also important nineteenth- and early twentieth-century contributions by Darwin on corals (1851) and earthworms (1881), Lewis Henry Morgan on the beaver (1868), Shaler's (1892) and Taylor's (1935) analyses of how animals affect soils, and of course Frederic Clements's classic (1916) work on plant ecological succession. However, these studies are perhaps better characterized as the idiosyncratic investigations by intellectual pioneers rather than reflective of a general consensus that the topic of niche construction was of central importance within biology. At the time of Lewontin's writings, in neither ecology nor evolutionary biology had that long but sparse history of investigation yet coalesced into a coherent program of research.

That began to change around the 1990s. The study of ecosystem engineering (a concept that overlaps with niche construction) took off in the mid-1990s (Jones, Lawton, and Shachak 1994, 1997; Jones and Lawton 1995) within ecosystem ecology. Prior to that, the prevailing approach within population ecology had been to treat ecosystems as food webs, dominated by trophic and competitive interactions (e.g., May 1973), with little consideration of how organisms create habitat or resources for themselves and other organisms. That organisms could influence the flows of energy and matter was widely appreciated among ecosystem ecologists and environmental scientists who studied, for instance, the nitrogen cycle, the oxygenation of the earth's atmosphere by photosynthetic organisms, and biogeochemical cycles, but these were primarily non-evolutionary approaches (O'Neill et al. 1986). Researchers lamented the failure to integrate the different branches of ecology, and in particular ecosystem ecology with evolutionary biology (Jones and Lawton 1995; Holt 1995). Likewise, it was not until the twenty-first century that the investigation of eco-evolutionary feedbacks and dynamics became a vigorous branch of evolutionary ecology (Pelletier, Garant, and Hendry 2009; Post and Palkovacs 2009; Hendry 2017).

When we began investigating niche construction in the late 1980s and early 1990s, that the activities of organisms could exert a widespread and general influence on the flows of energy and matter in ecosystems, and that these activities could accumulate across individuals and over time to significantly modify natural selection, were hypotheses that still remained to be proven. These proposals were viewed with skepticism by subsections of the biological community. Appreciation of the diversity and scale of niche construction had increased substantially since the 1940s and 50s, when Simpson (1949) and Dobzhansky (1955) had each claimed that humans alone substantially controlled and regulated their environments. But that niche construction was universal (as we characterize it) remained disputed, and there resided a common tendency to regard niche construction as only likely to be evolutionarily important if an adaptation (i.e., "extended phenotype"; Dawkins 1982). There were pockets of evolutionary and ecological theory that captured elements of niche construction, including evolutionary models of frequency and density dependent selection, habitat selection, coevolution, maternal effects, epistasis and indirect genetic effects, and ecological and demographic models of resource depletion (reviewed in Odling-Smee, Laland, and Feldman 2003). However, as we wrote in our book (Odling-Smee et al. 2003, 132), "while many separate theoretical domains investigate phenomena with features in common with niche construction, none of these captures all the pertinent characteristics." Likewise, although ecologists had been studying niche construction by organisms for many decades, there was, except in special cases, little recognition of the evolutionary significance of these activities.

Early in its development, NCT focused on documenting and categorizing the extent of niche construction in living organisms, devising methods by which its ecological and evolutionary consequences could be investigated, and deriving dedicated theory specifically designed to explore these questions (Laland, Odling-Smee, and Feldman 1996, 1999; Odling-Smee, Laland, and Feldman 1996; Odling-Smee et al. 2003; Silver and Di Paolo 2006; Kylafis and Loreau 2008; Lehmann 2008; see also Gurney and Lawton 1996). We defined niche construction as "*the process whereby organisms, through their metabolism, their activities, and their choices, modify their own and/or each other's niche,*" where a niche is "*the sum of all the natural selection pressures to which the population is exposed*" (Odling-Smee et al. 2003, 419). By this definition, niche construction encompasses animal artifacts that evolved as adaptations (e.g., birds' nests), environment-changing byproducts (e.g., plant litter), relocations (e.g., migrations), and learned activities (e.g., human agricultural practices). The breadth of our definition reflected the job we wanted the term to do. We wanted to draw scientific attention to the diverse ways in which organisms modify environmental states, and the myriad of important ecological and evolutionary consequences that follow from these activities. What the above activities have in common is that they modify selection, as well as flows of energy and matter. A narrower definition risked neglecting important processes and feedbacks that would fall outside it: for instance, it is now widely recognized that organisms open up new niches through their byproducts (e.g., San Roman and Wagner 2018).

By and large, these initiatives were uncontroversial and broadly welcomed. If NCT is contentious, it is not because it has encouraged the investigation of niche construction and/or its ecological and evolutionary ramifications. In spite of this, even prior to Lewontin's influential essays, as far back as Conrad Waddington's (1957, 1959) writings, niche construction theory, and its intellectual predecessors, generated contention. For instance, Waddington's evolutionist contemporaries frequently (and wrongly) associated his writings with Lamarckism (Peterson 2016). Over the years, NCT has been subject to a number of different classes of criticism, but we suggest that much of this criticism is derivative, and that the controversy to a significant degree revolves around the key issue of *agency*.[2] NCT is more than just a framework for studying niche construction: it is an attempt to capture the role of organismal agency in evolution, and this issue remains a major source of contention.[3]

Waddington (1957, 1959) believed that the neo-Darwinism of his day was an incomplete description of the evolutionary process. He maintained that there were important processes that were causally important in evolution, but whose significance had not been fully appreciated, one of which he called "the exploitive system." He wrote (1959, 1635–1636):

Animals... are usually surrounded by a much wider range of environmental conditions than they are willing to inhabit. They live in a highly heterogeneous "ambience," from which they themselves select the particular habitat in which their life will be passed. Thus the animal by its behaviour contributes in a most important way to determining the nature and intensity of the selective pressures which will be exerted on it. Natural selection is very far from being as external a force as the conventional picture might lead one at first sight to believe.

Waddington thought that the exploitive system applied primarily to animals, perhaps because it is obvious that animals choose components of their environments and construct artifacts such as nests, burrows, dams, and webs. In fact, the same reasoning applies to other organisms (Lewontin 1983; Odling-Smee et al. 2003; Sultan 2015). Plants, for instance, create shade, influence wind speeds, alter hydrological cycles, influence the cycling of nutrients, and even choose environments for their seeds by altering the timing of flowering or germination and thereby influencing the conditions to which their descendants are exposed (Donohue 2005; Sultan 2015).

Richard Lewontin (1982, 1983, 2000) brought this reasoning to prominence within the evolutionary biology community. His writings stressed how genes, organisms, and environments are in reciprocal interaction with each other in such a way that each is both cause and effect (Lewontin 1983, 281):

What is left out of [the] adaptive description of organism and environment is the fact, clear to all natural historians, that the environments of organisms are made by the organisms themselves as a consequence of their own life activities. How do I know that stones are part of the environment of thrushes? Because thrushes break snails on them. Those same stones are not part of the environment of juncos who will pass by them in their search for dry grass with which to make their nests. Organisms do not adapt to their environments: they construct them out of the bits and pieces of their external world. This construction process has a number of features:

(1) Organisms determine what is relevant. While stones are part of a thrush's environment, tree bark is part of a woodpecker's, and the underside of leaves part of a warbler's. It is the life activities of these birds that determine which parts of the world, physically accessible to all of them, are actually parts of their environments. ...

(2) Organisms alter the external world as it becomes part of their environments. All organisms consume resources by taking up minerals, by eating. But they may also create the resources for their own consumption, as when ants make fungus farms, or trees spread out leaves to catch sunlight. ...

(3) Organisms transduce the physical signals of the external world. Changes in external temperature are not perceived by my liver as thermal changes but as alterations in the concentrations of certain hormones and ions. ...

(4) Organisms create a statistical pattern of environment different from the pattern in the external world. Organisms, by their life activities, can damp oscillations, for example in food supply by storage, or in temperature by changing their orientation or moving. They can, on the contrary, magnify differences by using small changes in abundance of food types as a cue for switching search images. They can also integrate and differentiate. Plants may flower only when a sufficient number of days above a certain temperature have been accumulated.

Both Waddington and Lewontin, then, objected to any portrayal of organisms as passive victims of selection. They viewed organisms as co-directors, or active agents, in their own evolution, with their activities a major determinant of fitness differences and hence of natural selection.

2.   Agency and Process

Let us be clear what we mean by "agency," as it is a much misunderstood term, for which there exist numerous and diverse definitions (Walsh 2015). By agency we mean *the intrinsic capacity of individual living organisms to act on, and in, their world, and thereby to modify their experience of it, including in ways that are neither predetermined, nor random.* For reasons described in section 5 that derive from consideration of the properties of life and the principles of thermodynamics (Schrödinger 1944), we view agency to be an essential and inescapable aspect of nature. However, for us, agency is a property of individual organisms, not of sub-organismal components such as tissues or organs. We emphasize that living organisms are not just passively pushed around by external forces, but rather they act on their world according to intrinsically generated but historically informed capabilities. Organisms are self-building, self-regulating, highly integrated, functioning, and (crucially) "purposive" wholes, which through wholly natural processes exert a distinctive influence and a degree of control over their own activities, outputs, and local environments. Indeed, organisms *must* have these properties in order to be alive (Schrödinger 1944; Odling-Smee 1988; Odling-Smee et al. 2003; Turner 2016). How, through exclusively natural processes, a purposeless universe could give rise to the evolution of purposive living systems is both an important and tractable question for evolutionary biology.

Our use of agency does not imply conscious, sentient, or deliberate action, nor vitalism (or any mystical power that imbues living tissue), nor a rejection of mechanistic explanation, nor the belief that living organisms possess any desire to evolve, or achieve some final state. Of course, human niche construction frequently is conscious, intentional, and deliberate, and sometimes does set out to achieve longer-term goals, but these are neither necessary nor defining characteristics of niche construction. When we assert that organisms are "purposive" we mean nothing more than that organisms exhibit goal-directed activities, such as foraging, courtship, or phototaxis, which are entirely natural tendencies with short-term local objectives, and that have themselves evolved. The "goals" and "purposes" to which we refer can be defined with respect to general aspects of biological function, such as resource acquisition, stress avoidance, and reproduction. Hence organismic effects such as decomposition and resource depletion are included as niche construction.

Niche construction is often the product of multiple individuals' collective and sometimes coordinated activities, including from multiple species (see Chiu, this volume). Our characterization of agency as a property of individual organisms raises familiar questions concerning how "individuals" are to be defined (Clarke 2010), which we will not dwell on here. Suffice to say that, for us, biological "agents" are not always bounded by a continuous skin or an integrated nervous system, and that we recognize agency in the activities of many clonal (fungal or plant), colonial (invertebrate), or social (humans) organisms.

Yet, in acknowledging the legacy of past natural selection in shaping how contemporary organisms manifest their agency in the world, we reject the suggestion that the actions of organisms are fully and satisfactorily explained by prior selection, such that organismal activities contribute nothing to explanations of their consequences. From the outset, and particularly as manifest in Lewontin's writings (e.g., Lewontin 1982, 1983, 2000), central to the niche-construction perspective has always been a view of developmental processes as open and constructive through self-assembly, and a corresponding rejection of the idea that organisms and their activities are fully specified by genetic programs. Organisms are regarded as influenced, but not determined, by their genes, and their activities as shaped by developmental information-gaining processes as well as natural selection acting on genetic variation (Odling-Smee et al. 2003; Laland and Sterelny 2006; see Stotz, this volume).

This stance is significant because for researchers sympathetic to a "constructive development" mindset, but not necessarily believers in "programmed development" (see Laland et al. 2015 for a more detailed account of these terms), there exists an *explanatory gap* between the ancestral natural selection that shaped the niche-constructing capability and the consequences of their environmental modification. Phenotypes are complex interactions between genotype, epigenotype, maternal (and sometimes other ancestral) environment, current environment, learned behavior, other ontogenetic exploratory and selective processes, as well as developmental noise (Gerhart and Kirschner 1997; Lewontin 2000; Odling-Smee et al. 2003; Kirschner and Gerhardt 2005; Sultan 2017, this volume). Our characterization of niche construction as an evolutionary process (i.e., as a nontrivial cause and co-director of evolutionary change) can be construed as a label for the explanatory gap linking the ancestral natural selection that is partly responsible for the evolution of niche construction and the subsequent natural selection that is partly a consequence of niche construction. Organisms are not merely objects through which the causal explanatory power of natural selection flows; rather they are active agents that transduce and filter genetic inputs that derive from prior selection, as well as environmental inputs. They also exert a biased and directional[4] influence on the natural selection that follows directly or indirectly from their own activities.

We believe that NCT has proven contentious primarily because of its claim that niche construction should be treated as an evolutionary process, a deliberate and conscious framing designed to recognize organismal agency as a putative evolutionary cause, and which was implicit in Waddington and Lewontin's writings. Other issues, such as the contribution of ecological legacies to widening views of inheritance (Danchin et al. 2011; Bonduriansky 2012), the issue of whether NCT presents a challenge to proximate-ultimate dichotomous conceptions of biological causation (Laland et al. 2011), and the association of NCT with the extended evolutionary synthesis (EES; Laland et al. 2014; Laland et al. 2015) are also subject to much debate. However, each of these controversies, in different ways, is connected to the central and fundamental issue of which biological phenomena should be accepted as fundamental causes of evolutionary change, and thereby incorporated into explanations as to why and how evolution occurs. In particular, each is related to the issue of whether there is a directing evolutionary role for the activities of organisms.

In this light, NCT can be viewed as having inherited contention that can be traced back at least a century (Peterson 2016), and probably much longer (Riskin 2016). The forging of the modern evolutionary synthesis in the early part of the twentieth century comprised, in large part, a rejection of internalist explanations for evolutionary change, such as Lamarckism, vitalism, and orthogenesis (Peterson 2016; Riskin 2016). Since that time, scientific attempts to recognize an evolutionary role for organism-derived internal causes have been strongly out of favor (Gould 2002; Allen 2014; Love 2015). Historically, an emphasis on agency has been associated with vitalism and mysticism (Riskin 2016). Right up to the present, a stress on organismal agency has been somewhat ungenerously portrayed as reflecting a personal failing to accept what is depicted as the "true" but bleak externalist account of adaptive evolution which results from nothing more than the selection of randomly generated genetic variation (e.g., Futuyma 2017; Welch 2017). Seemingly, the fact that the traditional evolutionary framework can give an agent-less evolutionary explanation of the same phenomena is sufficient for many researchers to conclude, on grounds of parsimony, that there is no need for an agential explanation.

Yet there is nothing mystical about the fact that living organisms exhibit activities, outputs, and choices that modify the selection to which they, and other organisms, experience. Nor is there anything vitalist about the possibility that such modifications could bias the course of evolution. As stressed above, an emphasis on agency need not imply commitment to the notion that organismal activities are conscious or deliberate. The possibility remains that an evolutionary framework that recognized organismal agency could prove more productive than one that dismissed this possibility.

3. Conceptual Frameworks and the Reception of NCT

It is important to recognize that the issues raised above have wider implications than the specific debates over agency, and apply equally to long-standing controversies over developmental constraints and bias. To quote Gould (2002, 1028, italics in original):

orthodox Darwinians have not balked at *negative* connotations of constraint as limits and impediments to the power of natural selection in certain definable situations. But they have been far less willing to embrace *positive* meanings of constraint as *promoters*, *suppliers*, and *causes* of evolutionary *direction* and *change* ... the admission of a potent and positive version of constraint would compromise the fundamental principle that variation (the structuralist and internalist component of evolution) only proposes, while selection (the functionalist and externalist force) disposes as the only effective cause of change.

Almost the exact same claim could be made about niche construction: "orthodox Darwinians" are seemingly entirely comfortable with the ideas that niche construction evolves and modifies selective environments, but they are manifestly far less willing to accept it as a cause of evolutionary direction and change.

While most scientists would agree that their field should investigate all credible causes of the phenomena they wish to understand, inevitably there will be disagreement as to which causal factors are credible. This would seem to be the case for organismal agency, which some evolutionary biologists clearly struggle to acknowledge as an explanatory factor in evolution. Differences arise in part because scientists possess variant ways of understanding the world—different conceptual frameworks. To make progress, scientists must specify the phenomena that require explanation, and the methods, data and analyses required. In the process they inadvertently create (or, inherit) a way of thinking for their field (Kuhn 1962; Lakatos 1978). Such conceptual frameworks inevitably encourage some lines of research to the detriment of others. For this reason, alternative conceptual frameworks can be valuable because they draw attention to constructive new ways of thinking, additional causal influences, alternative predictions, or new lines of enquiry (Lakatos 1978). Science thrives on diversity; without it there could be no scientific progress. Specifying alternatives is often a necessary first step toward testing between them through experimentation. Generative frameworks that allow alternative hypotheses to be derived, and/or that encourage the opening of new lines of scientific enquiry should generally be welcomed, as science is a sorting process and typically only those avenues perceived by scientists to possess explanatory potential will flourish. Lewontin (Levins and Lewontin 1979) was explicit in embracing a dialectical stance. While we offer no such justification, our claim is no less based upon the belief that proposing an alternative scientifically motivated and empirically tractable way of thinking, as a means to highlight key issues and stimulate innovation and progress by specifying where and how expectations differ, is often a useful exercise (at least, in the context of exploration and discovery).

It is in this spirit that NCT was presented as an alternative conceptual framework, explicitly so in Odling-Smee et al.'s (2003) book, and more implicitly, as another way of thinking about the issues, in the writings of Waddington (1957, 1959) and Lewontin (1982, 1983). NCT starts from the premise that niche construction is an evolutionary process, and then explores where this leads. This is no act of faith. We are entirely persuaded that there are good theoretical reasons for thinking that niche construction must be an evolutionary process, as we describe in section 5 below. However, in our judgment, NCT should be evaluated, not on whether the premise is right or wrong, but on whether it is, or turns out to be, productive to accept this premise (i.e., whether it facilitates good science). Given the many researchers that have drawn on NCT in their investigations, the prominent journals in which their work has been published, and the positive manner in which the bulk of this work has been received within their respective disciplines, we submit that the productivity of NCT ought to be beyond dispute. NCT has been widely deployed across multiple fields to inspire empirical research and shed new evolutionary light on diverse topics, ranging from paleontology (Erwin and Valentine 2013), evolutionary ecology (Schwilk and Ackerly 2001; Post and Palkovacs 2009), evolutionary developmental biology (Schwab, Casasa, and Moczek 2017), biodiversity (Boivin et al. 2016), human evolution (Fuentes 2009; Kendal, Tehrani, and Odling-Smee 2011), evolutionary archaeology (Smith 2007; Zeder 2015; Piperno 2017), and the origin of language (Bickerton 2009). The rapidly growing literature on niche construction now encompasses well over a thousand articles and many tens of books. Not all of this work explicitly endorses NCT (as specified by Odling-Smee et al. 2003), but virtually all can be viewed as part of an intellectual tradition that through conceptual development and citation can be traced back to the seminal writings of Lewontin, and before that Waddington. Our strong impression is that Lewontin's and Waddington's emphasis on the active agency of the organism is a major attraction of NCT to many and diverse researchers. It is an emphasis that resonates with their own assessments of the role that agents play in their science. For whatever reasons, numerous researchers in diverse fields have found it is helpful to use the NCT framework with its emphasis on niche construction as process in advancing their science.

Yet, for all of its impact in other fields, within evolutionary and population genetics NCT has been received more circumspectly. Nevertheless, NCT achieved interdisciplinary significance for emphasizing the importance of eco-evolutionary dynamics and ecological legacies before these topics became widely studied, and for stressing the generality of feedbacks from organismal activities at a time when investigation of such feedbacks were largely restricted to social interactions (i.e., sexual selection), or other direct interactions between biota (i.e,. coevolution). It is surely not entirely coincidental that research in these topics has expanded substantially over the last 20 years. Yet, even if this more diffuse influence is acknowledged, it raises the question of why some evolutionary biologists should feel skeptical about niche construction theory when researchers in many other fields do not. We submit that the likely answer is, at least in part, that many evolutionary biologists feel uncomfortable recognizing organismal agency. Certainly, where niche construction is investigated within evolutionary biology, it has typically been stripped of its agency implications. To many evolutionary biologists, the claim that "niche construction is a process" makes NCT appear unnecessarily baroque, even misguided, particularly when other feedbacks from organisms' activities have been investigated productively without this claim (Scott-Phillips et al. 2013).

We note that in other fields, however, most obviously the human sciences, such as archaeology (e.g., Smith 2007; Zeder 2015), biological anthropology (e.g., Fuentes 2009, 2017), linguistics (e.g., Bickerton 2009), physiology (e.g., Turner 2016), and developmental psychology (Flynn et al. 2013), but also ecology (Ellis 2015; Alberti 2015), the emphasis that NCT places on (local, short-term) agency has been welcomed by some practitioners. To the extent that such researchers are broadly representative of their fields, they raise the possibility that differences between academic fields in attitudes toward agency and, more generally, in their overarching conceptual frameworks, might explain some of the interdisciplinary variability in the reception of NCT. Plausibly, as a legacy of the fractious history of attempts to incorporate agency into evolutionary accounts (Riskin 2016), the traditional evolutionary framework has become highly sensitive and resistant to attempts to incorporate agency into its ontology, while other fields of study, with different histories and attitudes toward agency, experience no such tension.

Differences between fields in how researchers understand development (ontogeny) are also highly relevant; indeed, these issues are intertwined. While evolutionary biologists are very conscious of environmental influences on phenotypes, and hence eschew a simple notion of genetic determinism, nonetheless a programmed view of development as controlled by genes, in the form of genetically specified switches or reaction norms that respond to environmental inputs, remains extremely common (E. Keller 2010). If the manner in which organisms engage in niche construction is perceived to be fully and satisfactorily explained by earlier natural selection, then niche construction would appear to do no additional explanatory work, and its influence can be characterized as the current proximate manifestation of an earlier evolutionary cause, namely natural selection. From this perspective, which is perhaps representative of a substantial proportion of contemporary evolutionary biologists, the aforementioned "explanatory gap" between ancestral selection and niche-constructing outputs may appear small, and perhaps, entirely invisible.

Conversely, human scientists (Flynn et al. 2013; Ellis 2015; Fuentes 2017), and some researchers in other fields, such as ecological developmental biology (Gilbert and Epel 2015), physiology (Noble 2017) and other subfields of developmental biology (Kirschner and Gerhardt 2005), like Lewontin (1982, 1983), are often unwilling to accept genetic programming and commonly conceive of development as a much more constructive, open-ended process, in which some very general "knowledge-gaining" adaptations, such as learning, and other exploratory processes (Gerhart and Kirschner 1997; Kirschner and Gerhardt 2005) play important roles. The constructive development perspective recognizes that the evolutionary process itself evolves (Laland et al. 2015; Watson and Szathmary 2016). In particularly, multicellular life has evolved ontogenetic processes that mimic the Darwinian algorithm of natural selection, and generate variation (perhaps at random), select or retain functional variants while removing unnecessary or dysfunctional solutions, and then regenerate in an iterative manner (Gerhart and Kirschner 1997; Kirschner and Gerhardt 2005). This confers on developmental systems, such as the adaptive immune system, the nervous system, microtubule assembly in eukaryotes, the vertebrate vascular system, the insect trachea system, or behavior influenced by learning, an adaptability to respond flexibly to novel challenges, be they environmental stressors or internal events such as mutation, and to generate novel, coherent, functional phenotypes.

Under this perspective, it would be a vulgar and naïve simplification to characterize behavior as genetically controlled. The distinction between programmed and constructive development is important, because in the former phenotypes are viewed as prescreened by prior selection, while for the latter developmental processes can produce genuinely novel phenotypes that subsequently trigger genetic accommodation (West-Eberhard 2003; Laland et al. 2015). Researchers who view development as constructive are thus predisposed to recognize the explanatory gap filled by niche construction as a process. If the adaptability of developmental processes can generate novel functional phenotypes, then it follows that niche construction (a product of such developmental processes) can lead to functional forms of environmental modification that were not prescripted by prior natural selection. For the same reasons, constructive or programmed views of development may help to explain whether or not researchers assign important evolutionary roles to developmental bias, developmental plasticity, and extra-genetic inheritance in evolution.

No less relevant to these debates is the causal interdependence of the processes that generate phenotypic variation, differential fitness, and inheritance, which are Lewontin's (1970) three conditions for evolution by natural selection (Odling-Smee et al. 2003; Walsh 2015; Uller and Helanterä 2017; Pocheville, this volume). For illustration, an orthodox stance would not recognize earthworm niche construction as an evolutionary cause but rather as the proximate outcome of earlier natural selection that favored earthworm soil-processing capabilities (figure 7.1, top). Implicit in this reasoning is the separation of the processes that generate variation (e.g., mutations in soil-processing activity) and the causes of fitness differences (e.g., the alternative environmental conditions that favored those mutations). This allows the explanation for the adaptive fit between earthworms and their soil environment to begin with that change in ancestral environmental conditions that generated selection for earthworm soil processing. In reality, because ancestral earthworm activity is itself the cause of the soil environment that favors mutations in soil processing, the causes of variation and fitness differences are not independent (figure 7.1, bottom). The explanation for the adaptive fit between earthworms and their soil environment cannot begin with those ancestral external environmental conditions that elicited selection for earthworm soil processing, since those conditions were themselves products of earlier earthworm niche construction. There is no longer a clean separation of selection and variation: these processes are causally intertwined. This confound is not resolved by pushing back the explanation to commence with an early episode of selection, as the selective environment is always partly constructed by organismal activities. Worse, the processes underlying inheritance are also not independent of the causes of variation and fitness. The genetically specified propensity for soil processing that contemporary earthworms inherit only functions effectively as the source of earthworm adaptation to a soil environment because contemporary earthworms also inherit the ecological legacy of a modified soil environment that is the product of ancestral earthworm niche construction. Recognizing niche construction as an evolutionary process is an attempt to build this causal interdependence into evolutionary explanation (Uller and Helanterä 2017). Neglecting or downplaying this causal interdependence is the manner in which it is more commonly treated.

Whether it is because of differential attitudes toward agency, different conceptions of development, alternative assumptions regarding the causal independence of evolutionary processes, or other factors, there clearly exist different, and often hidden, assumptions between academic fields, and between individuals within fields, which contribute to the

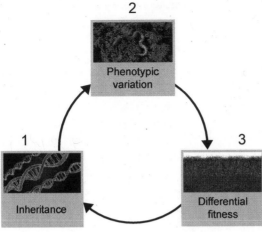

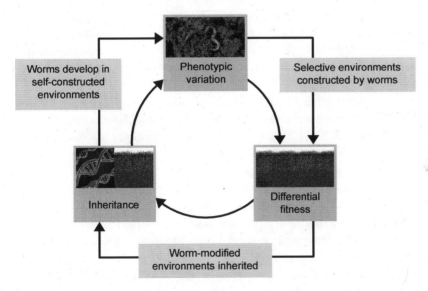

Figure 7.1
The assumption of the quasi-independence of processes responsible for inheritance, phenotypic variation, and differential fitness (Walsh 2015) is manifest in the top panel, illustrated with an earthworm soil-processing example. Here, a random mutation (1) in the earthworm population (2) changes the worm's soil-processing capability in a way that (3) has beneficial effects on fitness in the soil environment, and consequently is inherited by the next generation. On this perspective, the characterization of niche construction as an evolutionary process appears to be a conceptual error, since it mislabels a proximate cause (the earthworm's modification of the environment) as an evolutionary cause. However, in reality the situation is more complex, as illustrated in the lower panel. Differential fitness is not independent of phenotypic variation as soil conditions are partly constructed by earthworms, inheritance is not independent of selection because worm-modified environments persist for long periods of time and are inherited by descendants, and phenotypic variation is not independent of inheritance since worms develop in a self-constructed environment. Here causality is intertwined rather than quasi-independent, and there is no more reason to explain away niche construction as a product of past selection than there is to explain away natural selection as a product of earlier niche construction (Walsh 2015; Uller and Helanterä 2017).

variable reception of NCT. Seemingly, NCT is framed in a way that resonates with and excites some researchers, but it touches a nerve with others. When we published our monograph on the topic (Odling-Smee et al. 2003), the review in *Science* by ecologist John Vandermeer (2003) described it as "a major breakthrough" while the corresponding review in *Nature* by evolutionary biologist Laurent Keller (2003) stated: "it is unfortunate that the authors attempt to oversell the significance of niche construction." Strong endorsements by Richard Lewontin, Robert May, and David Hull were counterbalanced by forceful criticism from Richard Dawkins (2004), who went so far as to recommend that the term "niche construction" be abandoned forthwith. The primary concern of Keller, Dawkins, and others (Brodie 2005; Scott-Phillips et al. 2013) is that, in asserting that niche construction is an evolutionary process, NCT overstates its significance.

We agree that from an orthodox perspective it does not make sense to characterize niche construction as an evolutionary process. However, there is more than one way to think, and each way of thinking has associated with it a specific set of assumptions, with specific processes taking on greater or lesser significance according to the perspective from which they are viewed. From the developmental perspective championed by Gould, developmental constraints have far greater significance compared to the mainstream perspective, since they can be active and thereby co-direct evolutionary change. So it is from the constructivist perspective championed by Lewontin that niche construction takes on greater significance than is conventional.

All too often, such debates have been mischaracterized as revolving around whether particular processes (plasticity, constraint, extra-genetic inheritance) have or have not been "neglected" (e.g., Wray et al. 2014). This characterization is a distortion. As Gould (2002) makes clear, the issue is not whether constraints have been neglected—surely not, evolutionary biologists study constraints all the time—but whether the manner in which mainstream evolutionary biology comprehends and studies constraints hinders recognition of their true significance in evolution. Again, the parallels with niche construction are strong: the issue is not whether niche construction has been neglected—it hasn't—but whether the manner in which the mainstream comprehends and studies niche construction hinders recognition of its true significance in evolution.

We can surely all agree that there is merit in scientists being open to considering the possibility that particular processes might be more significant than is immediately apparent from their current standpoint. In an ideal world, researchers would take the trouble to try to understand the perspective of others. Then it would be perfectly reasonable for a researcher to express a preference for one conceptual framework over another, on the grounds that it appears to be more useful or productive. However, criticizing the postulates of one conceptual framework by the standards of another amounts to a failure of imagination— an inability to take the other's perspective. It would be no less a mistake to criticize neo-Darwinism for "overselling" natural selection (e.g., Ho 1988).

4. Organismal Agency and Evolutionary Process

Traditional evolutionary approaches largely disregard the agency of organisms. To quote Walsh (2015, 18):

The purposive complexity of organisms is seen as simply a consequence of evolution, which, in turn, according to currently orthodox Modern Synthesis thinking is basically a molecular process.

Historically, population genetics has investigated evolution assuming relatively simple relationships between genotypes and phenotypes (e.g., Hartl and Clark 1989). Commonly, for mathematical convenience evolutionary models allocate fitness terms directly to genotypes, ignoring phenotypes altogether. This seems perfectly reasonable given the mysteries and complexity of the genotype-to-phenotype translation. In conventional accounts organisms are credited with playing some roles in evolution (e.g., as the result of gene products that determine differential survival and reproduction), but they are not widely thought of as *causes* of evolutionary change.

In contrast, NCT phenotypes have an evolutionary agency not captured by traditional evolutionary approaches and not fully determined by naturally selected genes: namely, the capacity to modify environmental conditions and consequently natural selection and to generate ecological inheritances (including modifying developmental conditions for their descendants) in ways that bias the development and subsequent evolution of both themselves and other organisms (Odling-Smee 1988; Odling-Smee et al. 2003). NCT emphasizes that feedback from organisms' activities can be important for their survival and reproduction. However, it is the emphasis on organismal agency as a directing influence in evolution that is the defining characteristic of NCT.

What constitutes an evolutionary process is not universally accepted by evolutionary biologists, who have operated under a wide variety of definitions and assumptions. Most textbooks recognize as evolutionary processes those phenomena that directly change gene frequencies (e.g., selection, drift, mutation, gene flow), although other processes are sometimes included as affecting evolution (Laland et al. 2015). Niche construction is not included among these. Thus it is fair to say that niche construction is not currently accepted as an evolutionary process, nor even recognized as a candidate. Futuyma (2017, 4) represents this position well in asserting:

niche construction can influence or even cause the evolutionary process of natural selection, but it is not itself an evolutionary process, any more than a changing environment is. ... The sources of ... selection are not processes.

What this standpoint omits, and what potentially differentiates niche construction from environmental change more generally, is that niche construction initiates and modifies the patterns of natural selection directly affecting the constructor (and other species that share its environment) *in an orderly, directed and sustained manner* (Odling-Smee 1988;

Odling-Smee et al. 2003; Laland, Odling-Smee, and Endler 2017). As a consequence, niche construction directs adaptive evolution in a manner that other sources of environmental change, which do not bear the imprint of purposive organisms, do not. Niche construction, as an evolutionary process, imposes a statistical bias on the direction and mode of selection, and hence on the speed and direction of evolution. By systematically creating and reinforcing specific environmental states, niche construction directs evolution along particular trajectories. In this respect, the postulate that niche construction is an evolutionary process parallels the (no less contested) postulate that developmental processes bias the course of adaptive evolution (Arthur 2004; Brakefield 2011). Since Darwin, it has been well-recognized that not all phenotypic variants are likely to be generated (Darwin 1859; Maynard Smith et al. 1985). The bias imposed on the distribution of phenotypic variation arising from the structure, character, composition, or dynamics of the developmental system is known as "developmental bias" (Maynard Smith et al. 1985; Uller et al. 2018). We suggest that niche construction can be viewed as a form of developmental bias, unusual only in the respect that it is expressed outside of the organism.

For niche construction to be recognized as an evolutionary process, then, a broadening of current conceptions of evolutionary process is required. Endler (Endler 1986; Endler and McLellan 1988) develops a broad classification scheme, which includes categories of evolutionary processes, such as "adaptive processes," "rate-determining processes," and "direction-determining processes." It is in these capacities that niche construction has important, or even central, roles to play in evolution. If niche construction operates as an evolutionary process it does so in a manner qualitatively different from the standardly recognized evolutionary processes.

Futuyma (2017, 5) claims:

So far, no new, general theoretical principles that promise to guide novel empirical research have been articulated by proponents of niche construction.

This claim is evocative of that made by Lynch (2007, 8598) concerning evo devo:

No principle of population genetics has been overturned by an observation in molecular, cellular, or developmental biology, nor has any novel mechanism of evolution been revealed by such fields.

Stoltfus (2017, this volume) takes issue with Lynch, pointing out that, consistent with the claims deriving from evo devo that developmental bias is an important evolutionary process, Yampolsky and Stoltzfus (2001) used a population-genetic model to show that mutational or developmental biases in the introduction of variation can be a true cause of evolutionary bias. In our view, Lynch's claim is neither fair (as the science of other fields is not designed to shed light on population genetics processes, or primarily to address evolutionary mechanisms) nor true: findings from other fields have revealed clear evidence for transgenerational epigenetic inheritance (Jablonka 2017), developmental bias (e.g., Kavanagh, Evans, and Jernvall 2007; Wagner 2011), horizontal gene transfer (Syvanen and Kado 2002), facilitated variation (Watson et al. 2014), and numerous other challenges to conventional population-genetic principles. Futuyma's and Lynch's claims seemingly ignore many hundreds of studies that use alternative approaches productively, and suggest that the onus is on others to demonstrate that the author's views need to be changed. At a time when biology is moving away from privileging null hypotheses to a more even-handed model comparison approach, this stance appears questionable. A willingness to think outside of one's own conceptual framework is surely healthy.

We are nonetheless happy to take up Futuyma's challenge and to lay out an overarching general principal that guides and inspires niche construction research:

The bias that niche construction imposes on selection will be manifest as qualitative or quantitative differences between organism-constructed sources of selection in environments and autonomous sources of selection (i.e., those without a clear signal of organismal agency), with the former associated with greater consistency and predictability in the rate and direction of selection than other sources.

We elaborate on this general principal below. See also Odling-Smee et al. (2003) and Laland, Odling-Smee, and Endler (2017) for detailed expositions.

5. Why Does NCT Expect Niche Construction to Be Systematic and Directional?

The theoretical argument that lies at the heart of NCT's claim that niche construction is an evolutionary process derives from the expectation that niche construction must be purposive and goal directed, which in turn follows from consideration of some basics of thermodynamics (Schrödinger 1944).

The great physicist, Erwin Schrödinger (1944) first pointed out that living organisms are far-from-equilibrium systems relative to their physical or abiotic surroundings. To preserve their out-of-equilibrium status, while creating order (structure, organization) in their bodies and their immediate surrounds, organisms must do work on their environment. Living organisms can only survive by constantly exchanging energy and matter with their environments. Organisms feed on molecules rich in free energy, using the harvested energy to do work, and such work is necessary to allow organisms to produce and maintain a specified orderliness, be it inside their bodies or in their immediate external environments. Schrödinger explains why this work cannot be random; it must be directional. Random niche-constructing acts could not provide organisms with a basis for sustained life. It follows that organisms must change environments in systematic ways (Schrödinger 1944; Odling-Smee et al. 2003). Constructed environments must bear the signature of purposive agents.

Over and above these thermodynamic considerations, that the capability of organisms to engage in niche construction has itself evolved also leads us to expect it to be orderly.

Niche construction generates environmental states that are coherent and integrated with the organism's phenotype, and that are adaptive for the constructor or its descendants largely because the organism's capability to do so has evolved through earlier natural selection. Organisms in the same species possess broadly similar evolved capabilities for niche construction, so they are expected to modify environmental states in broadly similar ways, while the niche construction by offspring is expected to resemble that by their parents. In addition, Odling-Smee et al. (2003) emphasize how one important class of niche construction, termed "counteractive," functions to regulate change and variability in the external environment and to ensure that relevant conditions remain within tolerable bounds. For instance, by building nests, animals typically ensure a reliable, safe environment for both themselves and their relatives, in which temperature, humidity, gas levels, and other parameters can be regulated. Natural selection arising from animals' interactions with their nests is thereby expected to be unusually consistent because these niche-constructing activities themselves generate dependable conditions (Odling-Smee et al. 2003; Laland, Odling-Smee, and Endler 2017). Even the niche construction that arises as byproducts of metabolism is likely to be consistent and orderly, since the biological processes from which they derive are reliably produced, generation after generation (e.g., San Roman and Wagner 2018). For these reasons, NCT expects niche construction to scale up across individuals in populations and over time to generate stable modification of environmental components, and as a result to generate consistent and sustained forms of natural selection.

Why, if niche construction has itself evolved, can researchers not simply attribute any directionality that niche construction imposes on the evolutionary process to earlier selection? In fact, that attribution is probably the default assumption within evolutionary biology, a stance that allows any adaptive directionality in evolution produced by niche construction to be explained by past selection (a similar attribution is made for developmental bias). There are two reasons why we regard such an attribution to be illegitimate.

First, from the standpoint of constructive development, the niche-constructing capabilities of living organisms are only partly explained by earlier selection (i.e., the aforementioned "explanatory gap"). Generative and selective processes occurring during ontogeny, which are also information gathering and which are also capable of producing adaptive phenotypes in ways that mimic a Darwinian algorithm, are no less responsible for niche-constructing phenotypes. For illustration, the fact that humans possess an evolved capacity to learn does not provide a complete and satisfactory explanation for why some humans populations happen to take up dairy farming, and hence why those particular populations evolved adult lactose tolerance through self-imposed selection on LCT alleles (Tishkoff et al. 2007). The existence of this explanatory gap is what leads us to view living organisms as evolutionary agents.

The second reason is the aforementioned "causal intertwining" of evolutionary processes. Niche construction (and again, developmental bias) can only be reduced to past natural selection in a model of evolution in which the causes of Lewontin's (1970) three conditions of "phenotypic variation," "differential fitness," and "inheritance" are treated as quasi-independent: conversely, where these processes are causally intertwined, this interpretation does not follow (Walsh 2015; Uller and Helanterä 2017; figure 7.1).

6. Empirical and Theoretical Approaches to Demonstrating That Niche Construction Generates an Evolutionary Bias

There is ample empirical evidence that organisms engage in niche construction and modify selection (e.g., Kerr et al. 1999; Schwilk 2003; Post and Palkovacs 2009; Buser et al. 2014; Sultan 2015), and that they do so in consistent and reliable ways (reviewed in Odling-Smee et al. 2003; Sultan 2015). However, these empirical data do not decisively cement niche construction's status as an evolutionary process. That is because such data are equally consistent with niche construction being viewed as a background condition, or just another source of environmental change. Investigations of a single episode of niche construction, even those demonstrating a corresponding response to selection, do not demonstrate that niche construction is capable of generating a systematic evolutionary bias.

NCT may well have stimulated empirical investigations of niche construction, ecological inheritance, and eco-evolutionary dynamics, as well as mathematical analyses of these phenomena using standard population-genetic and population ecology tools, but that is not the same as demonstrating that niche construction is an evolutionary process. How could this be done? Niche construction is undoubtedly systematic and directional in some cases, but is it always? Or even typically? How could this be tested?

There are practical difficulties here (for instance, it is hard to measure all of the myriad different ways an organism modifies its environment). There are also conceptual difficulties: for instance, can the way in which niche construction modifies environments be predicted a priori? If organisms modify environments in different species-specific ways, then perhaps there are no laws of niche construction?

Laland, Odling-Smee, and Endler (2017) suggest that these difficulties can largely be circumvented by focusing on responses to selection. A priori prediction of the likely properties of the response to selection resulting from niche construction across multiple niche-constructing episodes in multiple populations potentially could provide evidence of such a bias. A key issue is the expectation, derived from NCT, of differential responses to constructed verses non-constructed elements of the environment. Recently, Laland, Odling-Smee, and Endler (2017) generated a series of predictions for how responses to constructed verses nonconstructed environmental components should differ. These include that (i) niche construction will typically generate more consistent selection, both in time and space, manifest as reduced temporal and spatial variance in selection gradients relative to non-constructed environments; (ii) well-established environment buffering (counteractive) niche construction will typically reduce the rate of response to selection relative to autonomous sources of selection, as manifest in reduced directional, stabilizing and correlational selection magnitudes; and (iii) novel (inceptive) niche construction activities will initially on average generate unusually strong selection, as manifest in larger selection gradients.

These and other predictions follow from the aforementioned observation that organisms control elements in their environment (e.g., nests, burrows), often pushing them into states that they would not otherwise occupy, thereby imposing an order or regularity on a subset of the selection that they experience, and reliably triggering adaptive responses, or buffering such responses, depending on circumstances. The suggestion here is that in some respects niche construction resembles the artificial selection carried out by the practices of the breeder/experimentalist. Like artificial selection, the selection pressures that niche construction generates will typically be reliable, directed, orderly, and often highly consistent across diverse organisms that perform similar niche-constructing activities, relative to selection arising from autonomous or nonconstructed sources. These expectations can potentially be evaluated through meta-analyses of published selection gradients. Confirmation of the predictions would show that constructed environments differ systematically from non-constructed environments in at least some respects. This does not prove that niche construction is a process, although the findings are consistent with this hypothesis; nonetheless, a failure to find these differences would have undermined the case that niche construction is a process.

A second approach, also specified in Laland, Odling-Smee, and Endler (2017), is to use comparative phylogenetic tools to try to predict sequences of trait evolution. This is based on the hypothesis that constructed elements of the environment, by virtue of being reliably produced, generate reliable evolutionary responses that can be anticipated. Constructed features should support longer more predictable trait-evolution sequences than non-constructed features, while similar constructions in independent taxa should elicit similar responses, and thereby generate parallel trait evolution (see Laland, Odling-Smee, and Endler 2017 for further detail).

The bottom line is that if NCT can be used to make evolution more predictable it potentially demonstrates its heuristic value to the field of evolutionary biology. Beyond this, that greater predictability can potentially be construed as evidence that niche construction operates as an evolutionary bias. Plausibly, niche construction could impose evolutionary biases, and organismal agency could affect evolution, in ways other than those we have considered, but for these too the general approach that we advocate here would hold: as long as those other ways have signatures that can be tested for, then specifying and testing for those signatures is surely the way to move the debate forward. Once again, the logic is identical to developmental bias, where a consideration of developmental mechanism is helping to make evolutionary biology a more predictable science (Kavanagh, Evans, and Jernvall 2007).

Theoretical approaches to the problem are also conceivable. Existing mathematical models of niche construction are designed to explore the consequences of niche construction but not to demonstrate that niche construction acts as a systematic evolutionary bias. While demonstrating bias using standard population-genetic approaches may be possible, such methods typically operate at the wrong (microevolutionary) scale to demonstrate that a *systematic* bias is operating. Patterns of bias manifest across multiple evolutionary episodes may only show up at meso- or macro-evolutionary scales. However, if the consistent, reliable, and predictable properties of niche construction are to be shown to evolve rather than be artificially imposed on the system, micro-evolutionary events need to be incorporated in models and analyses. The problem is potentially theoretically tractable, but new modeling approaches may be required to address this issue.

Conclusion

Evolutionary biology largely ignores the agency of organisms as an explanatory factor in evolution. Niche construction theory seeks to compensate for this by treating niche construction as an evolutionary process. This stance has generated contention, as the theoretical rationale, which derives from thermodynamics, is neither well-known nor well-appreciated. There appear to be strong differences between academic fields in responses to NCT, which may reflect different core assumptions and attitudes toward agency, the processes of development, and whether or not evolutionary processes are recognized to be causally intertwined. We suggest nonetheless that this issue is one that is potentially resolvable through empirical and/or theoretical investigation. Until such time as there is clear evidence that niche construction operates as an evolutionary bias, NCT is likely to remain contentious within mainstream evolutionary biology.

Acknowledgments

This chapter was originally drafted as a discussion paper for the Konrad Lorenz Institute workshop *Cause and Process in Evolution* in Vienna May 15–18, 2017. Research was supported in part by a grant from the John Templeton Foundation entitled "Putting the Extended Evolutionary Synthesis to the Test" (grant no. 60501). We are grateful to Lynn Chiu, Sonia Sultan, and Tobias Uller for helpful comments.

Notes

1. The term "niche construction theory" is commonly used in two senses in the literature: (i) as a body of conceptual theory with its own distinctive set of assumptions, and (ii) the formal mathematical theory of niche construction, its evolution and its evolutionary consequences. When deployed in the first sense, it is perhaps debatable as to whether NCT meets formal specifications of a theory as most scientists would understand the term (i.e., a particular explanation or prediction for an aspect of nature), and arguably it is better characterized as a standpoint, or way of thinking (i.e., "the niche construction perspective").

2. This is not to suggest that agency is widely recognized to lie at the heart of the contention. In the main, those stressing the evolutionary significance of niche construction have not explicitly singled out "agency" as something "missing" from mainstream evolutionary explanations, while non-sympathizers and critics of the niche construction perspective have objected to features of constructivist writing other than their emphasis on agency. Nonetheless, we regard the issue of agency as a hidden antagonist in these debates. The issues to which critics of NCT and its intellectual forebears object, such as the characterization of niche construction as an evolutionary process, or the suggestion that organisms can impose direction on evolution, were embraced by NCT precisely as a means to recognize organismal agency.

3. For instance, in what could be construed as a recent critique of NCT, Welch (2017), who defends orthodox thinking, attributes a need to include "imaginary agency" in evolutionary accounts as responsible for the "mistaken charges of narrowness and oversimplification" that he believes have been leveled against the field of evolutionary biology.

4. As for our use of the term "agency," our references to organisms "directing" evolutionary episodes should not be read as implying deliberate or conscious guidance, but rather as a statistical bias.

References

Alberti, M. 2015. "Eco-evolutionary Dynamics in an Urbanizing Planet." *Trends in Ecology and Evolution* 30 (2): 114–126.

Allen, G. E. 2014. "The History of Evolutionary Thought." In *The Princeton Guide to Evolution*, edited by J. B. Losos. Princeton, NJ: Princeton University Press.

Arthur, W. 2004. *Biased Embryo and Evolution*. Cambridge: Cambridge University Press.

Bickerton, D. 2009. *Adam's Tongue: How Humans Made Language, How Language Made Humans*. New York: Hill and Wang.

Boivin, N. L., M. A. Zeder, D. Q. Fuller, A. Crowther, G. Larson, J. M. Erlandson, T. Denham, and M. D. Petraglia. 2016. "Ecological Consequences of Human Niche Construction: Examining Long-term Anthropogenic Shaping of Global Species Distributions." *Proceedings of the National Academy of Sciences USA* 113 (23): 6388–6396.

Bonduriansky, R. 2012. "Rethinking Heredity, Again." *Trends in Ecology and Evolution* 27:330–336.

Brakefield, P. M. 2011. "Evo-devo and Accounting for Darwin's Endless Forms." *Philosophical Transactions of the Royal Society B* 366:2069–2075.

Brodie, D. E. 2005. "Caution: Niche Construction Ahead." *Evolution* 59:249–251.

Buser, C. C., R. D. Newcomb, A. C. Gaskett, and M. R. Goddard. 2014. "Niche Construction Initiates the Evolution of Mutualistic Interactions." *Ecology Letters* 17:1257–1264

Clarke, E. 2010. "The Problem of Biological Individuality." *Biological Theory* 5:312–325.

Clements, F. E. 1916. *Plant Succession: An Analysis of the Development of Vegetation*. Publication 242. Washington, DC: Carnegie Institute.

Danchin, E., A. Charmantier, F. A. Champagne, A. Mesoudi, and B. Pujol. 2011. "Beyond DNA: Integrating Inclusive Inheritance into an Extended Theory of Evolution." *Nature Reviews Genetics* 12:475–486.

Darwin, C. 1851. *The Structure and Distribution of Coral Reefs*. Oakland: University of California Press.

Darwin, C. 1881. *The Formation of Vegetable Mould through the Action of Worms, with Observations on their Habits*. London: Murray.

Dawkins, R. 1982. *The Extended Phenotype*. Oxford: Oxford University Press.

Dawkins, R. 2004. "Extended Phenotype—But Not Too Extended. A Reply to Laland, Turner and Jablonka." *Biology and Physiology* 19:377–396.

Dobzhansky, T. 1955. *Evolution, Genetics, and Man*. London: John Wiley and Sons.

Donohue, K. 2005. "Niche Construction through Phonological Plasticity: Life History Dynamics and Ecological Consequences." *New Phytologist* 166:83–92.

Ellis, E. C. 2015. "Ecology in an Anthropogenic Biosphere." *Ecological Monographs* 85 (3): 287–331.

Endler, J. A. 1986. "The Newer Synthesis? Some Conceptual Problems in Evolutionary Biology." *Oxford Surveys in Evolutionary Biology* 3:224–243.

Endler, J. A., and T. McLellan. 1988. "The Processes of Evolution: Towards a Newer Synthesis." *Annual Review of Ecology* 19:395–421.

Erwin, D. H., and J. W. Valentine. 2013. *The Cambrian Explosion: The Reconstruction of Animal Biodiversity.* Greenwood Village, CO: Roberts & Co.

Feldman, M. W., and L. L. Cavalli-Sforza. 1976. "Cultural and Biological Evolutionary Processes, Selection for a Trait under Complex Transmission." *Theoretical Population Biology* 9:238–259.

Flynn, E. G., K. N. Laland, R. L. Kendal, and J. R. Kendal. 2013. "Developmental Niche Construction." *Developmental Science* 16 (2): 296–313.

Fuentes, A. 2009. *Evolution of Human Behavior.* Oxford: Oxford University Press.

Fuentes, A. 2017. *The Creative Spark.* New York: Dutton.

Futuyma, D. J. 2017. "Evolutionary Biology Today and the Call for an Extended Synthesis." *Interface Focus.* http://dx.doi.org/10.1098/rsfs.2016.0145.

Futuyma, D. J., and M. Slatkin. 1983. *Coevolution.* Sunderland, MA: Sinauer.

Gerhart, J., and M. Kirschner. 1997. *Cells, Embryos, and Evolution.* Malden, MA: Blackwell Science.

Gilbert, S. F., and D. Epel. 2015. *Ecological Developmental Biology*, 2nd ed. Sunderland, MA: Sinauer.

Gould, S. J. 2000|2002. *The Structure of Evolutionary Theory.* Cambridge, MA: Belknap Press.

Gurney, W. S. C., and J. H. Lawton. 1996. "The Population Dynamics of Ecosystem Engineers." *Oikos* 76:273–283.

Hartl, D. L., and A. G. Clark. 1989. *Principles of Population Genetics.* Sunderland, MA: Sinauer.

Hendry, A. 2017. *Eco-Evolutionary Dynamics.* Princeton, NJ: Princeton University Press.

Ho, M-W. 1988. *Evolutionary Processes and Metaphors.* Chichester, UK: Wiley.

Holt, R. D. 1995. "Linking Species and Ecosystems: Where's Darwin?" In *Linking Species and Ecosystems,* edited by C. G. Jones and J. H. Lawton, 273–279. New York: Chapman and Hall.

Jablonka, E. 2017. "The Evolutionary Implications of Epigenetic Inheritance." *Interface Focus.* http://dx.doi.org/10.1098/rsfs.2016.0135.

Jones, C. G., and J. H. Lawton. 1995. *Linking Species and Ecosystems.* New York: Chapman and Hall.

Jones, C. G., J. H. Lawton, and M. Shachak. 1994. "Organisms as Ecosystem Engineers." *Oikos* 69:373–386.

Jones, C. G., J. H. Lawton, and M. Shachak. 1997. "Positive and Negative Effects of Organisms as Physical Ecosystem Engineers." *Ecology* 78:1946–1957.

Kavanagh, K. D., A. R. Evans, and J. Jernvall. 2007. "Predicting Evolutionary Patterns of Mammalian Teeth from Development." *Nature* 449:427–432.

Keller, E. 2010. *The Mirage of a Space between Nature and Nurture.* Durham, NC: Duke University Press.

Keller, L. 2003. "Changing the World." *Nature* 425:769–770.

Kendal, J., J. J. Tehrani, and F. J. Odling-Smee. 2011. "Human Niche Construction in Interdisciplinary Focus." *Philosophical Transactions of the Royal Society B* 366:785–792.

Kerr, B., D. W. Schwilk, A. Bergman, and N. W. Feldman. 1999. "Rekindling an Old Flame: A Haploid Model for the Evolution and Impact of Flammability in Resprouting Plants." *Evolutionary Ecology Research* 1:807–833.

Kirkpatrick, M. 1982. "Sexual Selection and the Evolution of Female Choice." *Evolution* 36:1–12.

Kirschner, M. W., and J. C. Gerhart. 2005. *The Plausibility of Life: Resolving Darwin's Dilemma.* New Haven, CT: Yale University Press.

Kuhn, T. 1962. *The Structure of Scientific Revolutions.* Chicago: University of Chicago Press.

Kylafis, G., and M. Loreau. 2008. "Ecological and Evolutionary Consequences of Niche Construction for Its Agent." *Ecology Letters* 11:1072–1081.

Lakatos, I. 1978. *The Methodology of Scientific Research Programmes.* Philosophical Papers, Vol. 1. Cambridge: Cambridge University Press.

Laland, K. N., F. J. Odling-Smee, and J. Endler. 2017. "Niche Construction, Sources of Selection and Trait Coevolution." *Interface Focus*. http://dx.doi.org/10.1098/rsfs.2016.0147.

Laland, K. N., F. J. Odling-Smee, and M. W. Feldman. 1996. "On the Evolutionary Consequences of Niche Construction." *Journal of Evolutionary Biology* 9:293–316.

Laland, K. N., F. J. Odling-Smee, and M. W. Feldman. 1999. "Evolutionary Consequences of Niche Construction and Their Implications for Ecology." *Proceedings of the National Academy of Sciences USA* 96:10242–10247.

Laland, K. N., and K. Sterelny. 2006. "Seven Reasons (Not) to Neglect Niche Construction." *Evolution* 60:1751–1762.

Laland, K. N., K. Sterelny, J. Odling-Smee, W. Hoppitt, and T. Uller. 2011. "Cause and Effect in Biology Revisited: Is Mayr's Proximate-Ultimate Dichotomy Still Useful?" *Science* 334:1512–1516.

Laland, K. N., T. Uller, M. W. Feldman, K. Sterelny, G. B. Müller, A. Moczek, E. Jablonka, and F. J. Odling-Smee. 2014. "Does Evolutionary Theory Need a Rethink? Yes, Urgently." *Nature* 514:161–164.

Laland, K. N., T. Uller, M. W. Feldman, K. Sterelny, G. B. Müller, A. Moczek, E. Jablonka, and F. J. Odling-Smee. 2015. "The Extended Evolutionary Synthesis: Its Structure, Assumptions and Predictions." *Philosophical Transactions of the Royal Society B*. doi: 10.1098/rspb.2015.10.

Lande, R. 1980. "Sexual Dimorphism, Sexual Selection, and Adaptation in Polygenic Characters." *Evolution* 34:292–307.

Lehmann, L. 2008. "The Adaptive Dynamics of Niche Constructing Traits in Spatially Subdivided Populations: Evolving Posthumous Extended Phenotypes." *Evolution* 62:549–566.

Levins, R., and R. C. Lewontin. 1985. *The Dialectical Biologist*. Cambridge, MA: Harvard University Press.

Lewontin, R. C. 1970. "The Units of Selection." *Annual Review of Ecology and Systematics* 1:1–18.

Lewontin, R. C. 1982. "Organism and Environment." In *Learning, Development and Culture*, edited by H. C. Plotkin. New York: Wiley.

Lewontin, R. C. 1983. "Gene, Organism, and Environment." In *Evolution from Molecules to Men*, edited by D. S. Bendall, 273–285. Cambridge: Cambridge University Press.

Lewontin, R. C. 2000. *The Triple Helix: Gene, Organism, and Environment*. Cambridge, MA: Harvard University Press.

Love, A. C. 2015. *Conceptual Change in Biology: Scientific and Philosophical Perspectives on Evolution and Development*. New York: Springer.

Lynch, M. 2007. "The Frailty of Adaptive Hypotheses for the Origins of Organismal Complexity." *PNAS* 104:8597–8604.

May, R. M. 1973. *Stability and Complexity in Model Ecosystems*. Princeton, NJ: Princeton University Press.

Maynard Smith, J. 1982. *Evolution and the Theory of Games*. Cambridge: Cambridge University Press.

Maynard Smith, J., R. Burian, S. Kauffman, P. Alberch, J. Campbell, B. Goodwin, R. Lande, D. Raup, and L. Wolpert. 1985. "Developmental Constraints and Evolution." *Quarterly Review of Biology* 60 (3): 265–287.

Morgan, L. H. 1868. *The American Beaver and His Works*. Philadelphia: J. B. Lippincott.

Noble, D. 2017. *Dance to the Tune of Life: Biological Relativity*. Cambridge: Cambridge University Press.

Odling-Smee, F. J. 1988. "Niche-constructing Phenotypes." In *The Role of Behavior in Evolution*, edited by H. C. Plotkin, 73–132. Cambridge, MA: MIT Press.

Odling-Smee, F. J., K. N. Laland, and M. W. Feldman. 1996. "Niche Construction." *The American Naturalist* 147:641–648.

Odling-Smee, F. J., K. N. Laland, and M. W. Feldman. 2003. *Niche Construction: The Neglected Process in Evolution*. Princeton, NJ: Princeton University Press.

O'Donald, P. 1980. *Genetic Models of Sexual Selection*. Cambridge: Cambridge University Press.

O'Neill, R. V., D. L. DeAngelis, J. B. Waide, and T. F. H. Allen. 1986. *A Hierarchical Concept of Ecosystem*. Princeton, NJ: Princeton University Press.

Pelletier, F., D. Garant, and A. P. Hendry. 2009. "Eco-evolutionary Dynamics." *Philosophical Transactions of the Royal Society B* 364:1483–1489.

Peterson, E.L. 2016. *The Life Organic: The Theoretical Biology Club and the Roots of Epigenetics.* Pittsburgh: University of Pittsburgh Press.

Piperno, D. R. 2017. "Assessing Elements of an Extended Evolutionary Synthesis for Plant Domestication and Agricultural Origin Research." *Proceedings of the National Academy of Sciences USA* 114 (25): 6429–6437.

Post, D. M., and E. P. Palkovacs. 2009. "Eco-evolutionary Feedbacks in Community and Ecosystem Ecology: Interactions between the Ecological Theatre and the Evolutionary Play." *Philosophical Transactions of the Royal Society B* 364:1629–1640.

Riskin, J. 2016. *The Restless Clock: A History of the Centuries-Long Argument over What Makes Living Things Tick.* Chicago: University of Chicago Press.

San Roman, M., and A. Wagner. 2018. "An Enormous Potential for Niche Construction through Bacterial Cross-feeding in a Homogeneous Environment." *PLoS Computational Biology* 14 (7): e1006340.

Schrödinger, E. 1944. *What Is Life? The Physical Aspect of the Living Cell.* Cambridge: Cambridge University Press.

Schwab, D. B., S. Casasa, and A. P. Moczek. 2017. "Evidence of Developmental Niche Construction in Dung Beetles: Effects on Growth, Scaling and Reproductive Success." *Ecology Letters* 20 (11): 1353–1363. doi:10.1111/ele.12830.

Schwilk, D. 2003. "Flammability Is a Niche Construction Trait: Canopy Architecture Affects Fire Intensity." *The American Naturalist* 162:725–733.

Schwilk, D. W., and D. D. Ackerly. 2001. "Flammability and Serotiny as Strategies: Correlated Evolution in Pines." *Oikos* 94:326–336.

Scott-Phillips, T. C., K. N. Laland, D. M. Shuker, T. E. Dickins, and S. A. West. 2013. "The Niche Construction Perspective: A Critical Appraisal." *Evolution.* doi: 10.1111/evo.12332.

Shaler, N. 1892. "Effects of Animals and Plants on Soils." In *Origin and Nature of Soils*, 12th Annual Report, Director US Geol. Survey, Part J. Geology Annual Report, Sector of the Interior. Washington, DC: Government Printing Office.

Silver, M., and E. A. Di Paolo. 2006. "Spatial Effects Favour the Evolution of Niche Construction." *Theoretical Population Biology* 70:387–400.

Simpson, G. G. 1949. *The Meaning of Evolution.* New Haven, CT: Yale University Press.

Slatkin, M. 1979. "The Evolutionary Response to Frequency- and Density-Dependent Interactions." *The American Naturalist* 114:384–398.

Smith, B. 2007. "Human Niche Construction and the Behavioral Context of Plant and Animal Domestication." *Evolutionary Anthropology* 16:188–199.

Stoltfus, A. 2017. "A New Evolutionary Cause and some Implications." Extended Evolutionary Synthesis (blog post). Accessed September 25, 2017. http://extendedevolutionarysynthesis.com/a-new-evolutionary-cause-and-some-implications/.

Sultan, S. E. 2015. *Organism and Environment.* Oxford: Oxford University Press.

Sultan, S. E. 2017. "Developmental Plasticity: Re-conceiving the Genotype." *Interface Focus* 7 (5): 20170009

Syvanen, M., and C. I. Kado. 2002. *Horizontal Gene Transfer*, 2nd ed. London: Academic Press.

Taylor, W. P. 1935. "Some Animal Relations to Soils." *Ecology* 16:127–136.

Tishkoff, S. A., F. A. Reed, A. Ranciaro, B. F. Voight, C. C. Babbitt, J. S. Silverman, K. Powell, et al. 2007. "Convergent Adaptation of Human Lactase Persistence in Africa and Europe." *Nature Genetics* 39:31–40.

Turner, J. S. 2016. "Homeostasis and the Physiological Dimension of Niche Construction Theory in Ecology and Evolution." *Evolutionary Ecology* 30:203–219.

Uller, T., and H. Helanterä. 2017. "Niche Construction and Conceptual Change in Evolutionary Biology." *British Journal for the Philosophy of Science*, axx050. doi: org/10.1093/bjps/axx050.

Uller, T., A. Moczek, R. A. Watson, P. M. Brakefield, and K. N. Laland. 2018. "Developmental Bias and Evolution: A Regulatory Network Perspective." *Genetics* 209:1–18.

Vandermeer, J. 2003. "The Importance of a Constructivist View." *Science* 303:472–474.

Waddington, C. H. 1957. *The Strategy of the Genes: A Discussion of Some Aspects of Theoretical Biology.* New York: Routledge.

Waddington, C. H. 1959. "Evolutionary Systems—Animal and Human." *Nature* 183:1634–1638.

Wagner, A. 2011. *The Origins of Evolutionary Innovations.* Oxford: Oxford University Press.

Walsh, D. 2015. *Organisms, Agency and Evolution.* Cambridge: Cambridge University Press.

Watson, R. A., and E. Szathmary. 2016. "How Can Evolution Learn?" *Trends in Ecology and Evolution* 31:147–157.

Watson, R. A., G. P. Wagner, M. Pavlicev, D. M. Weinreich, and R. Mills. 2014. "The Evolution of Phenotypic Correlations and 'Developmental Memory.'" *Evolution* 68:1124–1138.

Welch, J. 2017. "What's Wrong with Evolutionary Biology?" *Biology and Philosophy* 32 (2): 263–279.

West-Eberhard, M. J. 2003. *Developmental Plasticity and Evolution.* New York: Oxford University Press.

Wray, G. A., H. E. Hoekstra, D. J. Futuyma, R. E. Lenski, T. F. Mackay, D. Schluter, and J. E. Strassmann. 2014. "Does Evolutionary Theory Need a Rethink? No, All Is Well." *Nature* 514:161–164.

Wright, S. 1984. *Evolution and the Genetics of Populations.* Vol. 2: *The Theory of Gene Frequencies.* Chicago: University of Chicago Press.

Yampolsky, L. Y., and A. Stoltzfus. 2001. "Bias in the Introduction of Variation as an Orienting Factor in Evolution." *Evolution and Development* 3:73–83.

Zeder, M. A. 2015. "Core Questions in Domestication Research." *Proceedings of the National Academy of Sciences USA* 112:3191–3198.

8 Biological Dynamics and Evolutionary Causation

Renée A. Duckworth

Introduction

Why has understanding causation been so difficult in evolution? Perhaps it is because evolutionary biology has the unique role among biological disciplines of interpreting patterns at all levels of biological organization. Each subdiscipline is tasked with investigating how various systems currently function: molecular and cellular biologists at the cellular level, developmental biologists and physiologists at the organismal level, and ecologists at the community and ecosystem level. Evolutionary biologists not only need to understand how these systems function in current time, but they also need to understand how their functioning has changed over time. This requires investigating causal mechanisms that simultaneously span multiple timescales and levels of biological organization.

Studying both current and historical function of biological systems across multiple levels is a truly daunting task, and it is no wonder that a focus on genetic variation and natural selection, which are often measured as statistical summaries of the underlying dynamics of development and ecology, dominates the field. While such summaries of the underlying dynamics, reflected in patterns of genetic, phenotypic, and fitness variation, remain a core component of evolutionary research, they can only indicate correlations and thus should be viewed as a starting point for the determination of evolutionary mechanisms, not an end point. The central thesis of this chapter is that a deeper understanding of how evolution works requires examining system dynamics to determine how the behavior and interaction of biological entities at one scale influence patterns of variation at another scale.

In this chapter, I first discuss approaches to causation from two classic papers in ecology (Levin 1992) and evolutionary biology (Mayr 1961). I suggest that Mayr's proximate/ ultimate distinction touches on the fundamental problem of timescale in evolutionary change, but ignores the underlying dynamics of biological systems. In contrast, Levin's perspective, which focuses mainly on ecological timescales, iterates a clear view of how to dissect and understand causation in dynamic systems across scales. I suggest that integrating these two perspectives can provide novel insight into outstanding questions in evolution and illustrate this with discussion of the mechanisms of micro- and macroevolutionary change. Finally, I draw on nearly two decades of my own work on the dynamics of competition among cavity nesting birds to illustrate how patterns of dynamic stability and robustness in ecological systems are underlain by factors such as niche construction, developmental plasticity, and habitat selection. I suggest more explicit incorporation of these factors in evolutionary biology may be key to explaining patterns of species stasis.

Proximate and Ultimate: A Dichotomy of Timescale and System Dynamics

Levin (1992) argued that the problem of scale is at the crux of how scientists gain a mechanistic understanding of patterns in nature and suggested that it is as fundamental to understanding ecological and evolutionary patterns as it is to understanding the nature and behavior of matter in physics and chemistry. However, despite the widespread importance of the problem of scale across disciplines, explicit discussion of it has been rare in evolution and so it has been relegated mainly to philosophy of science with limited impact on empirical studies of evolution. The one major exception is Ernst Mayr's delineation of proximate and ultimate causation (Mayr 1961), which persists as an important dichotomy of causation in evolutionary biology.

Mayr defined proximate causes as *current* causes of phenotypic variation, and in this category he placed physiology and development; he defined ultimate causes as *historical* causes, and in this category he placed genes and natural selection. Mayr's goal was not to explicitly address the problem of scale in evolution—instead he was more concerned with explaining why functional and evolutionary biologists were often talking past one another and arguing over causation of the phenotype. However, a close reading of Mayr's explanation of this conceptual distinction (Mayr 1961; Mayr 1993) makes it clear that he was using these terms to distinguish between distinct timescales of phenotypic change (Haig 2013). The persistence of this dichotomy, despite many criticisms of it (Francis 1990; Thierry 2005; Laland et al. 2011; Calcott 2013), emphasizes how deeply entrenched it remains in current biological thought.

While the idea of distinguishing between timescales of phenotypic change is central to evolutionary theory, Mayr's placement of genes and natural selection only in the category of "historical causes" burdens his timescale dichotomy with an additional separation of dynamic versus static causes of evolution. This is evident in his treatment of genes as solely a historical cause ignoring the fact that within an individual, gene expression is also a proximate cause of developmental and physiological processes. Mayr put genes only in the category of historical cause because he was focused on inheritance; however, by treating genes as inanimate particles he essentially decoupled the active role of gene expression in current time from evolutionary explanations of phenotypic variation. Similarly, natural selection as a historical cause is often portrayed as a "force" external to the organism (Endler 1986), but, in reality, it is a summary of the behavioral interactions and resulting ecological dynamics of organisms. Thus, the main problem with the proximate/ultimate dichotomy is not that it delineates distinct timescales for the study of the phenotype, but that it treats the dynamics of biological systems as only relevant for studying current (within generation) timescales of phenotype change and makes inanimate genes and external forces of natural selection the only proper evolutionary causes. Yet, proximate causes of development and physiology in conjunction with ecological interactions are the dynamic processes that unfold each generation that are the building blocks of evolutionary change.

This separation of the dynamics of biological entities from evolutionary causation is problematic because it is the behavior and interactions of biological entities that are key to understanding mechanism (Craver and Bechtel 2006). Levin (1992) writes that "mechanisms operate at different scales than those on which the patterns are observed." They either emerge "from the collective behavior of large ensembles of smaller scale units … [Or] the pattern is imposed by larger scale constraints." The coordinated behavior and interactions of entities at lower scales often lead to emergent properties at higher scales that are not easily predicted from the behavior of an individual entity. This is not because the properties cannot be explained mechanistically, but instead it is because the properties at a higher scale are a statistical summary of the behavior of a group of entities at a lower scale. Constraints place boundaries on the system and limit in some way the variation of the lower scale entities (Bishop 2012). Thus, a complete understanding of causation requires integrating information about the (usually higher scale) constraints, how the component parts of a system (the lower scale entities) interact to produce emergent properties at a higher scale, and how external factors (that act on a different scale) may change these interactions.

Chave (2013), on the 20th anniversary of the publication of Levin's paper, writes of the impact it has had on the field of ecology: "Few research articles have been more influential to our discipline. … It has introduced a generation of ecologists to interdisciplinary thinking, and to two crucial concepts for ecology, pattern and scale." Is it possible to apply Levin's approach to pattern and scale to evolutionary causation and usher in a similar era of interdisciplinary thinking? If so, what are the "smaller scale units" in evolution whose behavior scales up to produce emergent properties at a higher scale? What are the "larger scale" constraints in the system? There are two important considerations in answering these questions. First, all life is organized hierarchically, where the behavior and interaction of lower-level entities scale up to produce emergent properties at a higher level. While such hierarchical organization of nature makes intuitive sense, working out a theoretical framework that captures the importance of hierarchical thinking for evolution has been challenging. Scale and level are typically treated as distinct concepts, yet they are intimately related (DiFrisco 2016). To be at the same hierarchical level, entities must interact to some extent (Salthe 2012). To interact, they must be on the same scale in multiple dimensions—at the same spatial, temporal, and size scale. While the nature of the interactions that bind entities to a particular level of biological organization are a matter of debate among biologists and philosophers (see Okasha 2006 for a summary of the main points), Levin essentially sidesteps this debate by treating the whole question as an empirical matter. He writes, "the objective of a model should be to ask how much detail can be ignored without producing results that contradict specific sets of observations, on particular scales of interest. In such an analysis, natural scales and frequencies may emerge, and in these rests the essential nature of the system dynamics." For Levin, levels are simply natural scales and frequencies that emerge from study of the system. In essence, Levin does not treat biological hierarchies as any different than hierarchies from other disciplines such as what is modeled in "statistical mechanics" and "interacting particle systems." Most importantly, he makes the point that, "there is no single 'correct' scale on which to describe populations or ecosystems" because the importance of a particular scale will vary depending on the focal organism or ecosystem. Similarly, there is no single correct scale at which to study evolution. Evolutionary patterns occur at every scale of biological organization and time—from molecular evolution to evolution of ecological communities to macroevolutionary dynamics—and the most powerful insight into causal mechanisms comes from identifying the constraints and mechanisms acting at one scale above and below the focal scale. The farther one gets from the focal scale, the less direct is the link between cause and effect. This is why the link between genotype and phenotype is such a persistent problem in evolution—it requires crossing multiple levels in the hierarchical organization of life from the transcription of genes that influence processes of cellular metabolism, which in turn influence cell behavior and morphology, to the role of tissue function and size on overall organismal phenotype. As one gets further from the molecular scale, additional system inputs and interactions occur that dilute the direct effects of genes on phenotypes. Thus, the overall organismal phenotype is typically too far removed in scale from the genotype to draw any direct lines of causation.

The second important consideration when applying Levin's approach to evolutionary patterns is that inheritance complicates assigning a scale to particular entities in evolution. This is because the genome can be either a lower scale entity whose behavior influences higher scale cellular processes or it can be a higher scale constraint in the system. Which of these roles it takes depends on the timescale of interest. Within a generation, the genome directly influences many organismal processes in the current generation by providing a template for cells to produce the raw materials needed for construction and maintenance of an individual organism (Nijhout 1990). In this way, the genome plays an important role in maintaining cellular and organismal stability by internalizing production of a subset of key resources to make the organism less dependent on variable and less predictable external resources. However, the inheritance of a genome across generations acts as a higher scale constraint because it restricts the range of phenotypes possible at the start of each new organism. Thus, genomes have a double role: within a generation, they are a smaller scale unit whose behavior and interaction with other molecular components influences the phenotype at the cellular and organismal level and across generations they are a larger scale constraint that limits the range of phenotypes possible.

If inheritance is a larger scale constraint in evolution, what then is the role of natural selection? And what are the lower scale entities whose behavior scales up to influence higher scale evolutionary patterns? The most basic unit of evolution is population change each generation, which occurs through processes of birth and death. If births and deaths vary with phenotype, then we may see population-level changes in mean trait values from one generation to the next. It is the interaction and behavior of individuals within populations that cause variation in births and deaths. Thus, patterns of variation in individual behavior are the basis of variable ecological dynamics, which are the mechanisms of natural selection. The idea that ecology is the cause of natural selection has been often made (Eldredge 1985; Van Valen 1989; Okasha 2006). However, there has been less emphasis on the behavior and interaction of individuals in evolutionary processes despite the fact this is what drives ecological dynamics in the first place (Lewontin 1982; Wcislo 1989; Werner 1992; Fryxell and Lundberg 1997; Duckworth 2009).

By behavior, I do not mean only cognitive behavior of animals with a nervous system. Instead, I use behavior in the sense of physics: to describe the movement and action of matter, including its inaction. Behavior is what biological entities (or any entities) do and applies to all biological entities, from the movement of molecules within a cell to the restructuring of the gut after a meal, from the opening of stomatal cells to the growth of branches toward a source of light, from the aggressive defense of a territorial boundary to the construction of a home to live in. In this sense, the behavior of individuals in a population is causally analogous to the behavior of individual water molecules in a pool of water. Just as by determining the principles by which water molecules interact, it is possible to understand the properties of water that influence wave dynamics, by determining the principles by which individuals interact with each other and the environment, it is possible to determine the properties of ecological interactions that influence evolutionary dynamics. This view is not reductionist because it does not suggest that we use the same rules that govern the behavior of molecules or other particles to understand the behavior of individuals. Instead, it requires that we figure out the rules that govern the behavior of individuals (by actually studying the behavior of individuals) to understand the mechanisms of population and community dynamics.

In Levin's view, being able to navigate across scales deftly not only provides insight into causation, but also enables prediction:

… by changing the scale of description, we move from unpredictable, unrepeatable individual cases to collections of cases whose behavior is regular enough to allow generalizations to be made. In so doing, we trade off the loss of detail or heterogeneity within a group for the gain of predictability; we thereby extract and abstract those fine-scale features that have relevance for the phenomena observed on other scales. (Levin 1992, 1947)

Gaining predictability necessitates ignoring some details and focusing on the larger patterns that emerge at a higher scale. For example, natural selection is a statistical summary of the outcome of the ecological dynamics each generation. By tallying the number of individuals that persist, are born and die each generation and seeing whether they differ in phenotype, it is possible to extract and abstract the higher scale pattern of biased survival or reproduction that results from all the detailed actions and interactions that produce ecological dynamics. Such abstraction is useful in pinpointing which traits might be expected to change from one generation to the next. However, it provides little insight into the mechanism by which trait variation is linked to fitness. And without an understanding of mechanism, the outcome of selection cannot be predicted beyond a few generations.

Such lack of understanding of the mechanisms underlying patterns of natural selection may be why it is often difficult to predict when and why natural selection will produce evolutionary change. Even though selection can fluctuate strongly across generations (Grant and Grant 2002; Dingemanse et al. 2004; Siepielski, DiBattista, and Carlson 2009) and can lead to large phenotypic responses on short timescales (Reznick, Rodd, and Nunney 2004), when summed across many empirical examples it is often found to be weak overall (Kingsolver et al. 2001). Such generally weak selection seems at odds with large spatial and temporal environmental variation often observed in nature. Resolution of this paradox comes from the acknowledgment that organisms are not passive entities in evolutionary processes (Lewontin 1983; Bateson 1988; Plotkin 1988; Duckworth 2009). Instead, it is the active strategies and responses of organisms to environmental variation that enables them to survive and reproduce in a variety of circumstances and that makes populations robust to environmental variation.

In essence, a complete understanding of evolutionary causation requires integrating information about inheritance (the higher scale constraint), how the behavior and interaction of individuals (the lower scale units) produce predictable ecological dynamics, and how external environmental factors (often the source of selection) may sometimes disrupt these predictable dynamics. This latter point—the disruption of otherwise predictable ecological dynamics—is the main source of natural selection in populations (Badyaev 2011). Thus, detecting natural selection is easy enough to do statistically, but interpreting its significance in order to predict evolutionary change requires understanding the ecological dynamics of a system enough to know what constitutes a stable and long-term disruption of the system. As an example of the insights that can be gained from applying this perspective, I turn to a long-standing problem of evolution, the link between micro- and macroevolution.

The Problem of Integrating Micro- and Macroevolutionary Dynamics

On the one hand, patterns at the macro and microevolutionary timescales are hypothesized to be causally linked (Charlesworth, Lande, and Slatkin 1982). On the other hand, their dynamics appear to be distinct (Jablonski 2017a). The patterns we observe at the macroevolutionary timescale are many orders of magnitude removed from the timescale of microevolutionary change that can be observed in populations over just a few generations (Ellner, Geber, and Hairston 2011). These same rapid evolutionary changes in phenotype, when viewed over a century or more, may appear as nothing more than stochastic noise. Larger climate-driven divergence among populations, such as those driven by glacial cycles, might produce cycles of local adaptation among populations within a species (Phillimore et al. 2007; Hunt, Hopkins, and Lidgard 2015), but resilient species-wide phenotypic divergence seems to require a timescale of one million years or more (Uyeda et al. 2011).

The idea of gradualism—that natural selection slowly and gradually shapes the phenotype of a species over many millennia or eons—is not supported by the fossil record (Hunt 2007; Eldredge 2016). Instead, species seem largely static in their morphology for the majority of their evolutionary history and this stasis is periodically punctuated by bouts of evolutionary change (Eldredge and Gould 1972; Gould and Eldredge 1977). Widely criticized when it was first proposed, a pattern of punctuated equilibria (if not the proposed mechanisms underlying it) is now largely accepted (Pennell, Harmon, and Uyeda 2014). A study by Uyeda et al. (2011) combined datasets that span different timescales, from field studies to fossil timeseries to phylogenetic datasets, to show that there is a million-year wait for macroevolutionary bursts in body size evolution. They call this the blunderbuss pattern, named for one of these old-fashioned guns with a long narrow barrel that flares out at the end. The Uyeda et al. (2011) study adds to a growing body of evidence that phenotypic change is limited for most of a taxa's evolutionary history (see also Hunt, Hopkins, and Lidgard 2015).

This macroevolutionary pattern of stasis often seems at odds with more recent studies of eco-evolutionary dynamics in which evolutionary changes in traits are observed on timescales of only a few generations (Hendry and Kinnison 1999; Hairston et al. 2005). Not only does significant evolutionary change occur on short timescales, but it is often quite strong (Hereford, Hansen, and Houle 2004). This is emphasized in a classic study by Gingerich (1983), which shows that rates of evolution measured over shorter periods of time are much higher than rates measured over long periods of time (figure 8.1A). One interpretation of this pattern is that evolution follows a weak zigzag pattern with little net directionality (Stanley and Yang 1987; figure 8.1B). In other words, when we scale out to a macroevolutionary timescale, rapid evolution measured in studies of eco-evolutionary dynamics is nothing more than stochastic noise. At moderate scales, perhaps glacial cycles or other climate changes push populations back and forth around a mean, but over very long timescales, there is little net evolutionary change.

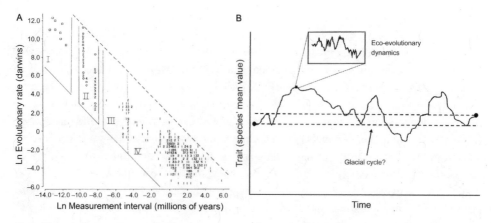

Figure 8.1
Reconciling evolutionary rates and net evolutionary change across diverse timescales. (A) Evolutionary
rates scale with measurement interval. Rates are from (i.) laboratory selection experiments (open squares),
(ii.) historical colonization events (open circles), (iii.) post-Pleistocene faunal recovery from glaciation, and
(iv.) fossil invertebrates and vertebrates. Numbers in (iii.) and (iv.) represent multiple cases falling at the same
point. Modified from Gingerich (1983). (B) Changes in a trait over hundreds of thousands of years in a hypothetical lineage. The inset shows trait changes over a randomly selected 50-year period that comprise rapid
microevolutionary changes due to eco-evolutionary dynamics. At the macroevolutionary scale these changes are
not even visible and resemble stochastic noise. General trends of increases or decreases in trait value on the
order of 1000s or 10,000s of years might correspond to glacial cycles or other long-term climate-driven changes.
The dots at the ends of the solid line show sampling points from a fossil timeseries and the dashed lines
correspond to the amount of net evolutionary change the lineage experiences at the macroevolutionary scale.
These types of dynamics at different timescales might produce the relationship observed in A.

In a follow-up to Uyeda et al. (2011), Arnold (2014) proposed that the blunderbuss
pattern is consistent with Simpson's idea of quantum evolution in which, for the majority
of a species' history, evolution takes place within relatively narrow and bounded adaptive
zones, but on rare occasions there are rapid transitions to a new adaptive zone (Simpson
1944). Arnold suggested that the long barrel of the blunderbuss is evolution occurring in
adaptive zones and that quantum evolution only occurs during rare events. He noted that
the amount of evolution that occurs in the adaptive zones is substantial, on the order of
±65 percent change in body size. Thus, Arnold referred to the pattern as "bounded evolution" rather than literal stasis. However, whatever the terminology, the pattern of limited
evolutionary change for the majority of a lineage's existence remains clear. The most
common mechanism proposed for species' adaptive zones is stabilizing selection (Charlesworth, Lande, and Slatkin 1982; Estes and Arnold 2007). But stabilizing selection as a
cause can only be half the story—the statistical summary of the underlying dynamics of
the system. To gain a mechanistic understanding of the pattern of evolutionary stasis, we
must also understand the collective behavior of the smaller scale units to try to discern
how lower scale interactions lead to this larger scale pattern.

Dynamic Stability and Robustness of Ecological Systems as Mechanisms of Evolutionary Stasis

The problem of evolutionary stasis is analogous in many ways to the problem of understanding robustness at the molecular, cellular, and organismal scales. Robustness is a ubiquitous property of biological systems (Wagner 2005). A system is robust when its outputs or state can remain stable despite variable inputs (Kitano 2004). Examples include cellular and organismal metabolism, the process of development, gene regulatory systems, and physiological homeostasis. Robustness does not mean that all components of a system stay unchanged in the face of variable inputs—it means the function or output of a system is stable. To achieve this stability at a higher level of organization, the lower scale entities must be in constant action—modifying their behavior and interactions in response to environmental changes. Hence, robust systems maintain a dynamic stability with built-in flexibility that tracks and responds to internal and external changes and maintains the same outputs as long as these changes are within the range of normal conditions (Nijhout and Reed 2014; Badyaev and Morrison 2018). Defining "normal" is the key to predicting how external conditions will influence functioning of the system. Biological systems can be adapted to function in variable environments as long as the variation experienced does not exceed the average or typical variation experienced in its evolutionary history. When it does, the system must either change or cease to function; in evolutionary terms this means evolve or go extinct.

Might Simpson's adaptive zones be produced by mechanisms of robustness at the population, community, or ecosystem scale? If so, then a species' adaptive zone is defined by the region of environmental or genetic variation that a species can experience without substantially changing the overall species' phenotype and large-scale evolutionary changes or extinction would only occur when a species encounters some extreme perturbation. There has long been an interest in the properties that account for stability of ecological communities (May 1974), and recent studies propose that similar properties of communities and ecosystems may account for their dynamic equilibrium state (Borrelli et al. 2015), making community ecology and coevolution perspectives essential to interpretation of patterns of evolutionary stasis (Stenseth and Maynard Smith 1984; Eldredge 1985; Morris et al. 1995; Lieberman, Miller, and Eldredge 2007; Voje et al. 2015). Evolutionary changes that are stable enough to appear in the fossil record may only occur with a major, usually catastrophic, event that pushes entire communities out of their equilibrium (Eldredge 2007). Such events are rare and only occur on the order of a million years or more, and when they occur, species either go extinct or change in profound ways (Gould and Eldredge 1977). Thus, macroevolutionary patterns cannot be understood by simply scaling up microevolutionary changes over a long period of time, because the important evolutionary events, such as those that precipitate mass extinctions, do not occur on a constant gradual basis (Jablonski 2017b). This point is only obvious by explicitly comparing patterns and mechanisms across multiple timescales (Badyaev 2018). Moreover, given that the clearest pattern in the history of life is one of stasis, this emphasizes that, in order to understand evolutionary processes, it is as important to explain how and why populations don't change as it is to explain how and why they do change.

Fortunately, the general mechanisms that produce robustness at the cellular, developmental, and physiological levels are well-known. These include negative and positive feedbacks, source–sink gradients, redundancy, saturation kinetics, modularity, and the ability of components to switch between bi-stable states (Eldar, Shilo, and Barkai 2004; Kitano 2004; Whitacre and Bender 2010; Nijhout and Reed 2014). There are ecological equivalents of all of these mechanisms. For example, one of the most common negative feedback processes in populations is density dependence. Source–sink gradients at a cellular level seem remarkably like source–sink population dynamics at a larger scale. Saturation kinetics refers to the interaction between enzyme and substrate where the reaction rate is limited by the availability of the substrate. These dynamics are similar to one of the most important relationships in ecology, the population growth curve where growth is limited by the availability of resources. Metapopulation structure of many species might be analogous to modularity within organisms, which confers robustness because it allows changes in one component without disrupting the whole system. This idea is consistent with the observation that species with larger ranges and more subspecies are more resistant to extinction (Vrba 1992; Eldredge et al. 2005). Finally, many robust systems display bi-stability in their components, where system components can switch between different alternative states while maintaining overall system functionality (Badyaev 2018). This might be analogous to alternative strategies at the organismal level where plastic responses enable tracking of variable environments and thus enable species persistence. In sum, any mechanism that allows species to be buffered from environmental changes will contribute to robustness of populations.

Robustness Mechanisms for Ecological Scales: An Empirical Example of Dynamic Stability

Discerning robustness mechanisms in ecological interactions provides a novel way of interpreting the dynamic stability that is often observed in communities and ecosystems. By applying concepts from molecular biology and physiology to higher scale processes, we may begin to observe similarities in the dynamics of biological systems across scales and thus gain deeper insight into evolutionary processes. To show how such dynamics might be studied and how a body of work can be reinterpreted with this novel approach, I draw on my work on the population and community dynamics of cavity nesting birds. Cavity nesting bird populations have been studied extensively in the forest ecosystems of the Northwestern United States. This community of birds depends on habitat that is created by forest fire, which has a long history of management in the United States, making the ecology of post-fire communities extremely well studied (Hutto 2006). My work has focused on the dynamic interplay of two species that depend on this post-fire habitat.

Mountain and western bluebirds (*S. currucoides* and *S. mexicana*) are two passerine species that compete for territories in areas where their ranges overlap (Duckworth 2014). For successful reproduction and survival, they have two main habitat requirements: open meadows for foraging and nest cavities for breeding. Forest fires generate suitable habitat by opening up understory vegetation and creating dead snags. Primary cavity nesters, such as woodpeckers, excavate nest cavities in these snags that are eventually used by secondary cavity nesters, such as bluebirds. This habitat can last for 20–30 years until regrowth of the forest eliminates the open meadows bluebirds require for insect prey. The successional nature of post-fire habitat means that populations of bluebirds (and the entire post-fire community) have a discrete beginning and end and, at any given time, there are numerous populations of varying ages and successional stages connected by dispersal in a temporally and spatially varying metapopulation structure.

In the bluebird system, the successional dynamics mean that the system is in constant change, but both species of bluebirds have different strategies for maintaining population persistence on a regional scale. Mountain bluebirds are more dispersive than western bluebirds and are among the earliest colonizers of new habitat patches, whereas western bluebirds often show delayed patterns of colonization (Kotliar, Kennedy, and Ferree 2007; Saab, Russell, and Dudley 2007; Duckworth et al. 2017) but are better interference competitors (Duckworth 2014). Mountain bluebirds are eventually displaced by the slower-arriving western bluebirds within a habitat patch, but because they are more dispersive and show a greater flexibility in their nest site and habitat choice (Duckworth et al. 2017; Johnson and Dawson 2019), they can quickly move to new areas when western bluebirds arrive. Western bluebirds, on the other hand, have evolved two distinct strategies that track the rapidly changing environmental conditions across successional stages enabling them to maintain high fitness in distinct competitive environments. Highly aggressive and dispersive western bluebirds are the first to arrive to a new habitat patch and easily displace earlier-arriving heterospecific competitors (Duckworth and Badyaev 2007). These aggressive dispersers excel in populations with low conspecific densities because they acquire large resource-rich territories and thus are able to maintain high fitness (Duckworth 2008; Duckworth 2014); however, they are eventually replaced by a nonaggressive philopatric type that excels in high-density populations because they cooperate with their relatives in territory defense and invest highly in parental care (Duckworth 2006; Aguillon and Duckworth 2015). Eventually, forest succession makes a habitat patch no longer suitable for bluebirds to breed in and the whole cycle—mountain bluebird to aggressive to nonaggressive western bluebird—repeats itself in a new patch. Western bluebird populations display many of the dynamics that characterize robust systems, such as density-dependent fitness leading to negative and positive feedbacks on each behavioral type driven by logistic growth of newly colonized populations (Duckworth and Aguillon 2015). The predictable increases in western bluebird breeding density over time leads to a predictable decline in the number of available nest cavities (Duckworth et al. 2017), and the switch between the two behavioral types closely tracks changes in population density and resources (Duckworth and Aguillon 2015).

The switch between the two behavioral types is adaptive and is maternally induced via differential allocation of testosterone to clutches. Females in late stage, crowded populations have small territories with few extra nest cavities and experience heightened competition. This induces them to allocate more testosterone to their clutches and produce aggressive sons that colonize new areas. In contrast, aggressive females that colonize new populations acquire large resource-rich territories and have lower stress levels (Duckworth unpub data). As a consequence, they allocate less testosterone to their clutches and immediately start producing nonaggressive offspring that bud off the parental territory and reap the benefits of nest cavity resources their parents acquired (Duckworth, Belloni, and Anderson 2015). The differences in competitive ability of the two types of males produce a type of competitive habitat selection that sets the stage for rapid changes in the cue that induces the maternal effect on offspring aggression allowing a match between behavioral type and the ecological context where they will perform best. Thus, this system provides a clear example of how active responses of organisms to environmental variation can allow for maintenance of high fitness across spatially and temporally varying populations.

Western and mountain bluebirds defend exclusive territories, and the majority of competitive interactions among bluebirds occur in March and April, well before the onset of oogenesis, making them rare during the period when egg production actually occurs. However, bluebirds defend their primary cavity throughout the breeding period from other species, and the peak of these non-bluebird intrusions overlaps with oogenesis (Duckworth, Belloni, and Anderson 2015). Females with more nest cavities on their territory experience fewer intrusions by these other species because one of them typically occupies the extra nest cavity on a bluebird territory and keeps others of its own kind away, effectively buffering the focal bluebird female from further competition. As a result, females on territories with extra nest cavities experience less competition from nest site competitors and produce sons that remain philopatric. Thus, competitive differences in aggressive and nonaggressive birds during settlement of territories lead to a sort of inadvertent niche construction where the outcome of competition over territories has cascading effects on territory quality, maternal effects, and offspring phenotype. These cascading effects are predictable because the extreme limitation of nest cavities for the entire community of secondary cavity nesters means there will always be competitor species intruding on bluebird territories, making competitive interactions over nest sites a reliable cue of the availability of parental resources.

This example shows how predictable higher scale community dynamics can occur despite the contingent nature of local-scale behavioral interactions. While there are many elements interacting in this system, many of the parts are interchangeable. For example, as long as females on small territories are harassed by competitors for their nest cavity, the particular competitor species doesn't matter. There is a suite of competitor species that can fulfill this role: tree swallows, house sparrows, mountain chickadees, and house wrens are all actively searching for nest cavities while female bluebirds are laying eggs. Thus, species identity of the competitors is a detail that can be extracted and abstracted to understand the larger scale link between population density and maternally induced switches in behavioral phenotype. Another example of interchangeability are the primary cavity nesters, which are a keystone species in this habitat (Martin, Aitken, and Wiebe 2004). They excavate the nest cavities that are then used by numerous species of mammals, birds, reptiles, amphibians, and insects (Bunnell, Kremaster, and Wind 1999). As long as nest cavities are being produced—whether by black-backed woodpeckers, northern flickers, or hairy woodpeckers—this resource will become available to the multitude of secondary cavity nesters that depend on it, fundamentally shaping the community composition of post-fire forests. Such redundancy of functional parts is a common feature of robust systems (Kitano 2004) and of ecological communities (Eldredge 1985).

Other mechanisms of robustness are also evident in this system. The developmental plasticity of dispersal phenotypes in western bluebirds is key to maintaining robustness of fitness across populations and is analogous to the switch between two bistable states of lower scale components that is often observed in robust systems. Moreover, density-dependent fitness of these types (Duckworth and Aguillon 2015) is a common feature of many disturbance-dependent species and is a type of feedback mechanism at the population level that is similar in dynamics to negative feedback mechanisms in molecular and organismal pathways. Thus, this case study shows many features at the population level that are typical of robust systems at the molecular, cellular and organismal levels. However, this is one case study and we can only assess the general importance of such mechanisms by comparing across enough such case studies to determine their prevalence. In other words, we need to change the scale of description and "move from unpredictable, unrepeatable individual cases to collections of cases whose behavior is regular enough for generalizations to be made" (Levin 1992).

One example of such extrapolation is the observation that many plants and animals evolve maternally induced dispersal strategies similar to those expressed in western bluebirds (Harrison 1980; Donohue 1999; Larios and Venable 2015; Duckworth, Potticary, and Badyaev 2018). The similarity across taxa is not in the specific traits linked to dispersal—seed polymorphisms in plants, wing dimorphisms in insects, and distinct personality types in vertebrates—but in the function of the suite of traits that enable the species to escape habitats declining in quality (Harrison 1980; Roff 1994; Cote et al. 2010; Duckworth 2012). Often such dispersal polymorphisms evolve in species that depend on successional or disturbance-prone habitat (Zera and Denno 1997; Armsworth and Roughgarden 2005). In this case, the details that can be extracted and abstracted across species are the specific traits that enable individuals to stay or go, and the generality is that in successional or disturbance-prone habitat dispersal polymorphisms frequently evolve. Thus, by looking across a wide variety of taxa that experience a similar problem for maintaining meta-population persistence in a constantly changing environment, we can begin to extract and abstract the key repeatable pattern that unites all of these examples.

Given that western bluebirds live in a world that is in constant flux, it seems that they would be resistant to most environmental perturbations. How do we even know what constitutes an environmental perturbation in a system that is constantly changing? The answer is to know enough about the ecology of a species or system to understand which resources and interactions, though constantly changing, are predictable for the species and whose disappearance would constitute a significant perturbation. For western bluebirds, their high competitive ability means that, as long as forests burn, they will have plenty of nest cavities to fight for and win, but take away this key resource and this dominant species becomes vulnerable. When the United States put in place a policy of fire suppression that largely eliminated post-fire habitat (Arno 1980) western bluebirds went extinct from the northern part of their range (Duckworth and Badyaev 2007). In contrast, mountain blue-birds, the subdominant species that evolved greater flexibility in habitat and nest cavity preferences, were able to persist (Duckworth and Badyaev 2007). When humans began to put up man-made nest boxes and fire policy changed in the 1980s to allow fires to burn in wilderness areas once again, western bluebirds quickly returned to the parts of their range where they were formerly extirpated. This human-induced large-scale experiment showed that removal of fire from the landscape was too large of a disruption for western bluebird populations and the species range contracted; if this disruption of their main resource had occurred range-wide, this species would surely have been in danger of extinction.

Conclusion

Applying the robustness concept to ecological systems raises a number of novel questions that have the promise of uniting currently disparate studies in ecology and evolution. What is the role of coevolutionary interactions in the maintenance of community and ecosystem robustness? Do such interactions make systems more or less resilient in the face of envi-ronmental change? How is the breadth of environments that a species is buffered against determined? Clearly, species are buffered against the sort of environmental changes that organisms are likely to experience during a lifetime, such as seasonal changes and weather variation across years. But how common and predictable does an environmental change need to be for robustness mechanisms to evolve? Does it depend on the temporal or spatial variability that a species experiences or both? Hunt (2007) found that only 5 percent of fossil timeseries showed directional change with the rest fitting a model of stasis or a random walk with no net directional change. Is there something different about the taxa that do show directional change? Is there something different about the communities they are a part of? Are they invaders? Do they have a unique role in the community or are they more (or less) likely to be species that show coevolutionary dynamics? All of these questions are at the crux of understanding patterns of stasis—one of the clearest macroevolutionary patterns. To answer them requires greatly expanding the role of ecology, paleontology, and ethology in modern evolutionary biology.

Calls for an Extended Evolutionary Synthesis have often pointed out several factors, such as developmental plasticity, niche construction, and habitat selection, that were left out of the Modern Synthesis but are key to a complete causal understanding of evolution (Laland et al. 2015). What all of these factors have in common is that they encompass the active role of the organism in evolution, and I suggest they are the key mechanisms underlying patterns of stasis. Every time we see organisms taking an active role in responding to or shaping their environment, we are seeing the adaptations that buffer organisms from perturbations—from wing dimorphisms in insects, to maternally induced defensive responses in plants, to helmet formation in daphnia, to habitat selection in birds, to the formation of a biofilm in bacteria, to the construction of elaborate homes in beavers. I suggest that the reasons for one of the most well-documented patterns in evolution—long-term stasis—are the interactions that many organismal biologists and ecologists have been studying all along. But without a unifying principle to tie them together, these studies seem merely a collection of individual case studies. What is now needed is to determine whether the mechanisms conferring robustness at the cellular, physiological, and developmental scale really do work similarly at the population, community, and ecosystem scale.

Each of the subdisciplines within biology—molecular biology, physiology, developmental biology, and ecology—is tasked with figuring out how various biological systems work at different levels of organization. The task of evolutionary biologists is to take this knowledge and use it to determine how these systems change over time so that we can understand how evolution works. Thus, the first step to understanding what processes lead to evolutionary change may be to learn what processes allow populations to stay the same.

Acknowledgments

I am grateful to Kevin Laland and Tobias Uller for inviting me to participate in the workshop that led to this chapter. In working out my ideas for this chapter, I greatly benefited from the general discussions and conversations I had there, particularly with Samir Okasha and Richard Watson. I am grateful to Alex Badyaev, Per Lundberg, and an anonymous reviewer for providing insightful comments on a draft of the manuscript. This work was supported by NSF DEB-1350107.

References

Aguillon, S. M., and R. A. Duckworth. 2015. "Kin Aggression and Resource Availability Influence Phenotype-dependent Dispersal in a Passerine Bird." *Behavioral Ecology and Sociobiology* 69 (4): 625–633. doi: 10.1007/s00265-015-1873-5.

Armsworth, P. R., and J. E. Roughgarden. 2005. "Disturbance Induces the Contrasting Evolution of Reinforcement and Dispersiveness in Directed and Random Movers." *Evolution* 59:2083–2096.

Arno, S. F. 1980. "Forest Fire History in the Northern Rockies." *Journal of Forestry* 78:460–465.

Arnold, S. J. 2014. "Phenotypic Evolution: The Ongoing Synthesis." *The American Naturalist* 183 (6): 729–746. doi: 10.1086/675304.

Badyaev, A. V. 2011. "Origin of the Fittest: Link between Emergent Variation and Evolutionary Change as a Critical Question in Evolutionary Biology." *Proceedings of the Royal Society B: Biological Sciences* 278 (1714): 1921–1929. doi: 10.1098/rspb.2011.0548.

Badyaev, A. V. 2018. "Evolutionary Transitions in Network Controllability Reconcile Adaptation with Continuity of Evolution." *Seminars in Cell and Developmental Biology.* https://doi.org/10.1016/j.semcdb.2018.05.014.

Badyaev, A. V., and E. S. Morrison. 2018. "Emergent Buffering Balances Evolvability and Robustness in the Evolution of Phenotypic Flexibility." *Evolution.* In press.

Bateson, P. 1988. "The Active Role of Behaviour in Evolution." In *Evolutionary Processes and Metaphors*, edited by M.-W. Ho and S. W. Fox. New York: John Wiley and Sons.

Bishop, R. C. 2012. "Fluid Convection, Constraint and Causation." *Interface Focus* 2 (1): 4–12. doi: 10.1098/rsfs.2011.0065.

Borrelli, J. J., S. Allesina, P. Amarasekare, R. Arditi, I. Chase, J. Damuth, R. D. Holt, et al. 2015. "Selection on Stability across Ecological Scales." *Trends in Ecology and Evolution* 30 (7): 417–425. doi: 10.1016/j.tree.2015.05.001.

Bunnell, F. L., L. L. Kremaster, and E. Wind. 1999. "Managing to Sustain Vertebrate Richness in Forests of the Pacific Northwest: Relationships within Stands." *Environmental Reviews* 7:97–146.

Calcott, B. 2013. "Why How and Why Aren't Enough: More Problems with Mayr's Proximate-Ultimate Distinction." *Biology and Philosophy* 28 (767–780). doi: 10.1007/s10539-013-9367-1.

Charlesworth, B., R. Lande, and M. Slatkin. 1982. "A Neo-Darwinian Commentary on Macroevolution." *Evolution* 36:474–498.

Chave, J. 2013. "The Problem of Pattern and Scale in Ecology: What Have We Learned in 20 Years?" *Ecology Letters* 16 Suppl 1:4–16. doi: 10.1111/ele.12048.

Cote, J., J. Clobert, T. Brodin, S. Fogarty, and A. Sih. 2010. "Personality-dependent Dispersal: Characterization, Ontogeny and Consequences for Spatially Structured Populations." *Philosophical Transactions of the Royal Society B: Biological Sciences* 365 (1560): 4065–4076. doi: 10.1098/rstb.2010.0176.

Craver, C. F., and W. Bechtel. 2006. "Top-Down Causation without Top-Down Causes." *Biology and Philosophy* 22 (4): 547–563. doi: 10.1007/s10539-006-9028-8.

DiFrisco, J. 2016. "Time Scales and Levels of Organization." *Erkenn.* doi: 10.1007/s10670-016-9844-4.

Dingemanse, N. J., C. Both, P. J. Drent, and J. M. Tinbergen. 2004. "Fitness Consequences of Avian Personalities in a Fluctuating Environment." *Proceedings of the Royal Society B: Biological Sciences* 271 (1541): 847–852. doi: 10.1098/rspb.2004.2680.

Donohue, K. 1999. "Seed Dispersal as a Maternally Influenced Character: Mechanistic Basis of Maternal Effects and Selection on Maternal Characters in an Annual Plant." *The American Naturalist* 154:674–689.

Duckworth, R. A. 2006. "Behavioral Correlations across Breeding Contexts Provide a Mechanism for a Cost of Aggression." *Behavioral Ecology* 17:1011–1019.

Duckworth, R. A. 2008. "Adaptive Dispersal Strategies and the Dynamics of a Range Expansion." *The American Naturalist* 172:S4–S17.

Duckworth, R. A. 2009. "The Role of Behavior in Evolution: A Search for Mechanism." *Evolutionary Ecology* 23:513–531.

Duckworth, R. A. 2012. "Evolution of Genetically Integrated Dispersal Strategies." In *Dispersal Ecology and Evolution*, edited by J. Clobert, M. Baguette, T. G. Benton, and J. M. Bullock, 83–94. Oxford: Oxford University Press.

Duckworth, R. A. 2014. "Human-induced Changes in the Dynamics of Species Coexistence: An Example with Two Sister Species." In *Avian Urban Ecology: Behavioural and Physiological Adaptations*, edited by D. Gil and H. Brumm, 181–191. Oxford: Oxford University Press.

Duckworth, R. A., and S. M. Aguillon. 2015. "Eco-evolutionary Dynamics: Investigating Multiple Causal Pathways Linking Changes in Behavior, Population Density and Natural Selection." *Journal of Ornithology* 156:115–124. doi: 10.1007/s10336-015-1239-9.

Duckworth, R. A., and A. V. Badyaev. 2007. "Coupling of Dispersal and Aggression Facilitates the Rapid Range Expansion of a Passerine Bird." *Proceedings of the National Academy of Sciences USA* 104 (38): 15017–15022. doi: 10.1073/pnas.0706174104.

Duckworth, R. A., V. Belloni, and S. R. Anderson. 2015. "Cycles of Species Replacement Emerge from Locally Induced Maternal Effects on Offspring Behavior in a Passerine Bird." *Science* 374:875–877.

Duckworth, R. A., K. K. Hallinger, N. Hall, and A. L. Potticary. 2017. "Switch to a Novel Breeding Resource Influences Coexistence of Two Passerine Birds." *Frontiers in Ecology and Evolution* 5. doi: 10.3389/fevo.2017.00072.

Duckworth, R. A., A. L. Potticary, and A. V. Badyaev. 2018. "On the Origins of Adaptive Behavioral Complexity: Developmental Channeling of Structural Trade-Offs." *Advances in the Study of Behavior* 50:1–36. doi: 10.1016/bs.asb.2017.10.001.

Eldar, A., B. Z. Shilo, and N. Barkai. 2004. "Elucidating Mechanisms Underlying Robustness of Morphogen Gradients." *Current Opinion in Genetics and Development* 14 (4): 435–439. doi: 10.1016/j.gde.2004.06.009.

Eldredge, N. 1985. *Unfinished Synthesis: Biological Hierarchies and Modern Evolutionary Thought*. Oxford: Oxford University Press.

Eldredge, N. 2007. "Hierarchies and the Sloshing Bucket: Toward the Unification of Evolutionary Biology." *Evolution: Education and Outreach* 1 (1): 10–15. doi: 10.1007/s12052-007-0007-6.

Eldredge, N. 2016. "The Checkered Career of Hierarchical Thinking in Evolutionary Biology." In *Evolutionary Theory: A Hierarchical Perspective*, edited by N. Eldredge, T. Pievani, E. Serrelli, and I. Tëmkin. Chicago: University of Chicago Press.

Eldredge, N., and S. J. Gould. 1972. "Punctuated Equilibria: An Alternative to Phyletic Gradualism." In *Models in Paleobiology*, edited by T. J. M. Schopf, 82–115. San Francisco: Freeman, Cooper and Company.

Eldredge, N., J. N. Thompson, P. M. Brakefield, S. Gavrilets, D. Jablonski, J. B. C. Jackson, R. E. Lenski, B. S. Lieberman, M. A. McPeek, and W. Miller, III. 2005. "The Dynamics of Evolutionary Stasis." *Paleobiology* 31:133–145.

Ellner, S. P., M. A. Geber, and N. G. Hairston. 2011. "Does Rapid Evolution Matter? Measuring the Rate of Contemporary Evolution and Its Impacts on Ecological Dynamics." *Ecology Letters* 14 (6): 603–614. doi: 10.1111/j.1461-0248.2011.01616.x.

Endler, J. A. 1986. *Natural Selection in the Wild*. Princeton, NJ: Princeton University Press.

Estes, S., and S. J. Arnold. 2007. "Resolving the Paradox of Stasis: Models with Stabilizing Selection Explain Evolutionary Divergence on All Timescales." *The American Naturalist* 169:227–244.

Francis, R. C. 1990. "Causes, Proximate and Ultimate." *Biology and Philosophy* 5:401–415.

Fryxell, J., and P. Lundberg. 1997. *Individual Behavior and Community Dynamics*. Vol. 20. New York: Chapman & Hall.

Gingerich, P. D. 1983. "Rates of Evolution: Effects of Time and Temporal Scaling." *Science* 222:159–161.

Gould, S. J., and N. Eldredge. 1977. "Punctuated Equilibria: The Tempo and Mode of Evolution Reconsidered." *Paleobiology* 3:115–151.

Grant, P. R., and B. Rosemary Grant. 2002. "Unpredictable Evolution in a 30-year Study of Darwin's Finches." *Science* 296:707–711.

Haig, D. 2013. "Proximate and Ultimate Causes: How Come? And What For?" *Biology and Philosophy* 28:781–786. doi: 10.1007/s10539-013-9369-z.

Hairston, N. G., Jr., S. P. Ellner, M. A. Geber, T. Yoshida, and J. A. Fox. 2005. "Rapid Evolution and the Convergence of Ecological and Evolutionary Time." *Ecology Letters* 8:1114–1127.

Harrison, R. G. 1980. "Dispersal Polymorphisms in Insects." *Annual Review of Ecology and Systematics* 11:95–118.

Hendry, A. P., and M. T. Kinnison. 1999. "The Pace of Modern Life: Measuring Rates of Contemporary Microevolution." *Evolution* 53:1637–1653.

Hereford, J., T. F. Hansen, and D. Houle. 2004. "Comparing Strengths of Directional Selection: How Strong Is Strong?" *Evolution* 58:2133–2143.

Hunt, G. 2007. "The Relative Importance of Directional Change, Random Walks, and Stasis in the Evolution of Fossil Lineages." *Proceedings of the National Academy of Sciences USA* 104 (47): 18404–18408. doi: 10.1073/pnas.0704088104.

Hunt, G., M. J. Hopkins, and S. Lidgard. 2015. "Simple versus Complex Models of Trait Evolution and Stasis as a Response to Environmental Change." *Proceedings of the National Academy of Sciences USA* 112 (16): 4885–4890. doi: 10.1073/pnas.1403662111.

Hutto, R. L. 2006. "Toward Meaningful Snag-Management Guidelines for Postfire Salvage Logging in North American Conifer Forests." *Conservation Biology* 20 (4): 984–993. doi: 10.1111/j.1523–1739.2006.00494.x.

Jablonski, D. 2017a. "Approaches to Macroevolution: 1. General Concepts and Origin of Variation." *Evolutionary Biology* 44 (4): 427–450. doi: 10.1007/s11692-017-9420-0.

Jablonski, D. 2017b. "Approaches to Macroevolution: 2. Sorting of Variation, Some Overarching Issues, and General Conclusions." *Evolutionary Biology* 44 (4): 451–475. doi: 10.1007/s11692-017-9434-7.

Johnson, L. S., and R. D. Dawson. 2018. "Mountain Bluebird (*Sialia currucoides*)," version 2.0. In *The Birds of North America*, edited by P. G. Rodewald. Ithaca, NY: Cornell Lab of Ornithology.

Kingsolver, J. G., H. E. Hoekstra, J. M. Hoekstra, David Berrigan, S. N. Vignieri, C. E. Hill, A. Hoang, P. Gibert, and P. Beerli. 2001. "The Strength of Phenotypic Selection in Natural Populations." *The American Naturalist* 157:245–261.

Kitano, H. 2004. "Biological Robustness." *Nature Reviews Genetics* 5 (11): 826–837. doi: 10.1038/nrg1471.

Kotliar, N. B., P. L. Kennedy, and K. Ferree. 2007. "Avifaunal Responses to Fire in Southwestern Montane Forests along a Burn Severity Gradient." *Ecological Applications* 17:491–507.

Laland, K. N., K. Sterelny, J. Odling-Smee, W. Hoppitt, and T. Uller. 2011. "Cause and Effect in Biology Revisited: Is Mayr's Proximate-Ultimate Dichotomy Still Useful?" *Science* 334 (6062): 1512–1516. doi: 10.1126/science.1210879.

Laland, K. N., T. Uller, M. W. Feldman, K. Sterelny, G. B. Muller, A. Moczek, E. Jablonka, and J. Odling-Smee. 2015. "The Extended Evolutionary Synthesis: Its Structure, Assumptions and Predictions." *Proceedings of the Royal Society B: Biological Sciences* 282 (1813). doi: 10.1098/rspb.2015.1019.

Larios, E., and D. L. Venable. 2015. "Maternal Adjustment of Offspring Provisioning and the Consequences for Dispersal." *Ecology* 96 (10): 2771–2780.

Levin, S. A. 1992. "The Problem of Pattern and Scale in Ecology." *Ecology* 73:1943–1967.

Lewontin, R. C. 1982. "Organism and Environment." In *Learning, Development, and Culture*, edited by H. C. Plotkin. New York: John Wiley and Sons.

Lewontin, R. C. 1983. "Gene, Organism and Environment." In *Evolution from Molecules to Men*, edited by D. S. Bendall. Cambridge: Cambridge University Press.

Lieberman, B. S., W. Miller, and N. Eldredge. 2007. "Paleontological Patterns, Macroecological Dynamics and the Evolutionary Process." *Evolutionary Biology* 34 (1–2): 28–48. doi: 10.1007/s11692-007-9005-4.

Martin, K., K. E. H. Aitken, and K. L. Wiebe. 2004. "Nest Sites and Nest Webs for Cavity-Nesting Communities in Interior British Columbia, Canada: Nest Characteristics and Niche Partitioning." *The Condor* 106:5–19.

May, R. M. 1974. *Stability and Complexity in Model Ecosystems*, Vol. 6. Princeton, NJ: Princeton University Press.

Mayr, E. 1961. "Cause and Effect in Biology: Kinds of Causes, Predictability, and Teleology Are Viewed by a Practicing Biologist." *Science* 134 (3489): 1501–1506.

Mayr, E. 1993. "Proximate and Ultimate Causations." *Biology and Philosophy* 8:93–94.

Morris, P. J., L. C. Ivany, K. M. Schopf, and C. E. Brett. 1995. "The Challenge of Paleoecological Stasis: Reassessing Sources of Evolutionary Stability." *Proceedings of the National Academy of Sciences USA* 92:11269–11273.

Nijhout, H. F. 1990. "Metaphors and the Role of Genes in Development." *BioEssays* 12 (9): 441–446.

Nijhout, H. F., and M. C. Reed. 2014. "Homeostasis and Dynamic Stability of the Phenotype Link Robustness and Plasticity." *Integrative and Comparative Biology* 54 (2): 264–275. doi: 10.1093/icb/icu010.

Okasha, S. 2006. *Evolution and the Levels of Selection*. Oxford: Clarendon Press.

Pennell, M. W., L. J. Harmon, and J. C. Uyeda. 2014. "Is There Room for Punctuated Equilibrium in Macroevolution?" *Trends in Ecology and Evolution* 29 (1): 23–32. doi: 10.1016/j.tree.2013.07.004.

Phillimore, A. B., C. D. Orme, R. G. Davies, J. D. Hadfield, W. J. Reed, K. J. Gaston, R. P. Freckleton, and I. P. Owens. 2007. "Biogeographical Basis of Recent Phenotypic Divergence among Birds: A Global Study of Subspecies Richness." *Evolution* 61 (4): 942–957. doi: 10.1111/j.1558–5646.2007.00068.x.

Plotkin, H. C. 1988. "Behavior and Evolution." In *The Role of Behavior in Evolution*, edited by H. C. Plotkin. Cambridge, MA: MIT Press.

Reznick, D. N., H. Rodd, and L. Nunney. 2004. "Empirical Evidence for Rapid Evolution." In *Evolutionary Conservation Biology*, edited by R. Ferrière, U. Dieckman, and D. Couvet, 101–118. Cambridge: Cambridge University Press.

Roff, D. A. 1994. "Habitat Persistence and the Evolution of Wing Dimorphism in Insects." *The American Naturalist* 144:772–798.

Saab, V. A., R. E. Russell, and J. G. Dudley. 2007. "Nest Densities of Cavity-Nesting Birds in Relation to Postfire Salvage Logging and Time since Wildfire." *The Condor* 109:97–108.

Salthe, S. N. 2012. "Hierarchical Structures." *Axiomathes* 22 (3): 355–383. doi: 10.1007/s10516-012-9185-0.

Siepielski, A. M., J. D. DiBattista, and S. M. Carlson. 2009. "It's About Time: The Temporal Dynamics of Phenotypic Selection in the Wild." *Ecology Letters* 12 (11): 1261–1276. doi: 10.1111/j.1461–0248.2009.01381.x.

Simpson, G. G. 1944. *Tempo and Mode in Evolution*. New York: Columbia University Press.

Stanley, S. M., and X. Yang. 1987. "Approximate Evolutionary Stasis for Bivalve Morphology over Millions of Years: A Multvariate, Multilineage Study." *Paleobiology* 13:113–139.

Stenseth, N. C., and J. Maynard Smith. 1984. "Coevolution in Ecosystems: Red Queen Evolution or Stasis?" *Evolution* 38:870–880.

Thierry, B. 2005. "Integrating Proximate and Ultimate Causation: Just One More Go!" *Current Science* 89 (7): 1180–1183.

Uyeda, J. C., T. F. Hansen, S. J. Arnold, and J. Pienaar. 2011. "The Million-Year Wait for Macroevolutionary Bursts." *Proceedings of the National Academy of Sciences USA* 108 (38): 15908.

Van Valen, L. 1989. "Three Paradigms of Evolution." *Evolutionary Theory* 9:1–17.

Voje, K. L., O. H. Holen, L. H. Liow, and N. C. Stenseth. 2015. "The Role of Biotic Forces in Driving Macroevolution: Beyond the Red Queen." *Proceedings of the Royal Society B: Biological Sciences* 282 (1808): 20150186. doi: 10.1098/rspb.2015.0186.

Vrba, Elisabeth S. 1992. "Mammals as a Key to Evolutionary Theory." *Journal of Mammalogy* 73:1–28.

Wagner, A. 2005. *Robustness and Evolvability in Living Systems*. Princeton, NJ: Princeton University Press.

Wcislo, William T. 1989. "Behavioral Environments and Evolutionary Change." *Annual Review of Ecology and Systematics* 20:137–169.

Werner, Earl E. 1992. "Individual Behavior and Higher-Order Species Interactions." *The American Naturalist* 140:S5–S32.

Whitacre, J. M., and A. Bender. 2010. "Networked Buffering: A Basic Mechanism for Distributed Robustness in Complex Adaptive Systems." *Theoretical Biology and Medical Modelling* 7:20. doi: 10.1186/1742-4682-7-20.

Zera, A. J., and R. F. Denno. 1997. "Physiology and Ecology of Dispersal Polymorphism in Insects." *Annual Review of Entomology* 42:207–230.

9 The Causes of a Major Transition: How Social Insects Traverse Darwinian Space

Heikki Helanterä and Tobias Uller

1. Introduction

Life is hierarchically organized. Cells come together to form tissues, organs, and organisms like ourselves, who in turn are members of social groups and populations. Given that organisms that are composed of many individual cells have the capacity for evolution by natural selection, it is natural to ask if social groups composed of multicellular individuals can also represent evolutionary units. Answers to this question are based on an understanding of how the evolutionary process can assemble populations of individuals into populations of collectives (Szathmáry and Maynard Smith 1995; Watson and Szathmáry 2016). The motivation for this chapter is whether or not the traditional view on evolutionary causation (Uller and Laland, this volume) is a satisfactorily framework for providing an answer.

A good starting point is the observation that evolution by natural selection can be applied to reproducing entities ("individuals") that fulfill three conditions: (i) variation in features among individuals (the principle of variation); (ii) variants differ in their survival or reproduction (the principle of differential fitness); and (iii) offspring resemble their parents (the principle of heredity) (Lewontin 1970). DNA, cells, and multicellular organisms all fulfill these criteria. This means that natural selection operates at multiple biological levels, often simultaneously. Considering at what level of organization that selection is operating—or operating most efficiently—has been used to assess the evolutionary potential of collectives (Rainey and Rainey 2003; Taylor, Zeyl, and Cooke 2002; Tsuji 1995; Wade 1977). However, this is unlikely to reveal the full capacity for adaptive evolution exhibited by a system, nor does it obviously address how new levels of organization originate and evolve.

This leads us to the two aims of this chapter. First, we illustrate how the major transitions in evolution can be described in terms of a heuristic tool known as the Darwinian space (Godfrey-Smith 2009, box 9.1). The axes of the Darwinian space are features of populations of entities that distinguish between instances of evolution by natural selection that merely fulfill the three necessary criteria from those instances that have the capacity to produce complex, cumulative adaptations. As explained below, making this distinction requires more than identifying reproducing entities that have differential fitness. Applying the Darwinian space to transitions in evolutionary individuality is not new (in fact, it was dedicated its own chapter in Godfrey-Smith 2009). However, as most of these discussions are concerned with transitions from unicellular to multicellular life (Clarke 2016b; Leslie, Shelton, and Michod 2017; Libby et al. 2016; Watson and Thies, this volume), we hope that applying it to social insects will make biologists more aware of its utility and inspire comparison across transitions.

Second, we ask what the Darwinian space framework can tell us about evolutionary causation. As the axes of the Darwinian space represent biological features that are not necessarily targets of selection (box 9.1), a focus on the inclusive fitness benefits of reproductive division of labor—a strong tradition in the evolutionary biology of social insects—appears to provide only a partial explanation for the acquisition of evolutionary potential of collectives. We suggest that a more inclusive view of evolutionary causation that focuses on the relationship between natural selection and "proximate" causes, such as plasticity and niche construction, will prove useful to a research program on the evolution of social insects.

2.   A Darwinian Space

Godfrey-Smith (2009; see also Godfrey-Smith [2013]) identified a range of features that characterize paradigmatic Darwinian populations (box 9.1), that is, populations with the capacity to produce cumulative complex adaptations through natural selection. The features of such populations, or axes of a "Darwinian space," are variation, fidelity of inheritance, intrinsicness of fitness, continuity of fitness, and competition. These terms are explained briefly in box 9.1. We focus on the first three. Firstly, cumulative adaptation requires that the supply of variation (V in box 9.1) from genetic (e.g., mutation, recombination) or environmental perturbation can, at least occasionally, produce new phenotypes without disrupting the overall function. Secondly, those phenotypes need to be heritable (H in box 9.1). If offspring do not resemble their parents, natural selection will struggle to produce adaptations because variation that has been selected cannot be retained down generations. Thirdly, the causes of fitness differences need to be intrinsic to individuals (S in box 9.1). If fitness differences are mainly determined by external objects, such as asteroids falling from the sky, the deaths of individuals do not amount to a persistent accumulation of particular features within the population. That is, in paradigmatic cases of evolution by natural selection, life and death depends on features that are intrinsic to individuals and reliably inherited by offspring. Most familiar organisms belong to this category.

In addition to such paradigm cases, there are entities that deviate more or less in these respects. Consider, for example, bacteria growing on tooth enamel and forming a biofilm. The bacteria within the biofilm fulfill the three principles for evolution by natural selection:

Box 9.1

A Darwinian Space

Peter Godfrey-Smith (2009) described a multidimensional space where populations of any kind can be placed to account for their capacity to evolve by natural selection. Paradigm cases of such "Darwinian populations" include familiar multicellular organisms like fish. These score high on all of the axes of the Darwinian space. Other entities, like sponges or bacteria films, score lower on at least one of the axes. Despite that all of these are, in some sense, collectives of individuals (fish, sponges, and biofilms are made up of individual cells), only some have evolutionary potential as populations of collectives. Other collectives, such as a school of fish, have virtually no capacity to undergo evolution by natural selection—the relevant process is taking place at the level of the fish, not the school.

Godfrey-Smith described five population features that support a high capacity for evolution by natural selection:

V: Variation. Without variation, natural selection has no raw material to work on. In addition to current standing variation, the long-term ability for adaptive evolution depends on an ability to generate new functional variants.

H: Fidelity of Inheritance. Adaptive evolution depends on parent–offspring similarity. Note that in Darwinian space the fidelity concept is not the same as the technical definition of heritability, but is a general measure of reliability of phenotype resemblance of parent and offspring.

S: Intrinsicness of Fitness. The extent to which fitness differences within a population depend on the intrinsic characters of the individuals rather than external features, such as their location. That is, fitness should be a function of the fit between characters that belong to the individual and its particular context, not context alone.

C: Continuity of the Fitness Landscape. Cumulative adaptive evolution is thwarted in a rugged fitness landscape, where small changes in the heritable material lead to very large changes in fitness.

α: Competition. Describes the extent to which reproductive success of one member of a population affects success of others or the extent to which reproduction is a zero-sum game.

Some life cycle features make collectives more likely to be members of a Darwinian population because they reduce the evolutionary potential of the individuals that make up collectives:

B: Bottleneck refers to the extent to which a life cycle features a stage with a limited material overlap and/or low genetic diversity, such as the zygote stage of a multicellular organism. This is important for ensuring high relatedness and low variation among group members.

G: Germline-Soma Separation, or Reproductive Division of Labor, refers to the extent to which group members vary in their chances of reproduction. This is important for which

(continued)

Box 9.1
(continued)

fitness differences are due to intrinsic versus relational properties. For example, both somatic
and germ cells have (long-term) fitness only as members of a group.

I: Integration refers to a composite of several features such as the extent of mutual dependency, complementary function and loss of totipotency of lower-level individuals, the extent
of boundary development of the higher-level individuals, and the nonreproductive division
of labor among individuals. For example, a multicellular body requires several different cell
types, both somatic and germline, in order to complete the life cycle. Thus, the capacity for
evolution by natural selection is better ascribed to the multicellular whole rather than to any
particular cell lineage.

(i) there is often variation in characters among bacteria, such as the ability to metabolize
particular nutrients; (ii) this variation underlies fitness differences in particular circumstances; and (iii) variation is often heritable, for example, because enzymes are encoded
in the DNA. However, the fitness of bacteria in a biofilm also depends on features of the
biofilm that can be attributed to the collective, such as properties of the jointly produced
extra-cellular matrix and synergistic community interactions within the biofilm (Clarke
2016b; Poltak and Cooper 2011). Among the competing bacteria growing on the enamel,
the dependence of reproductive differences on intrinsic features therefore appears reduced
compared to "free living" bacteria. What about biofilms—do they also form a Darwinian
population? There are arguably differences in survival and reproduction among biofilms,
and biofilms proliferate, for example, by breaking up into smaller pieces. However, competition for reproduction occurs primarily within the biofilm, such that fitness differences
among biofilms can largely be ascribed to fitness differences among individual bacteria
(Nadell, Xavier, and Foster 2009). Furthermore, although biofilms have features that are
reliably inherited to a new generation of biofilms (Kolenbrander et al. 2010), the fidelity
of inheritance of biofilm properties is limited by their aggregate life cycles and lack of
terminal reproductive differentiation (Clarke 2016b).

The important point of this example is that, at both levels of organization—the bacteria
and the multicellular biofilm—we can recognize the ingredients of evolution by natural
selection (i.e., variation, differential fitness, and heredity). But we can also identify features that make each level deviate from a paradigmatic Darwinian population, like the
populations of free-living bacteria or multicellular organisms like ourselves.[1] The Darwinian space helps to put the finger on what these differences are, thereby explaining why
some entities are more likely to evolve adaptations than other entities (Godfrey-Smith
2009; Sterelny 2011).

Another important feature of the Darwinian space perspective is that it emphasizes that
the ability to evolve by natural selection itself is evolving. That is, populations can move

from marginal or minimal cases toward paradigmatic cases (figure 9.1). A striking example is the association of individuals to form a new, higher level or more complex, individual (Bourke 2011; Buss 1987; Maynard Smith and Szathmáry 1995). The candidates for such events (often referred to as "major transitions" or "transitions in individuality") include the evolution of genomes, eukaryotic cells, multicellularity, and social insect colonies. Following a transition, we can still recognize lower-level units (e.g., individual cells of our bodies) that resemble a solitary ancestor (i.e., a free-living unicellular organism). These units may retain features that make them evolvable in the short term, but they have lost their long-term ability to reproduce, compete, and evolve independently of the new wholes. In contrast to the cells within the biofilm, the evolutionary prospect of individual cells within our bodies is rather bleak, despite that cells exhibit heritable variation and leave different numbers of descendants.[2] On the other hand, the collection of cells that make up bodies have properties that make multicellular organisms like ourselves capable of evolving by natural selection as paradigmatic Darwinian reproducing collectives.

A transition in individuality involves changes in the causal dependencies between biological entities that make lower-level entities move away from paradigm instances of Darwinian evolution (even if those entities may still exhibit differential fitness), while the higher-level entity gains properties that make it capable of evolution by natural selection (figure 9.1). This refers to two alternative "problem domains" of the transition: "de-Darwinization" (i.e., lower-level individuals lose their capacity for evolution by natural selection) and "Darwinization" (i.e., groups gain this capacity). The social insect literature on the major transition—with its emphasis on the evolution of altruism, suppression of conflicts, and reproductive division of labor (in particular worker sterility)—has mainly been concerned with the former (Boomsma 2009; Ratnieks, Foster, and Wenseleers 2006; Strassmann and Queller 2007; but see Johnson and Linksvayer 2010; Linksvayer and Wade 2005; Linksvayer et al. 2012; Seeley 1989). Philosophers of biology, on the other hand, have to a larger extent been concerned with the latter, in particular in the literature on the concept of individuality (Bouchard and Huneman 2013; Calcott and Sterelny 2011; Clarke 2014; Clarke 2016a; Griesemer 1999; Sterelny et al. 2013; Trestman 2012).

In what follows, we discuss these two problem domains for the evolutionary transition from solitary insects to the most extreme forms of insect sociality. After a brief overview of the biological steps toward complex sociality, exemplified by social Hymenoptera (i.e., ants, social bees, and wasps), we explain how individuals within societies lose their status as members of Darwinian populations and how this leads to a society life cycle. We then explore the extent to which de-Darwinization of individuals is associated with Darwinization of insect societies. We argue that an account that is primarily concerned with the loss of evolutionary capacity of individuals is insufficient to fully account for the extent to which social insect colonies acquire evolutionary individuality. This implies that something is missing from most contemporary accounts of the evolution of social insects. Specifically, we suggest that understanding evolutionary transitions in individuality

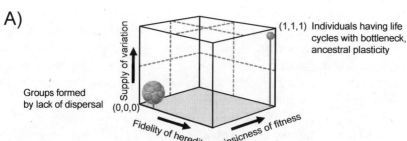

A)

Groups formed by lack of dispersal

Supply of variation

Fidelity of heredity   Intrinsicness of fitness

(0,0,0)

(1,1,1) Individuals having life cycles with bottleneck, ancestral plasticity

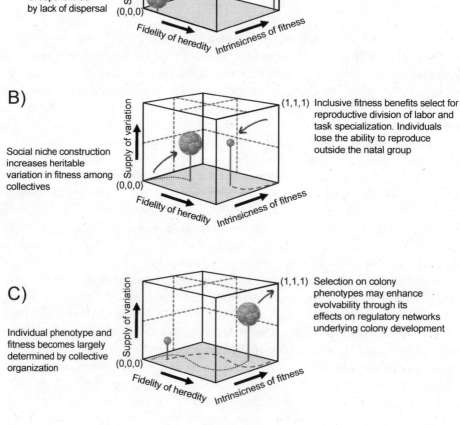

B)

Social niche construction increases heritable variation in fitness among collectives

Supply of variation

Fidelity of heredity   Intrinsicness of fitness

(0,0,0)

(1,1,1) Inclusive fitness benefits select for reproductive division of labor and task specialization. Individuals lose the ability to reproduce outside the natal group

C)

Individual phenotype and fitness becomes largely determined by collective organization

Supply of variation

Fidelity of heredity   Intrinsicness of fitness

(0,0,0)

(1,1,1) Selection on colony phenotypes may enhance evolvability through its effects on regulatory networks underlying colony development

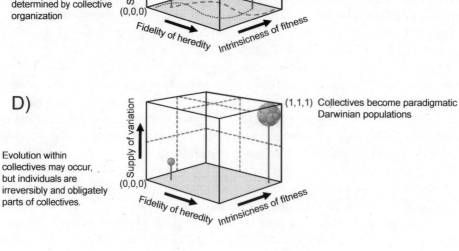

D)

Evolution within collectives may occur, but individuals are irreversibly and obligately parts of collectives.

Supply of variation

Fidelity of heredity   Intrinsicness of fitness

(0,0,0)

(1,1,1) Collectives become paradigmatic Darwinian populations

requires an account of how the "developmental" regulatory interactions within the higher-level reproducer (i.e., social insect colony) evolve from the social and environmental interactions of lower-level reproducers (i.e., individual insects). This puts the study of major transitions firmly at the interface between developmental biology, ecology, and evolution. We end with an outline of a research program dedicated to study if and how insect societies transform from mere collectives to evolutionary individuals.

3.   Evolution of Insect Societies

Advanced insect societies, such as those of termites, ants, and social bees and wasps, have been compared to "stereotypical" multicellular organisms for over a century (Huxley 1912; Weismann 1893; Wheeler 1911).[3] In the extreme cases, the interdependence of individual and group life and reproduction is obligate such that we can identify a colony life cycle. The steps by which these complex insect societies evolved are reasonably well understood, especially in the Hymenoptera (i.e., ants and social bees and wasps; figure 9.2). The ancestral solitary life cycle is a mated queen who rears offspring alone (Boomsma 2009; Cronin et al. 2013; Hughes et al. 2008). That is, she starts the nest, lays eggs, and forages and provisions for the brood. This solitary founding stage has been retained by many social ants, bees, and wasps, and the queen life cycle has effectively stayed the same throughout the transition to obligate, organismal social groups. Her male mate plays no other role than providing sperm that the queen stores for the rest of her life. The life cycle is completed when the sexual males and females that constitute the next generation are reared.

From an evolutionary perspective, the earliest social groups along the way to superorganisms consist of mothers and their offspring who forego or delay their own dispersal to help their mothers rear more of their siblings.[4] This required phenotypic plasticity (Page, Linksvayer, and Amdam 2009; Rehan and Toth 2015; Toth and Robinson 2007; West-Eberhard 1987), enabling facultative decisions to disperse or stay at the natal nest to provide help for the rearing of siblings. The role of the workers is typically either brood

---

Figure 9.1
Evolutionary transitions in individuality, exemplified by the evolution of insect societies, represented in a Darwinian space. (A) Before a transition is initiated populations of individual insects are paradigm cases of evolution by natural selection. Collectives may be reliably formed but lack capacity for evolution by natural selection. (B) Natural selection on individuals makes them move away from paradigmatic Darwinian populations. This may be accompanied by changes in the properties of colonies that make them acquire some capacity for evolution by natural selection in their own right. (C) Persistent selection—on both individuals and colonies—may result in further loss of evolutionary potential of the population individual insects, and increased evolutionary potential of colonies. (D) In the extreme case, the population of individuals lose their ability to evolve and most (or all) adaptive change now happens as a result of variation, differential fitness, and heredity attributed to properties of the colony.

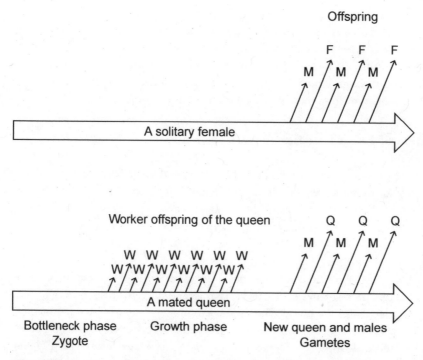

Figure 9.2
In a solitary insect life cycle, a female establishes a nest and rears her offspring, which disperse as reproductive males and females. In a social insect life cycle, the female (the queen) establishes a nest and rears her first offspring as workers who stay in the nest and help to rear the reproductive males and females. As a consequence, reproduction takes place in the context of a collective and the closing of the life cycle requires a phase of colony growth and maturation.

care and foraging or defense, and they help the queen rear the next cohort of foundress queens and males that disperse, mate, and form the next "zygote" generation. As a consequence, nonreproductive individuals proliferate (i.e., the growth phase of the colony) before new reproductive individuals develop (i.e., the reproductive phase of the colony). The presence of workers allows the reproductive individual to forego nonreproductive tasks, and is thus a primitive decoupling of the reproductive and nonreproductive phenotypes that constitute reproductive division of labor.

As the number of workers increased over evolutionary time, the growth phase of the colony life cycle increased in importance, allowing the colony to become more complex. Associated with this is a trend toward reproductive division of labor being determined increasingly early in development (Bourke 1999; Wheeler 1986). Once the development of individuals into queens and workers becomes controlled by the rearing environment (Johnson and Linksvayer 2010; Linksvayer and Wade 2005; Linksvayer et al. 2012), and irreversibly determined during larval stages, the transition to a superorganism is typically considered complete and irreversible (Boomsma and Gawne 2018). After this point, the developing individuals have lost independent breeding options and have fitness only as members of the group in which they are born as helpers.[5] Individuals have thus yielded the control of their phenotype development and group membership to the social environ- ment as it is the environment, and especially the nutrition provided, that sets the developing brood onto either a helper or a reproductive developmental trajectory (Wheeler 1986). Simultaneously, the group life cycle has become entirely dependent on the presence of the nonreproducing individuals, since no sexual offspring is produced until the colony has passed the growth or "ergonomic" stage (Oster and Wilson 1978) and reached a reproduc- tive size, sometimes requiring thousands of workers.

4.   Natural Selection and the De-Darwinization of Individuals within Collectives

One obvious approach to construct an account of evolutionary transitions in individuality is on the basis of natural selection. The paradox lies in that group features cannot be selected for before they exist, and so the challenge is to explain how selection on individu- als can give rise to collective reproducers capable of evolution by natural selection (Clarke 2014; Griesemer 1999; Trestman 2012). Indeed, a major hurdle for the evolution of highly coordinated groups, like those of social insects, is that evolution within collectives may compromise collective integrity and functionality. If collectives are to acquire evo- lutionary individuality, the evolutionary potential of its members must be relinquished. From the perspective of the Darwinian space, this means that there must be inclusive fitness benefits that make individuals move away from paradigm cases of evolution by natural selection (figure 9.1).

The evolution of insect societies is associated with several features that reduce the evolutionary potential of populations of individuals while promoting the stability and recurrence of the colony life cycle. The most conspicuous is that most of the colony members lose their ability to produce offspring. This obviously reduces the potential for evolution within collectives and is therefore a powerful mechanism of de-Darwinization. Selection against individual reproduction, causing reproductive division of labor, is pos- sible because of the consistently high relatedness among colony members (Boomsma 2009, 2013). In such a case, it is not necessarily sterility per se that is the target of selection, but a phenotype that is a good worker, with sterility arising as a consequence of functional trade-offs (the familiar "jack of all trades is a master of none").[6] Early commitment to a specific role, and release from pleiotropic constraints of having to carry out multiple tasks, allows evolution of morphological specialization, a hallmark of organismal insect societies (Gadagkar 1997). As the group becomes more complex, the indirect fitness benefits of staying increase while the expected success when leaving the natal group (i.e., the direct fitness cost of staying) or joining another group decreases.

A key feature that allows this evolutionary process to get started and proceed is the bottleneck stage in the colony life cycle (*B* in box 9.1). In social insects, the solitary nest founding by mothers is the ancestral state and the bottleneck did not need to evolve *de novo*, it only needed to be retained.[7] Thus, there is no need to invoke selection to explain the evolution of the bottleneck. A strong bottleneck, combined with group formation through nondispersal of offspring, promotes high relatedness to other group members (Boomsma 2009, 2013) and recurrence of social and ecological conditions. These ensure reliable cues for indirect fitness benefits of commitment to a helper and facilitates the evolution of co-adapted genomes and phenotypes (Linksvayer and Wade 2005). Natural selection for social determination of task specialization thus creates a positive feedback loop between loss of individual ability to reproduce outside of the collective and increased integration of the colony life cycle with an obligate phase of collective growth and differentiation.

5. Social Insect Colonies as Darwinian Populations

The de-Darwinization of colony members make colonies candidates for the kind of entities that undergo evolution by natural selection. But is it sufficient to capture how social insect colonies traverse the Darwinian space? We will restrict our discussion to three of the dimensions—the intrinsicness of fitness, the capacity to generate variation, and heredity (box 9.1). If social insect colonies have acquired evolutionary individuality, they must score high on these axes, while individuals within societies should score low on at least some of them (figure 9.1). But since intrinsicness of fitness and the capacity to generate heritable variation are features that are not typically considered targets of selection, it is not obvious how they evolve. In the following two sections, we briefly explain why one cannot assume that collectives with reproductive division of labor, task specialization, and high genetic relatedness will behave like paradigm cases of evolution by natural selection. The third section provides a clue toward what is missing.

Intrinsicness of Fitness

That a particular character is under natural selection implies that it is causally responsible for the realization of fitness differences between individuals (i.e., there is "selection for" the character; Sober 1984). That is, in a paradigm Darwinian population, differences in reproductive output between individuals depend on characters that are intrinsic to those individuals (*S* in box 9.1). Intrinsic characters of one object are those that "do not depend on the existence and arrangement of other objects. [...] It measures the degree to which the difference makers with respect to reproductive success approximate being dependent on the intrinsic character of the members of the population, instead of extrinsic or relational features" (Godfrey-Smith 2009, 53–55). Some causes of mortality depend on extrinsic characters. Individuals that die in an asteroid crash, for example, do not die because they have particular features. They die because they happened to be in the wrong place at the wrong time. In contrast, carriers of a missense mutation in the melanocortin-1 receptor tend to develop pale skin or fur no matter what they do or where they are. Thus, although carriers and noncarriers may have different fitness in only some environments, those fitness differences do depend on features that are intrinsic to individuals.

The evolutionary changes that occur to individual colony members during their de-Darwinization make differences in individual reproductive fitness depend primarily on the properties and arrangement of other objects, in particular the complementary phenotypes of other colony members (Watson and Thies, this volume).[8] Consider the pheromone trail system an ant colony uses to forage. It is a result of large numbers of individuals processing and responding to environmental (food availability, distance from nest) and social information (pheromone concentration; traffic levels; signals from nurse workers, brood, or queen) (Czaczkes, Grüter, and Ratnieks 2015). Even characters that were responsible for intrinsic fitness differences between individuals in the solitary ancestor, like the size of mandibles, power of flight muscles, or ovarian activity, lose their causal efficacy for fitness variation because of reproductive division of labor and functional integration among colony members. As a result, the fitness of individuals largely depends on the combination of phenotypes within the colony (e.g., complementarity of interacting phenotypes; Clarke 2014; Watson and Thies, this volume).

It seems reasonable to assume that this loss of intrinsicness of fitness at the level of individuals should be accompanied by a gain in intrinsicness of fitness at the colony level. However, this is not necessarily the case. Differences in fitness among colonies may primarily arise from extrinsic features, such as location. As fitness differences among individuals become located *between colonies*, those fitness differences must also derive from the distribution and organization of individuals *within colonies*. The de-Darwinization of individuals via selection for functional specialization alone is not sufficient to explain how difference makers of fitness become intrinsic properties of colonies.

Variation and Heredity

At first sight, the move away from a paradigmatic Darwinian population does not appear to involve a reduction in the capacity of individuals to generate heritable phenotypic variation. However, since individual phenotypes are largely determined by plastic responses to the collective organization, the parent–offspring resemblance is actually rather low. Queens do, after all, produce workers as well as new queens.

What about features of the colony? In social insects, variation among colonies and fidelity of colony phenotypes is facilitated by the bottleneck, reliable group formation, and reproductive division of labor. These features limit the effects of mutational diversity across generations, create genetic uniformity within groups and variation among groups, select against overt conflict among individuals, and increase genetic similarity between parental and offspring groups. These may be sufficient to account for the inheritance of some collective phenotypes, such as the responses to environmental and social cues that determine reproductive division of labor (Johnson and Linksvayer 2010). However, genetic similarity alone does not necessarily guarantee inheritance of phenotypes that rely on complex interactions. Even the smallest organismal insect societies consist of tens of individuals interacting with their environment, with complex communication systems and resource flows, individual specialization, and learning. In extreme cases, such as in leaf-cutter—fungus systems, there are millions of individuals, genetic diversity, several morphological sub-castes, and multiple mutualists. Like multicellular development, the recurrence of colony phenotypes down generations therefore likely capitalizes on self-organization, plasticity of lower-level entities, and exploratory processes that stabilize colony phenotypes that are functional (Kirschner and Gerhart 2005). As a consequence, inheritance of variable colony phenotypes cannot be solely assumed from genetic inheritance, despite that the latter is guaranteed by bottlenecks and reproductive division of labor.

Externalization and Internalization of Causation

Although brief, these considerations suggest that a focus on the fitness benefits of functional specialization of individuals, including reproductive division of labor, is not sufficient to explain if and how collectives become Darwinian populations. What appears to be missing is an account of how individuals and their phenotypes become organized relative to each other within collectives. Laubichler and Renn (2015) suggested that a transition in individuality can be conceptualized as changes in the causal dependencies between the parts of the collective and their surroundings, and by the acquisition of regulatory control over fitness and phenotype by the collective (figure 9.3).

First, as group living becomes obligate, the individual "externalizes" developmental control to features of its environment. This implies that individual insects are responding plastically to their social environment. One well-studied example is the regulatory network that determines food provisioning to the larvae and, hence, caste determination (Laubichler and Renn 2015; Linksvayer et al. 2012). The queen and brood signal to nurse workers who feed offspring and maintain them at right temperature, in the process directing the activity of forager workers (Linksvayer et al. 2012). Thus, even if we can still identify phenotypes that are very similar to those of solitary ancestors, the ontogeny of those phenotypes has evolved to become an integrated component of the collective life cycle.

Second, as the colonies grow in size and complexity, they increasingly construct their environments. As a consequence, societies become more robust to environmental perturbation

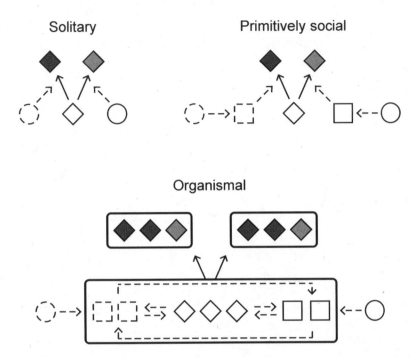

Figure 9.3
Schematic representation of how the major transition comprises acquisition of control over phenotype determination by the collective. Diamonds depict developing individuals facing a developmental switch point with two alternative phenotypes (dark vs. pale shading). Circles depict environmental states that affect development, and rectangles group members that mediate the effect of the environment. Top left: in solitary organisms, environmental inputs directly affect developmental outcomes. Top right: in primitively social organisms the environmental inputs may be mediated by group members. Bottom: in organismal societies, these interactions are mediated by the society, which responds to environmental cues through a collective organization. Determinants of development have been externalized from the perspective of the individual but internalized from the perspective of the group. The developmental outcomes are new group properties, that is, distributions of individual phenotypes.

due to homeostatic mechanisms such as thermoregulation, the context-dependence and plasticity of individual development and behavior, and modification of environments through mutualisms (e.g., aphid rearing and fungus growing). This represents an "internalization" of the regulatory interactions that determine features of the colony (Laubichler and Renn 2015), their inheritance, and fitness (Watson and Thies, this volume).

These two processes—externalization of control of individual development and internalization of control of colony development—can be coupled, such that the environmental features that determine individual development are products of the niche constructing activities of the colony. This should increase the extent to which colonies can vary in features that are heritable, and show fitness differences that are due to intrinsic, rather than extrinsic, features. Establishing and maintaining a feedback between the two processes may therefore be necessary if social insects are to traverse the Darwinian space. Studying the evolutionary dynamics of regulatory control of individual and colony phenotypes therefore becomes a central task in social insect evolutionary biology (Laubichler and Renn 2015). How can this be done?

6.   Toward a Research Program

Historically, researchers have defined and described evolution of social insects in terms of reproductive division of labor, usually in terms of the apparent evolutionary enigma of sterile workers, and the evolution of altruism (Boomsma and Gawne 2018). While this has been a very successful research program (Bourke 2011), it emphasizes only one of the two problem domains of evolutionary transitions—de-Darwinization of the lower level—and may even lose sight of the gain of evolutionary capacity at the higher level of biological organization, and thus fall short of a full picture of the evolution of social insects.

A research program aiming to address how social insects acquire evolutionary individuality would need to turn the considerations of the previous sections into a concrete theoretical and empirical enterprise. Here we emphasize three components of such a research program. Firstly, understanding the process by which societies become members of Darwinian populations requires moving beyond the fitness benefits that make individuals lose their Darwinian properties and toward an investigation of the evolution of collective organization. Second, the diversity of life cycle features in insect societies can point toward biological correlates of the position occupied in the Darwinian space. Thirdly, identifying what is different, and what is similar, between organismal insect societies and other transitions, such as multicellularity or obligate mutualistic symbioses, may establish key features of the evolution of individuality. Ultimately, these three components will contribute to understanding whether or not any social insect has, indeed could, become a paradigm Darwinian population.

Acquisition of Individuality through the Evolution of Regulatory Interactions

We pointed out above that the capacity of biological systems to generate heritable phenotypic variation is not typically believed to have immediate fitness consequences, which makes its evolution difficult to account for by referring to selection alone. The same might be said for the extent to which fitness differences between individuals depend on intrinsic versus extrinsic factors. These problems are well recognized in the literature on the evolution of evolvability, where a challenge is to account for developmental systems that are able to produce functional variation when perturbed (Brown 2014; Pigliucci 2008). Models that represent phenotypes in terms of evolving regulatory interactions, or networks, are helpful in this regard because they can reveal how selection on phenotypes can affect the potential for populations to generate adaptive heritable variation through regulatory change (reviewed in Uller et al. 2018).

To capture the movement of collectives in Darwinian space, such models must consider how the regulatory interactions that govern individual development and behavior evolve. Importantly, any variable and co-inherited properties of the system can usefully be seen as a part of the regulatory network (e.g., Salazar-Ciudad and Jernvall 2010). Multilayered regulatory networks should therefore be useful to explore how the evolution of interactions can concurrently externalize control of individual developmental to the social environment and internalize the control of the social environment by the colony (Johnson and Links-vayer 2010; Laubichler and Renn 2015; Linksvayer et al. 2012). For example, one may establish under what conditions selection will strengthen interactions between genes residing in different individuals, between individuals and recurrent components of the environment, or between the genomes of mutualist species (Linksvayer et al. 2012).

While these interactions initially evolve as a consequence of fitness differences among individuals, they come to influence the organization of collectives. This should make it possible to identify minimal conditions for the acquisition of evolutionary individuality by collectives (Laubichler and Renn 2015; Ryan, Powers, Watson 2016). Watson and Thies (this volume) have made a convincing case that group members must express plasticity, extra-genetic inheritance, and niche construction for collectives to acquire evolutionary individuality. Over evolutionary time, the increasing integration among multiple co-evolving phenotypes may transfer reproductive differences from properties of individual phenotypes to properties of interactions among phenotypes. The layers of regulatory control will therefore inevitably evolve in response to multilevel selection. This demonstrates the complexity of capturing the transition in regulatory networks. However, the recognition that connectionist models of learning can be put to use for similar problems may help to identify useful tools and to resolve conceptual issues (Watson and Szathmáry 2016).

Empirically, the evolution of collective organization can be reconstructed by assessing whether the topology of regulatory networks of organismal insect societies is qualitatively different to their more primitive counterparts, and documenting how networks have been rewired in the course of social evolution (Duncan, Hyink, and Dearden 2016; Hartfelder et al. 2015; Sobotka et al. 2016). This allows insights into the extent to which ancestral plasticity has been co-opted and refined in the course of evolution of social organization, and the extent to which individual phenotypes have relinquished control to the collective. As a first approximation, this may involve quantification of the responsiveness of individual and colony phenotypes to various aspects of social context (e.g., genotype composition of the colonies or behavioral variation). This has been done in the context of establishing indirect genetic effects on phenotypic variation (Linksvayer 2006, 2007), and how gene expression responds to social organization (Gempe et al. 2012; Malka et al. 2014; Nipitwattanaphon et al. 2013; Vojvodic et al. 2015). Particularly suitable taxa to address these issues may be the social bees and wasps. Both exhibit a continuum of different levels of sociality (Kapheim et al. 2015; Taylor, Bentley, and Sumner 2018; Woodard et al. 2011), and the genetic underpinnings of the early steps in social evolution (Amdam et al. 2006; Hunt et al. 2007; Johnson and Linksvayer 2010) and plasticity of social behavior of solitary organisms (Fewell and Page 1999) are reasonably well understood.

Finally, similar to how gene-regulatory networks can be used to study evolutionary innovations and developmental biases (e.g., Wagner 2011), recasting the social insect colony as a regulatory network makes it possible to study how new phenotypes can arise through changes in social organization and, for example, the role of modular organization and developmental biases in this process. Examples of such innovations include novel worker castes (Abouheif and Wray 2002; Rajakumar et al. 2012), inter-castes and loss of wings in queens (Molet, Wheeler, and Peeters 2012), novel physiological organs (e.g., glands; Jasper et al. 2015), and different forms of collective foraging and communication (Möglich, Maschwitz, and Hölldobler 1974; Seeley 1995).

Diversity of Darwinian-ness in Insect Societies

To establish the Darwinian status of social insects, it will also be necessary to gain a better empirical understanding of how colony phenotypes actually evolve. There are only a few decades-long studies of natural populations of social insects that allow tracking of parent–offspring relations and investigate fitness effects of colony phenotypes (Cole 2009; Gordon 2013; Vitikainen, Haag-Liautard, and Sundström 2015).[9] However, in the laboratory, breeding studies of domesticated honeybees have revealed that responses to selection involve a complex mixture of changes in direct and indirect (sib-social and offspring-maternal) interactions (Linksvayer, Fondrk, and Page 2009; Page and Fondrk 1995). Similarly, studies of genetic architecture of, for example, reproductive allocation decisions in *Temnothorax* ants have shown a complexity indicative of phenotypic control relinquished to properties of the collective (Linksvayer 2006, 2007, 2008).

One interesting aspect in this regard is that social insects have a diversity of life cycles. Bottlenecks have been lost as colony founding behaviors such as fission and budding have evolved, and iteroparous life cycles and perennial societies also exist (Helanterä 2016). Further, colonies with multiple queen and male lineages, novel castes, and caste determining systems have repeatedly evolved (Heinze 2008). This diversification is typically discussed in terms of its effects on internal conflicts (Ratnieks et al. 2006), but it also raises questions of how fidelity and heritable variation is maintained in genetically complex societies (Kennedy, Uller, and Helanterä 2014), and how inter-genome epistasis (Libbrecht and Keller 2013; Linksvayer 2007; Schwander and Keller 2008; Teseo et al. 2014), caste antagonistic fitness variation (Holman, Linksvayer, and d'Ettorre 2013; Pennell et al. 2018), and the low effective population sizes (Romiguier et al. 2014) of insect societies affect their ability to evolve through natural selection. A network perspective could prove useful in linking existing evolutionary genetic and statistical approaches to a more mechanistic understanding of how insect societies evolve.

How Similar Are Social Insects to Other Transitions in Individuality?

Maynard Smith and Szathmáry (1995) suggested that major transitions share certain fundamental features. But how comparable are different kinds of Darwinian populations? The biology of solitary insects evolving into social insects appears to be very different to single cells evolving multicellularity. Insect societies are not clonal, consist of highly mobile individuals, and the germline-soma separation of multicellular organisms is not strictly analogous with reproductive division of labor. This suggests that results from models of transitions to multicellularity (Michod 1999) cannot be directly applied to early evolution of insect societies. And this, in turn, begs the question how much the idiosyncratic, contingent features affect if and how collectives traverse Darwinian space. Comparisons of the early stages of the transition—and in particular the establishment of the necessary feedbacks and social niche construction (see above)—are likely to be informative (Ryan et al. 2016; Van Dyken and Wade 2012). Features that likely will vary across transitions include the coevolution of population structure, dispersal behavior, and reproductive division of labor (Gardner and West 2006; Powers, Penn, and Watson 2011), signals and helping decisions (Holman 2014; González-Forero 2015), and coevolution of social control and decisions on cooperation (Michod, Nedelcu, and Roze 2003; Michod and Roze 1997; Wenseleers et al. 2004).

We suggest that one particularly informative feature is that, considered as individuals, social insect colonies have fundamentally different developmental biology, life cycle, and reproductive biology to most multicellular organisms (Helanterä 2016; Kennedy, Uller, and Helanterä 2014; Yang 2007). For example, it has been proposed that there are five different mechanisms of cellular interactions in multicellular organisms (the inductive mechanism of cell communication, the morphogenetic mechanisms of cell division, cell adhesion, cell differentiation, and apoptosis; Salazar-Ciudad, Jernvall, and Newman 2003). As the "somatic" lineages of insect societies (i.e., workers) do not form lineages that divide and differentiate, but are constantly produced by the queen, not all of these mechanisms are available to insect societies. Other important features of colony development include the extreme mobility of workers and that colonies easily can grow and lose parts, thereby changing the dynamics of how the individuals interact with each other and their environment. Such features of the developmental biology of social insects should have implications for the capacity of colonies to produce novel variation and ultimately whether or not they are able to traverse the Darwinian space.

7. Conclusion

The considerations above suggest that to fully understand if and how insect societies acquire evolutionary individuality, research must expand beyond the reproductive division of labor and resolution of cooperative dilemmas that cause populations of individuals to lose their Darwinian status. As complex societies evolve from simple groups, those groups must acquire an integrated functional organization that is evolvable. This requires an explanation for the evolution of collective features that are not considered to be targets of selection, such as inheritance and the ability to produce variation. We predict that this explanation likely will include plasticity and social construction of both selective and developmental niches. This involves interactions between past natural selection and biological organization across levels, as well as lineage-specific ancestral features that facilitate or hinder transitions in individuality. Unraveling these processes in social insects can make them an illustrative model for the evolutionary emergence of evolvability.

Acknowledgments

We thank the organizers of the *Cause and Process in Evolution* Workshop at Konrad Lorenz Institute for the opportunity to present the first versions of these ideas, and the participants for discussions. This work was supported by a grant from the John Templeton Foundation (60501). H. H. was funded by the Kone Foundation and the Academy of Finland (Grant 284666 to Centre of Excellence in Biological Interactions), and T. U. was supported by the Knut and Alice Wallenberg Foundations. We are grateful to Charlie Cornwallis and Tim Linksvayer for comments on a draft of this chapter.

Notes

1. There is an important caveat; we would not survive without the micro-organisms in our bodies. Perhaps, as Scott Gilbert declared, "we were never individuals" (Gilbert, Sapp, and Tauber 2012; Sterelny 2011)? Although the status of the individual collective of its own cells and that of its micro-organisms—the holobiont—as a Darwinian individual is an important question (Queller and Strassmann 2016; Gilbert and Tauber 2016), we will not enter into this discussion here and assume that we can treat single-celled organisms like bacteria and multicellular organisms like ourselves as paradigm cases of Darwinian individuals (Godfrey-Smith 2009). There are pragmatic reasons for this assumption in this chapter since it makes it easier to see what it will take for social insects to make the transition from an evolving population of insects into an evolving population of insect societies.

2. That cells within our bodies retain some the capacity for evolution by natural selection is the reason why it can be useful to treat cancers as an evolving population of cells (Merlo et al. 2006). The facial tumor disease spreading among Tasmanian devils demonstrates that these cells sometimes may be able to escape their confinement to a particular individual and continue to evolve (Pearse and Swift 2006).

3. Some authors have emphasized reproductive division of labor (queens *versus* workers), while others have emphasized reproductive harmony and lack of within-group selection, individuation mechanisms or complex, emergent group level traits such as communication and task allocation systems (Clarke 2013; Gardner and Grafen 2009; Seeley 1989; Wilson and Sober 1989). In their review, Boomsma and Gawne (2018) point out that the historical literature consistently referred to societies as (super)organismal only when workers (i.e., nonreproductive individuals) have lost their totipotency and are incapable of leaving the nest, founding a nest, and reproducing on their own. In an unambiguous case, such as a colony of leafcutter ants, the female workers have lost the spermatheca and are physically unable to mate and produce diploid offspring (i.e., new workers and reproductive females). Similarly, in these societies the survival and reproduction of queens have evolved to rely on other group members. We will here concentrate on evolution of sociality in Hymenoptera, although many arguments we make apply to the evolution of organismal termite societies in Blattodea (Boomsma 2009, 2013).

4. In taxa where morphologically separated castes are absent, the diversity of founding modes and relatedness between helper and reproductives is high. However, organismal levels of sociality never seem to have evolved from groups of cooperative breeders of the same cohort where relatedness is less consistently high, even if such societies can be obligatory group living (Boomsma and Gawne 2017). We leave open the question whether they ever qualify as a major transition.

5. That fitness is achieved only as a member of the natal group does not mean there can be no selection within groups. Even in social insects, helpers may retain some reproductive potential. For example, workers of most Hymenoptera are able to lay unfertilized male-destined eggs, even if they are unable to found a colony and reproduce independently (Bourke 1988). Conflicts are also not completely absent after the loss of reproductive capacity, but have switched from conflicts over reproductive status to conflicts over allocation decisions such as sex and caste ratios (Boomsma and Gawne 2018; Ratnieks et al. 2006). Perfect harmony is a poor definition of a transition in individuality that does not reflect the diversity of collectives able to evolve by natural selection.

6. The existence of nonbreeding individuals, and the near-complete reproductive division of labor in social insects, is sometimes explained through coercion into sterility by the individuals rearing the offspring (Birch 2012; Frank 2003; Ratnieks and Wenseleers 2008). However, the origin of committed workers does not need to involve coercion. The determination of nonreproductive commitment early in development (and hence reproductive division of labor) can evolve from an ancestral plasticity if the larval environment, including signals from the mother and siblings, is a reliable cue to the fitness benefits of being a queen or a worker phenotype. Such responses are predicted to be especially likely when efficiency of helping behaviors is co-evolving with facultative dispersal and helping (González-Forero 2014, 2015).

7. Although some insect colonies have secondarily lost solitary founding stage (Cronin et al. 2013; Helanterä 2016), the queen's ability to facultatively produce workers and dispersing reproductive females that found new colonies maintains the dependency of reproduction on the collective.

8. The statistical framework of "indirect genetic effects" (IGE) is a powerful toolkit for assessing how the genetic architecture, fitness effects, and selective responses of social traits depend on the organization of genotypes within groups, and the correlations among their various traits (McGlothlin et al. 2010, 2014; Moore, Brodie, and Wolf 1997). We are not aiming to review this literature here but refer readers to the literature on the social insects that adopt this approach (Linksvayer 2006, 2008, 2015; Linksvayer et al. 2012).

9. Another unknown is the extent to which the reproductive success of one colony impacts the reproductive success of other colonies. The strength of competition (α in box 9.1) is singled out as an important feature of Darwinian populations by Godfrey-Smith (2009).

References

Abouheif, E., and G. A. Wray. 2002. "Evolution of the Gene Network Underlying Wing Polyphenism in Ants." *Science* 297 (5579): 249–252.

Amdam, G. V., A. Csondes, M. K. Fondrk, and R. E. Page. 2006. "Complex Social Behaviour Derived from Maternal Reproductive Traits." *Nature* 439 (7072): 76–78.

Birch, J. 2012. "Collective Action in the Fraternal Transitions." *Biology and Philosophy* 27:363–380.

Boomsma, J. J. 2009. "Lifetime Monogamy and the Evolution of Eusociality." *Philosophical Transactions of the Royal Society B: Biological Sciences* 364 (1533): 3191–3207.

Boomsma, J. J. 2013. "Beyond Promiscuity: Mate-Choice Commitments in Social Breeding." *Philosophical Transactions of the Royal Society of London B: Biological Sciences* 368 (1613): 20120050.

Boomsma, J. J., and R. Gawne. 2018. "Superorganismality and Caste Differentiation as Points of No Return: How the Major Evolutionary Transitions Were Lost in Translation." *Biological Reviews* 93:28–53.

Bouchard, F., and P. Huneman. 2013. *From Groups to Individuals: Evolution and Emerging Individuality.* Cambridge, MA: MIT Press.

Bourke, A. F. G. 1988. "Worker Reproduction in the Higher Eusocial Hymenoptera." *Quarterly Review of Biology* 63:291–311.

Bourke, A. F. G. 1999. "Colony Size, Social Complexity and Reproductive Conflict in Social Insects." *Journal of Evolutionary Biology* 12:245–257.

Bourke, A. F. G. 2011. *Principles of Social Evolution*. New York: Oxford University Press.

Brown, R. L. 2014. "What Evolvability Really Is." *British Journal for the Philosophy of Science* 65 (3): 549–572.

Buss, L. W. 1987. *The Evolution of Individuality*. Princeton, NJ: Princeton University Press.

Calcott, B., and K. Sterelny. 2011. *The Major Transitions in Evolution Revisited*. Cambridge, MA: MIT Press.

Clarke, E. 2013. "The Multiple Realizability of Biological Individuals." *Journal of Philosophy* 110:413–435.

Clarke, E. 2014. "Origins of Evolutionary Transitions." *Journal of Biosciences* 39 (2): 303–317.

Clarke, E. 2016a. "A Levels-of-Selection Approach to Evolutionary Individuality." *Biology and Philosophy* 31 (6): 893–911.

Clarke, E. 2016b. "Levels of Selection in Biofilms: Multispecies Biofilms Are Not Evolutionary Individuals." *Biology and Philosophy* 31 (2): 191–212.

Cole, B. J. 2009. "The Ecological Setting of Social Evolution: The Demography of Ant Populations." In *Organization of Insect Societies: From Genome to Sociocomplexity*, edited by Juergen R. Gadau and Jennifer H. Fewell. Cambridge, MA: Harvard University Press.

Cronin, A. L., M. Molet, C. Doums, T. Monnin, and C. Peeters. 2013. "Recurrent Evolution of Dependent Colony Foundation across Eusocial Insects." *Annual Review of Entomology* 58 (1): 37–55.

Czaczkes, T. J., C. Grüter, and F. L.W. Ratnieks. 2015. "Trail Pheromones: An Integrative View of Their Role in Social Insect Colony Organization." *Annual Review of Entomology* 60 (1): 581–599.

Duncan, E. J., O. Hyink, and P. K. Dearden. 2016. "Notch Signalling Mediates Reproductive Constraint in the Adult Worker Honeybee." *Nature Communications* 7 (August): 12427.

Fewell, J. H., and R. E. Page Jr. 1999. "The Emergence of Division of Labour in Forced Associations of Normally Solitary Ant Queens." *Evolutionary Ecology Research* 1 (5): 537–548.

Frank, S. A. 2003. "Repression of Competition and the Evolution of Cooperation." *Evolution* 57 (4): 693–705.

Gadagkar, R. 1997. "The Evolution of Caste Polymorphism in Social Insects: Genetic Release Followed by Diversifying Evolution." *Journal of Genetics* 76 (3): 167–179.

Gardner, A., and A. Grafen. 2009. "Capturing the Superorganism: A Formal Theory of Group Adaptation." *Journal of Evolutionary Biology* 22 (4): 659–671.

Gardner, A., and S. A. West. 2006. "Demography, Altruism, and the Benefits of Budding." *Journal of Evolutionary Biology* 19 (5): 1707–1716.

Gempe, T., S. Stach, K. Bienefeld, and M. Beye. 2012. "Mixing of Honeybees with Different Genotypes Affects Individual Worker Behavior and Transcription of Genes in the Neuronal Substrate." *PLoS ONE* 7 (2): e31653.

Gilbert, S. F., J. Sapp, and A. I. Tauber. 2012. "A Symbiotic View of Life: We Have Never Been Individuals." *Quarterly Review of Biology* 87 (4): 325–341.

Gilbert, S. F., and A. I. Tauber. 2016. "Rethinking Individuality: The Dialectics of the Holobiont." *Biology and Philosophy* 31 (6): 839–853.

Godfrey-Smith, P. 2009. *Darwinian Populations and Natural Selection*. New York: Oxford University Press.

Godfrey-Smith, P. 2013. "Darwinian Individuals." In *From Groups to Individuals: Evolution and Emerging Individuality*, edited by P. Bouchard and F. Huneman. Cambridge, MA: MIT Press.

González-Forero, M. 2014. "An Evolutionary Resolution of Manipulation Conflict." *Evolution* 68 (7): 2038–2051.

González-Forero, M. 2015. "Stable Eusociality via Maternal Manipulation When Resistance Is Costless." *Journal of Evolutionary Biology* 28 (12): 2208–2223.

Gordon, D. M. 2013. "The Rewards of Restraint in the Collective Regulation of Foraging by Harvester Ant Colonies." *Nature* 498 (7452): 91–93.

Griesemer, J. R. 1999. "Materials for the Study of Evolutionary Transition." *Biology and Philosophy* 14:127–142.

Hartfelder, K., K. R. Guidugli-Lazzarini, M. S. Cervoni, D. E. Santos, and F. C. Humann. 2015. "Old Threads Make New Tapestry—Rewiring of Signalling Pathways Underlies Caste Phenotypic Plasticity in the Honey Bee, Apis Mellifera L." *Advances in Insect Physiology* 48 (January): 1–36.

Heinze, J. 2008. "The Demise of the Standard Ant." *Myrmecological News* 11:9–20.

Helanterä, H. 2016. "An Organismal Perspective on the Evolution of Insect Societies." *Frontiers in Ecology and Evolution* 4 (February): 1–12.

Holman, L. 2014. "Conditional Helping and Evolutionary Transitions to Eusociality and Cooperative Breeding." *Behavioral Ecology* 25 (5): 1173–1182.

Holman, L., T. A. Linksvayer, and P. d'Ettorre. 2013. "Genetic Constraints on Dishonesty and Caste Dimorphism in an Ant." *The American Naturalist* 181 (2): 161–170.

Hughes, W. O. H., B. P. Oldroyd, M. Beekman, and F. L. W. Ratnieks. 2008. "Ancestral Monogamy Shows Kin Selection Is Key to the Evolution of Eusociality." *Science* 320 (5880): 1213–1216.

Hunt, J. H., B. J. Kensinger, J. A. Kossuth, M. T. Henshaw, K. Norberg, F. Wolschin, and G. V. Amdam. 2007. "A Diapause Pathway Underlies the Gyne Phenotype in Polistes Wasps, Revealing an Evolutionary Route to Caste-Containing Insect Societies." *Proceedings of the National Academy of Sciences USA* 104 (35): 14020–14025.

Huxley, J. 1912. *The Individual in the Animal Kingdom.* Cambridge: Cambridge University Press.

Jasper, W. C., T. A. Linksvayer, J. Atallah, D. Friedman, J. C. Chiu, and B. R. Johnson. 2015. "Large-Scale Coding Sequence Change Underlies the Evolution of Postdevelopmental Novelty in Honey Bees." *Molecular Biology and Evolution* 32 (2): 334–346.

Johnson, B. R., and T. A. Linksvayer. 2010. "Deconstructing the Superorganism: Social Physiology, Groundplans, and Sociogenomics." *Quarterly Review of Biology* 85 (1): 57–79.

Kapheim, K. M., H. Pan, C. Li, S. L. Salzberg, D. Puiu, T. Magoc, H. M. Robertson, et al. 2015. "Genomic Signatures of Evolutionary Transitions from Solitary to Group Living." *Science* 348 (6239): 1139–1143.

Kennedy, P., T. Uller, and H. Helanterä. 2014. "Are Ant Supercolonies Crucibles of a New Major Transition in Evolution?" *Journal of Evolutionary Biology* 27 (9): 1784–1796.

Kirschner, M. W., and J. C. Gerhart. 2005. *The Plausibility of Life: Resolving Darwin's Dilemma.* New Haven, CT: Yale University Press.

Kolenbrander, P. E., R. J. Palmer, S. Periasamy, and N. S. Jakubovics. 2010. "Oral Multispecies Biofilm Development and the Key Role of Cell–Cell Distance." *Nature Reviews Microbiology* 8 (7): 471–480.

Laubichler, M. D., and J. Renn. 2015. "Extended Evolution: A Conceptual Framework for Integrating Regulatory Networks and Niche Construction." *Journal of Experimental Zoology Part B: Molecular and Developmental Evolution* 324 (7): 565–577.

Leslie, M. P., D. E. Shelton, and R. E. Michod. 2017. "Generation Time and Fitness Tradeoffs during the Evolution of Multicellularity." *Journal of Theoretical Biology* 430 (October): 92–102.

Lewontin, R C. 1970. "The Units of Selection." *Annual Review of Ecology and Systematics* 1 (1): 1–18.

Libbrecht, R., and L. Keller. 2013. "Genetic Compatibility Affects Division of Labor in the Argentine Ant Linepithema Humile." *Evolution* 67 (2): 517–524.

Libby, E., P. L. Conlin, B. Kerr, and W. C. Ratcliff. 2016. "Stabilizing Multicellularity through Ratcheting." *Philosophical Transactions of the Royal Society B: Biological Sciences* 371 (1701): 20150444.

Linksvayer, T. A. 2006. "Direct, Maternal, and Sibsocial Genetic Effects on Individual and Colony Traits in an Ant." *Evolution* 60 (12): 2552–2561.

Linksvayer, T. A. 2007. "Ant Species Differences Determined by Epistasis between Brood and Worker Genomes." *PLoS ONE* 2:e994.

Linksvayer, T. A. 2008. "Queen-Worker-Brood Coadaptation rather than Conflict May Drive Colony Resource Allocation in the Ant Temnothorax Curvispinosus." *Behavioral Ecology and Sociobiology* 62 (5): 647–657.

Linksvayer, T. A. 2015. "The Molecular and Evolutionary Genetic Implications of Being Truly Social for the Social Insects." *Advances in Insect Physiology* 48:271–292.

Linksvayer, T. A., J. H. Fewell, J. Gadau, and M. D. Laubichler. 2012. "Developmental Evolution in Social Insects: Regulatory Networks from Genes to Societies." *Journal of Experimental Zoology Part B: Molecular and Developmental Evolution* 318 (3): 159–169.

Linksvayer, T.A., M. K. Fondrk, and R. E. Page. 2009. "Honeybee Social Regulatory Networks Are Shaped by Colony-Level Selection." *The American Naturalist* 173: E99–E2017.

Linksvayer, T. A., and M. J. Wade. 2005. "The Evolutionary Origin and Elaboration of Sociality in the Aculeate Hymenoptera: Maternal Effects, Sib-social Effects, and Heterochrony." *The Quarterly Review of Biology* 80 (3): 317–336.

Malka, O., E. L. Niño, C. M. Grozinger, and A. Hefetz. 2014. "Genomic Analysis of the Interactions between Social Environment and Social Communication Systems in Honey Bees (Apis Mellifera)." *Insect Biochemistry and Molecular Biology* 47 (April): 36–45.

Maynard Smith, J., and E. Szathmáry. 1995. *The Major Transitions in Evolution.* New York: Oxford University Press.

McGlothlin, J. W., A. J. Moore, J. B. Wolf, and E. D. Brodie III. 2010. "Interacting Phenotypes and the Evolutionary Process. Iii. Social Evolution." *Evolution* 64 (9): 2558–2574.

McGlothlin, J. W., J. B. Wolf, E. D. Brodie, and A. J. Moore. 2014. "Quantitative Genetic Versions of Hamilton's Rule with Empirical Applications." *Philosophical Transactions of the Royal Society of London. Series B, Biological Sciences* 369 (1642): 20130358.

Merlo, L. M. F., J. W. Pepper, B. J. Reid, and C. C. Maley. 2006. "Cancer as an Evolutionary and Ecological Process." *Nature Reviews Cancer* 6 (12): 924–935.

Michod, R. E. 1999. *Darwinian Dynamics: Evolutionary Transitions in Fitness and Individuality.* Princeton, NJ: Princeton University Press.

Michod, R. E., A. M. Nedelcu, and D. Roze. 2003. "Cooperation and Conflict in the Evolution of Individuality: IV. Conflict Mediation and Evolvability in Volvox Carteri." *BioSystems* 69 (2–3): 95–114.

Michod, R. E., and D. Roze. 1997. "Transitions in Individuality." *Proceedings of the Royal Society B: Biological Sciences* 264 (1383): 853–857.

Möglich, M., U. Maschwitz, and B. Hölldobler. 1974. "Tandem Calling: A New Kind of Signal in Ant Communication." *Science* 186 (4168): 1046–1047.

Molet, M., D. E. Wheeler, and C. Peeters. 2012. "Evolution of Novel Mosaic Castes in Ants: Modularity, Phenotypic Plasticity, and Colonial Buffering." *The American Naturalist* 180 (3): 328–341.

Moore, A. J., E. D. Brodie, and J. B. Wolf. 1997. "Interacting Phenotypes and the Evolutionary Process: I. Direct and Indirect Genetic Effects of Social Interactions." *Evolution* 51 (5): 1352–1362.

Nadell, C. D., J. B. Xavier, and K. R. Foster. 2009. "The Sociobiology of Biofilms." *FEMS Microbiology Reviews* 33 (1): 206–224.

Nipitwattanaphon, M., J. Wang, M. B. Dijkstra, and L. Keller. 2013. "A Simple Genetic Basis for Complex Social Behaviour Mediates Widespread Gene Expression Differences." *Molecular Ecology* 22 (14): 3797–3813.

Oster, G. F., and E. O. Wilson. 1978. *Caste and Ecology in the Social Insects.* Princeton, NJ: Princeton University Press.

Page, R. E., and M. K. Fondrk. 1995. "The Effects of Colony-Level Selection on the Social Organization of Honey Bee (Apis Mellifera L.) Colonies: Colony-Level Components of Pollen Hoarding." *Behavioral Ecology and Sociobiology* 36 (2): 135–144.

Page, R. E. Jr., T. A. Linksvayer, and G. V. Amdam. 2009. "Social Life from Solitary Regulatory Networks: A Paradigm for Insect Sociality." In *Organization of Insect Societies from Genome to Sociocomplexity,* edited by Juergen R Gadau and Jennifer H Fewell, 355–374. Cambridge, MA: Harvard University Press.

Pearse, A.-M., and K. Swift. 2006. "Transmission of Devil Facial-Tumour Disease." *Nature* 439 (7076): 549–549.

Pennell, T., L. Holman, E. H. Morrow, and J. Field. 2018. "Building a New Research Framework for Social Evolution: Intralocus Caste Antagonism." *Biological Reviews* 93:1251–1268.

Pigliucci, M. 2008. "Is Evolvability Evolvable?" *Nature Reviews Genetics* 9 (1): 75–82.

Poltak, S. R., and V. S. Cooper. 2011. "Ecological Succession in Long-Term Experimentally Evolved Biofilms Produces Synergistic Communities." *The ISME Journal* 5 (3): 369–378.

Powers, S. T., A. S. Penn, and R. A. Watson. 2011. "The Concurrent Evolution of Cooperation and the Population Structures That Support It." *Evolution* 65 (6): 1527–1543.

Queller, D. C., and J. E. Strassmann. 2016. "Problems of Multi-Species Organisms: Endosymbionts to Holobionts." *Biology and Philosophy* 31 (6): 855–873.

Rainey, P. B., and K. Rainey. 2003. "Evolution of Cooperation and Conflict in Experimental Bacterial Populations." *Nature* 425 (6953): 72–74.

Rajakumar, R., D. San Mauro, M. B. Dijkstra, M. H. Huang, D. E. Wheeler, F. Hiou-Tim, A. Khila, M. Cournoyea, and E. Abouheif. 2012. "Ancestral Developmental Potential Facilitates Parallel Evolution in Ants." *Science* 335:79–82.

Ratnieks, F. L. W., K. R. Foster, and T. Wenseleers. 2006. "Conflict Resolution in Insect Societies." *Annual Review of Entomology* 51 (1): 581–608.

Ratnieks, F. L. W., and T. Wenseleers. 2008. "Altruism in Insect Societies and Beyond: Voluntary or Enforced?" *Trends in Ecology and Evolution* 23 (1): 45–52.

Rehan, S. M., and A. L. Toth. 2015. "Climbing the Social Ladder: The Molecular Evolution of Sociality." *Trends in Ecology and Evolution* 30 (7): 426–433.

Romiguier, J., J. Lourenco, P. Gayral, N. Faivre, L. A. Weinert, S. Ravel, M. Ballenghien, et al. 2014. "Population Genomics of Eusocial Insects: The Costs of a Vertebrate-like Effective Population Size." *Journal of Evolutionary Biology* 27 (3): 593–603.

Ryan, P. A., S. T. Powers, and R. A. Watson. 2016. "Social Niche Construction and Evolutionary Transitions in Individuality." *Biology and Philosophy* 31 (1): 59–79.

Salazar-Ciudad, I., and J. Jernvall. 2010. "A Computational Model of Teeth and the Developmental Origins of Morphological Variation." *Nature* 464 (7288): 583.

Salazar-Ciudad, I., J. Jernvall, and S. A. Newman. 2003. "Mechanisms of Pattern Formation in Development and Evolution." *Development* 130 (10): 2027–2037.

Schwander, T., and L. Keller. 2008. "Genetic Compatibility Affects Queen and Worker Caste Determination." *Science* 322 (5901): 552

Seeley, T. D. 1989. "The Honey Bee Colony as a Superorganism." *American Scientist* 77:546–553.

Seeley, T. D. 1995. *The Wisdom of the Hive: The Social Physiology of Honey Bee Colonies*. Cambridge, MA: Harvard University Press.

Sober, E. 1984. *The Nature of Selection: Evolutionary Theory in Philosophical Focus*. Chicago: University of Chicago Press.

Sobotka, J. A., M. Daley, S. Chandrasekaran, B. D. Rubin, and G. J. Thompson. 2016. "Structure and Function of Gene Regulatory Networks Associated with Worker Sterility in Honeybees." *Ecology and Evolution* 6 (6): 1692–1701.

Sterelny, K. 2011. "Evolvability Reconsidered." In *The Major Transitions in Evolution Revisited*, edited by B. Calcott and K. Sterelny. Cambridge, MA: MIT Press.

Sterelny, K., R. Joyce, B. Calcott, and B. Fraser. 2013. *Cooperation and Its Evolution*. Cambridge, MA: MIT Press.

Strassmann, J. E., and D. C. Queller. 2007. "Insect Societies as Divided Organisms: The Complexities of Purpose and Cross-Purpose." *Proceedings of the National Academy of Sciences of the United States of America* 104 Suppl (June): 8619–8626.

Szathmáry, E., and J. Maynard Smith. 1995. "The Major Evolutionary Transitions." *Nature* 374 (6519): 227–232.

Taylor, D., M. A. Bentley, and S. Sumner. 2018. "Social Wasps as Models to Study the Major Evolutionary Transition to Superorganismality." *Current Opinion in Insect Science* 28 (August): 26–32.

Taylor, D. R., C. Zeyl, and E. Cooke. 2002. "Conflicting Levels of Selection in the Accumulation of Mitochondrial Defects in Saccharomyces Cerevisiae." *Proceedings of the National Academy of Sciences of the United States of America* 99 (6): 3690–3694.

Teseo, S., N. Châline, P. Jaisson, and D. J. C. Kronauer. 2014. "Epistasis between Adults and Larvae Underlies Caste Fate and Fitness in a Clonal Ant." *Nature Communications* 5 (February): 3363.

Toth, A. L., and G. E. Robinson. 2007. "Evo-Devo and the Evolution of Social Behavior." *Trends in Genetics* 23 (7): 334–341.

Trestman, M. 2012. "Which Comes First in Major Transitions: The Behavioral Chicken, or the Evolutionary Egg?" *Biological Theory* 7 (1): 48–55.

Tsuji, K. 1995. "Reproductive Conflicts and Levels of Selection in the Ant Pristomyrmex Pungens: Contextual Analysis and Partitioning of Covariance." *The American Naturalist* 146 (4): 586–607.

Uller, T., A. P. Moczek, R. A. Watson, P. M. Brakefield, and K. N. Laland. 2018. "Developmental Bias and Evolution: A Regulatory Network Perspective." *Genetics* 209 (4): 949–966.

Van Dyken, J. D., and M. J. Wade. 2012. "Origins of Altruism Diversity II: Runaway Coevolution of Altruistic Strategies via 'Reciprocal Niche Construction.'" *Evolution: International Journal of Organic Evolution* 66 (8): 2498–2513.

Vitikainen, E. I. K., C. Haag-Liautard, and L. Sundström. 2015. "Natal Dispersal, Mating Patterns, and Inbreeding in the Ant *Formica exsecta.*" *The American Naturalist* 186:716–727.

Vojvodic, S., B. R. Johnson, B. A. Harpur, C. F. Kent, A. Zayed, K. E. Anderson, and T. A. Linksvayer. 2015. "The Transcriptomic and Evolutionary Signature of Social Interactions Regulating Honey Bee Caste Development." *Ecology and Evolution* 5 (21): 4795–4807.

Wade, M. J. 1977. "An Experimental Study of Group Selection." *Evolution* 31 (1): 134–153.

Wagner, A. 2011. *The Origins of Evolutionary Innovations: A Theory of Transformative Change in Living Systems.* Oxford: Oxford University Press.

Watson, R. A., and E. Szathmáry. 2016. "How Can Evolution Learn?" *Trends in Ecology and Evolution* 31 (2): 147–157.

Weismann, A. 1893. "The All-Sufficiency of Natural Selection. A Reply to Herbert Spencer." *The Contemporary Review* 64:309–338.

Wenseleers, T., H. Helanterä, A. Hart, and F. L. W. Ratnieks. 2004. "Worker Reproduction and Policing in Insect Societies: An ESS Analysis." *Journal of Evolutionary Biology* 17 (5): 1035–1047.

West-Eberhard, M. J. 1987. "Flexible Strategy and Social Evolution." In *Animal Societies: Theories and Facts*, edited by Y. Ito, J. L. Brown, and J. Kikkawa, 35–51. Tokyo: Japan Scientific Societies Press.

Wheeler, D. E. 1986. "Developmental and Physiological Determinants of Caste in Social Hymenoptera: Evolutionary Implications." *The American Naturalist* 128 (1): 13–34.

Wheeler, W. M. 1911. "The Ant-Colony as an Organism." *Journal of Morphology* 22 (2): 307–325.

Wilson, D. S., and E. Sober. 1989. "Reviving the Superorganism." *Journal of Theoretical Biology* 136:337–356.

Woodard, S. H., B. J. Fischman, A. Venkat, M. E. Hudson, K. Varala, S. A. Cameron, A. G. Clark, and G. E. Robinson. 2011. "Genes Involved in Convergent Evolution of Eusociality in Bees." *Proceedings of the National Academy of Sciences USA* 108 (18): 7472–7477.

Yang, A. S. 2007. "Thinking Outside the Embryo: The Superorganism as a Model for EvoDevo Studies." *Biological Theory* 2 (4): 398–408.

10 Are Developmental Plasticity, Niche Construction, and Extended Inheritance Necessary for Evolution by Natural Selection? The Role of Active Phenotypes in the Minimal Criteria for Darwinian Individuality

Richard A. Watson and Christoph Thies

1. Evolution and "Extensions"

The process of evolution by natural selection can be characterized by three core mechanisms: variation, selection, and inheritance (Lewontin 1970; Godfrey-Smith 2007; Watson and Szathmáry 2016). Models of biological evolution often adopt several simplifying assumptions in their treatment of these mechanisms: that inheritance occurs by gene transmission only, that variation is genetic or genetically determined, and that selection is an exogenous filtering process. We refer to this set of simplifications as the "standard model" (table 10.1). In this simplification, variation, selection, and inheritance are treated as fixed and causally autonomous mechanisms—unaffected by the products (phenotypes) of the evolutionary process (Watson and Szathmáry 2016; Uller and Helanterä 2017; Walsh 2015).

Of course, biology is rarely, if ever, this simple. Organisms exhibit extended inheritance (e.g., epigenetic inheritance, parental effects, niche inheritance), environmentally sensitive phenotypic plasticity, and behaviors that modify or construct the selective conditions they experience. The standard model abstracts- and idealizes-away these influences of phenotypes on the mechanisms of evolution. This allows phenotypes to be backgrounded and play no active part in evolutionary change (except in so much as they mediate the fixed mapping between genotypes and fitness). Some researchers advocate the need for increased attention to mechanisms that operate outside these simplifying assumptions, where the phenotype and its interaction with the environment plays a more active and constructive role in the processes of variation, selection, and inheritance (table 10.1) (Pigliucci 2009; Laland et al. 2014; Laland et al. 2015, and references therein). We refer to the phenotypes in this extended model as "active phenotypes."

Although the biology involved in these extensions is not controversial, there is some considerable disagreement in how to interpret their significance (Laland et al. 2011; West, Mouden, and Gardner 2011; Dickins and Rahman 2012; Laland et al. 2013; Scott-Phillips et al. 2014; Laland et al. 2015; Charlesworth, Barton, and Charlesworth 2017; Futuyma 2017). It might be tempting to advocate the importance of the extensions, and the

Table 10.1.
Standard and extended models of variation, selection, and inheritance.

Standard model: passive phenotypes (Variational emphasis)*	**Extensions:** active phenotypes (Transformational emphasis)*
Variation: Variation that is evolutionarily relevant is genetic (or variation in phenotype is genetically determined).	**Phenotypic plasticity:** Phenotypic variation is environmentally sensitive.
Selection: Selection is a passive (genotype-) filtering process.	**Niche construction:** Salient selective criteria arise from how the phenotype interacts with, and modifies, the environment.
Inheritance: Inheritance is carried out by genetic transmission, unaffected by environment or phenotype.	**Extended inheritance:** Inheritance includes extra-genetic factors—epigenetics, parental effects, niche-inheritance, etc.

*The passive and active views parallel the variational and transformational accounts of evolutionary change (Lewontin 1983; Mayr 1989), respectively, as we will discuss later.

inadequacy of the standard model, simply on the grounds that the extensions violate the assumptions of the standard model. But it is also possible to view the standard model as a starting point that is robust to such "add-ons." Questions that lie outside those simplifying assumptions simply motivate more complicated treatments as and when required (Futuyma 2017). Examples exist where each of these complications is addressed in an otherwise standard theoretical approach (e.g., Cavalli-Sforza and Feldman 1981; Via and Lande 1985; Gomulkiewicz and Kirkpatrick 1992; Laland, Odling-Smee, and Feldman 1999; Lande 2009; Tal, Kisdi, and Jablonka 2010; Day and Bonduriansky 2011; Gomez-Mestre and Jovani 2013; Uller, English, and Pen 2015). After all, since the reality of natural biology is undeniably complicated it seems practical, as in any science, to treat some details as standard and others as optional extras or "add-ons." Thus, while they might be important in some specific cases, it is not clear that these extensions change how evolution works in any fundamental way (Dickins and Rahman 2012; Scott-Phillips et al. 2014; Wray et al. 2014; Futuyma 2017). Crucially, if the mechanisms of the standard model are necessary and sufficient conditions for evolution by natural selection then the extensions seem to be, at best, "optional extras." The mechanisms of the standard model are thus assumed to be causally prior, and extension mechanisms (if they are of any adaptive relevance) derivative products of the standard model.

In this chapter we question the validity of this view in light of the major evolutionary transitions (Maynard Smith and Szathmáry 1995). In the major evolutionary transitions, or evolutionary transitions in individuality, new evolutionary units are created at successively higher levels of biological organization (Maynard Smith and Szathmáry 1995; Michod 2000; Okasha 2006; Bourke 2011; Godfrey-Smith 2011, 2013; Buss 2014; Clarke 2014; Szathmáry 2015; West et al. 2015; Watson and Szathmáry 2016). Examples include the transitions from individual self-replicating molecules to chromosomes, from simple prokaryotes to eukaryotes with organelles, from single-celled life to multicellular organisms, and from solitary organisms to social colonies or super-organisms. In these transitions, the mechanisms of variation, selection, and inheritance at the lower level of organization can be different from those at the higher level. Consider the transition from unicellular to multicellular life. The inheritance, variation, and selection of the cells within the multicellular organisms are all nonstandard (table 10.1). Cells exhibit plasticity (e.g., cell differentiation and context-specific growth), extra-genetic inheritance (e.g., epigenetic inheritance of cell types), and niche construction (e.g., cells construct the local, intra-organismic, environment that they experience, which then determines differential cell reproduction or tissue growth). We want to explore whether this is a contingent fact of biology or logically essential for the origination and maintenance of Darwinian individuality at all levels of biological organization.

The transitions lead us to explore a line of argument arriving at a hypothesis contrary to the standard position: that active phenotypes are not (optional) extensions to the standard model but rather that they are necessary to instantiate the standard model in the first place (figure 10.1). This view sees the active roles of phenotypes as causally prior and the "standardization" of variation, selection, and inheritance mechanisms as a derivative, special case. This is clearly related to "phenotype-first" ideas (Müller and Newman 2003; Newman, Forgacs, and Müller 2003; West-Eberhard 2003; Jablonka 2006; Newman and Bhat 2008; Schwander and Leimar 2011; Uller and Helanterä 2011), but these previous works do not suggest that active phenotypes are necessary to instantiate Darwinian individuality.

We begin from the position that a necessary condition for Darwinian individuality is that an evolutionary unit exhibits heritable variation in fitness *over and above those of its component parts*. We describe the requirements for a putative individual to exhibit this "more than the sum of its parts" property in a formal sense. We find that it can only exhibit heritable fitness differences at the new level if its sub-units have active phenotypes. Thus, while the standard model is a description of a well-defined evolutionary unit, it cannot address how an evolutionary unit originated or how it is maintained. The evolutionary transitions therefore suggest a much more complicated and equitable relationship between the standard and active mechanisms, and their causal priority in evolutionary processes. This chapter is an informal exposition of an argument that is, at present, a work in progress (mathematical analyses are in development). We briefly discuss how learning theory might be used to formalize this relationship.

2. The Evolutionary Transitions in Individuality

In each of the evolutionary transitions in individuality "entities that were capable of independent reproduction before the transitions can replicate only as part of a larger whole after the transition" (Maynard Smith and Szathmáry 1995). In some cases, it is tempting

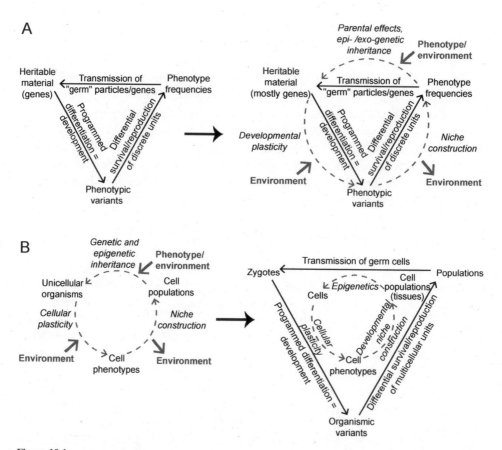

Figure 10.1
Views on the relationship between the mechanisms of the standard model and the roles of active phenotypes.
(A) The standard model supports the evolution of active phenotypes. Left: The standard model of evolution containing inheritance, variation, and selection (gene-only inheritance, genetically determined variation, passive selection on discrete units). Right: Prevailing view of active phenotypes as "extensions" or "add-ons" to the standard model, introducing interactions with phenotypes and the environment that complicate and possibly compromise the mechanisms of the standard model (e.g., hard inheritance is "spoiled" by parental effects/ ecological inheritance, programmed development "spoiled" by environmentally induced developmental plasticity, and selection viewed as a passive filter is "spoiled" by niche construction). (B) An alternative view: Active phenotypes at a lower level of organization support and enable the standard model at a higher level of organization. For example, Left: Unicellular organisms naturally inherit both genes and intra-cellular material, have phenotypes that modify and are modified by their local environment. Right: In multicellular organisms, each cell experiences and modifies the "environment" (i.e., the organism they are in, co-constructed with other cells). Epigenetic inheritance of cell states is necessary to support (nongenetic) inheritance of the pluripotent cell state enabling germ/soma separation, cellular plasticity is necessary for predictable cell differentiation enabling "programmed" development of the multicellular phenotype, and niche-construction of shared reproductive constraints binds the selective fate of cells together so that selection acts holistically at the level of the organism. In this view the active roles of phenotypes are necessary to originate and maintain the multicellular organism as an evolutionary unit. This "standardizes" the collective as an evolutionary unit (i.e., enabling it to exhibit heritable variation in fitness that is more than the sum of its parts).

to refer to "individuals" forming "groups" in the transitions, but to be more general, it is conventional to use the terms "particle" and "collective" (Okasha 2006).

Although the biology of transitions in individuality is well studied, and they are central in the origin of biological complexity (Maynard Smith and Szathmáry 1995), accommodating the transitions into conventional evolutionary theory is problematic (Okasha 2006; Watson and Szathamáry 2016). A key problem is that conventional theory takes the mechanisms of variation, selection, and inheritance as axiomatic assumptions—and thus assumes them to be fixed mechanisms that are not modified by the products of the evolutionary process. In the transitions, however, the mechanisms of variation, selection, and inheritance are repeatedly reinstantiated at a new level of biological organization (Maynard Smith and Szathmáry 1995; Watson and Szathmáry 2016).

In order for a new level of biological organization to take a meaningful causal role as an evolutionary unit, evolutionary outcomes of the collective must not be simply summary statistics of the lower-level units they contain. A crucial characteristic of the transitions is to consider the Darwinian individuality of the new unit (Clarke 2014), that is, the extent to which a collective exhibits heritable variation in fitness over and above the heritable variation in fitness of the particles it contains.

Godfrey-Smith (2011) advocates a continuous space of Darwinian individuality (rather than a categorical concept) that links marginal with paradigmatic instances of individuality along several dimensions. Three particular dimensions of this space, which feature prominently in his framework, are defined by:

a) The degree of functional integration of the components (e.g., functionally heterogeneous particles exploiting division of labor) that makes the function of the collective more than the sum of the parts it contains. This is important to create fitness benefits that belong to the collective level.

b) The degree or severity of a genetic bottleneck (e.g., as created by reproduction through a single-cell propagule, or a single individual in a social colony) creates genetically homogeneous collectives and thus removes fitness differences between the particles in the same collective.

c) The degree of reproductive specialization (e.g., early determination of the germ/soma separation in multicellular organisms). This is both a particularly profound type of functional integration (a specific type of division of labor) and a mechanism that involves a degree of genetic bottleneck (specific particles pass through the population bottleneck). Moreover, without reproductive specialization, individual particles can increase their likelihood of surviving the bottleneck "lottery" by proliferating within the collective. When the germline is determined, the incentive for particles to compete for positions in the bottleneck is removed. Consequently, particles can only increase their fitness by cooperating to increase the fitness of the collective (through the germline).

None of these dimensions is considered definitive for individuality and each may be present by degree (Godfrey-Smith 2011; see also Helanterä and Uller, this volume). While Lewontin's core mechanisms (variation, selection, and inheritance) describe the logical properties necessary for an evolutionary unit, Godfrey-Smith's dimensions describe biological properties that change the level of organization to which they apply. Specifically, movements in these dimensions constitute changes in the reproductive, developmental, and variational properties of the collective that enable it to express heritable variation in fitness at the collective level, referred to as the "Darwinization" of the collective (Godfrey-Smith 2011). This is accompanied by the "de-Darwinization" of the particles it contains—that is, the suppression or removal of heritable variation in fitness among particles of the same collective (e.g., by genetic homogeneity of cells in multicellular organisms).

These three dimensions are directly relevant to the argument we will develop. In each case notice that the biological property cannot be a property of a particle but changes in this dimension may nonetheless be motivated by benefits to particles. Specifically, functional integration implies that the fitness of a particle does not derive from its own properties but rather from the complementarity of its properties to those of the other particles with which it interacts. An increase in this dimension may nonetheless result from particle-level evolution if it is adaptive for a particle to control which other particles it interacts with, to influence the phenotypes of other particles or to adopt a phenotype that is sensitive to the phenotypes of others. Similarly, a genetic bottleneck is not a property of an individual particle, but an increase in the severity of the bottleneck may arise if it is adaptive for an individual particle to increase an ability to control population structure, such as propagule size or the severity of a repeated population bottleneck. Likewise, reproductive specialization is not a particle-level property, but an increase in reproductive specialization may arise if it is adaptive for a particle to have an increased ability to sustain a pluripotent state and induce a differentiated, nonreproductive state in other particles, or vice versa.

In the following sections we expand on these concepts, focusing on the minimal necessary and sufficient conditions for an evolutionary unit to be more than the sum of its parts in this Darwinian sense. We formalize these criteria for individuality following Clarke (2014) in seeking heritable variation in fitness at the collective level that is not attributable to the heritable variation in fitness exhibited by the particles it contains. To do so we start by formalizing Godfrey-Smith's notion of functional integration for a minimal (two particle) system using a game-theoretic approach: specifically, we analyze which specific two-player games create components of fitness that are not attributable to the fitnesses of the individual players. We then focus on the properties of particles required for them to exploit these components of fitness in this type of game.

3.   Necessary Properties of Particles to Support Individuality at the Collective Level

We aim to describe the minimal necessary and sufficient properties for an evolutionary unit to be more than the sum of its parts in a formal sense. We thus seek to identify the minimal properties of particles that support individuality at the collective level, and the relationship of these properties to plasticity, niche construction, and extended inheritance.

3.1   Collective Fitness Requires Complementary Particle Functions

We use the term *collective fitness* to refer to fitness at the collective level that is not attributable to fitness at the particle level (Clarke 2014). What is required to produce collective fitness?

To understand the problem, consider first a deficient example: suppose that the height of a collective is the average height of the particles it contains and that the fitness of a collective is proportional to its height. Such collectives cannot exhibit collective fitness because although collectives may have different compositions and thus different heights and different fitnesses, the variance in collective fitness is fully explained by variance in particle fitness (Okasha 2006).

Some sort of nonlinear interaction, or *synergy* (Corning and Szathmáry 2015), is necessary to avoid this. However, synergy is necessary but not sufficient to avoid a correlation between individual characters and collective fitness (Hammerschmidt et al. 2014; Pichugin 2015). If, for example, collective fitness was the square of the average height, this would be nonlinear but particle fitness is still positively correlated with particle character. Whenever increasing some character of the particle has the reliable effect of increasing the fitness of the collective, even if it is a non-additive effect, particle character will inevitably be correlated with particle fitness. In this case, the direction of change in character frequencies is still explained by the rankings of particle fitnesses (although the rate of change may vary). Therefore, heritable variation in the fitness of the collective is explanatorily redundant in determining which particle character is favored by selection (this is the problem of *cross-level by-products* [Okasha 2006]). Synergies that amount to *economies of scale*,[1] where collective fitness is a super-linear (but monotonic) function of average particle character, are thus not sufficient for generating collective fitness.

How can it be the case that the properties of collectives are entailed by the properties of the particles they contain and yet it not be the case that increasing some character of the particle has the reliable effect of increasing the fitness of the collective? Logically, this requires that increasing some character of the particle will sometimes increase the fitness of the collective and sometimes decrease the fitness of the collective. Whether it increases or decreases collective fitness must depend on what other particles are present.

In this way, and only in this way, the fitness of a collective will depend on the particle characters it contains (and may confer high fitness on the particles it contains) but there will not be any particle character that reliably confers increased particle fitness. There is therefore no particle character that explains particle fitness (independent of its context) but there is nonetheless a *collective* character that cannot be reduced to the character of individual particles, that confers (collective fitness and hence) particle fitness.

Intuitively, this requires fitness that arises from the interactions between particles, not from individual particle properties. Such a collective character must depend on the specific *combination* of particle characters it contains such as the heterogeneity or homogeneity, complementarity, coordination, or other more general spatial or functional *organization* of the particles within the collective. Changing the character of a particle will thus have not just a nonlinear effect on collective fitness but an effect on collective fitness that can change sign (otherwise particle character will inevitably carry a signal of particle fitness).[2] We refer to this as "sign synergy"—that is, a given change in particle character can either increase or decrease collective fitness (and hence particle fitness) depending on the context of other particle characters in the collective.[3]

Simple two-player, two-strategy games are sufficient to help us think about such interactions more clearly. Although the Prisoner's Dilemma is a popular choice for modeling social dilemmas, it does not exhibit sign synergy and therefore cannot be the basis of collective fitness. Games that exhibit sign synergy are one of only two types: coordination games and anti-coordination games. In coordination games high-fitness collectives contain homogeneous particles (e.g., AA or BB). Whereas, in anti-coordination games high-fitness collectives contain heterogeneous particles (e.g., AB or BA).[4] The latter type includes scenarios where a benefit is provided by avoiding unfavorable trade-offs between functions (Michod 2006; Ispolatov, Ackermann, and Doebeli 2012; Rueffler, Hermisson, and Wagner 2012) and can be formalized as *division of labor* games (Tudge, Watson, and Brede 2016). Both games can be described as the problem of finding the types that are *complementary;* meaning the types that "go well together." This concept of complementarity applies equally well to asymmetric (e.g., host–parasite) relationships as it does to symmetric relationships (e.g., non-trophic competition).[5]

There are several reasons to believe that heterogeneous collectives are more relevant to evolutionary individuality than homogeneous collectives. Although economies of scale (where high numbers of homogeneous particles are favored) are suggested as the selective impetus for some transitions (Queller 1997), they are not sufficient to create *sign* synergy, as noted above.[6] Homogeneous collectives cannot exhibit division of labor, central to the functional integration and reproductive specialization dimensions of Godfrey-Smith. Even though multicellular organisms may be *genetically* homogeneous, interesting examples have cell *phenotypes* and functions that are heterogeneous. Variability in multicellular morphology arises mostly from symmetry breaking in the placement, timing, abundance, or control of cells (Gilbert and Epel 2009), whereas a homogeneous collective cannot employ such mechanisms by definition. Finally, from a somewhat more pragmatic point of view, homogeneous collectives suffer the fact that collective selection, variation and inheritance can be easily conflated with particle selection, variation and inheritance.[7]

In sum, although the role of homogeneous collectives in individuality cannot be logically dismissed (and may be particularly relevant in the early stages of transitions) it seems to us that homogeneous collectives are limited in how far they can take us in a transition to individuality. Accordingly, in what follows we focus on anti-coordination or division of labor games to capture individuality of the collective.

3.2 Heritable Collective Fitness Requires Particle Plasticity

Solving a division of labor game, to create fitness differences at the collective level, requires heterogeneous collectives. However, fitness differences among the replicating particles that a collective contains (if persistent over collective-reproduction timescales) will change the composition of the collective and thus reduce its heritability (Okasha 2006). Accordingly, to preserve the heritability of the collective, particles must not be different from one another in fitness (Clarke 2014). This is the de-Darwinization of particles that coincides with the Darwinization of the collective (Godfrey-Smith 2011). Thus the different functional roles, or phenotypes, of particles (although necessary for collective fitness) must not result in the differential reproduction of those phenotypes in the long term.

This implies that it cannot be the functionally salient phenotype of the particles that determines their fitness—it must be something common to all particles within the collective. There appear to be two possibilities to solving this "heterogeneous function with homogeneous fitness" problem: (i) Something internal or intrinsic to the particles (not their phenotypes) determines their reproductive rate and this is the same for all of the particles in the collective[8]; or (ii) Something external or extrinsic to the particles (not their phenotypes) determines their reproductive rate and this is the same for all of the particles in the collective. Either way it is not the phenotypes of the particles that determines their fitness. These two logical possibilities are well represented in natural systems:

a) Fraternal transitions (Queller 1997), for example, transition to multicellularity. The reproductive rate of particles is controlled by their genotype (rather than phenotype), and all particles in the collective have the same genotype. In this case the particles of the collectives are phenotypically heterogeneous but genotypically homogeneous—as in, for example, many multi-cellular organisms. Particles therefore have the same fitness.[9]

b) Egalitarian transitions (Queller 1997), for example, transition to eukaryote cells with organelles. There are various mechanisms by which particle reproduction may be controlled by factors external to the particle (but internal to the collective). Examples include a shared public good or resource (fairly distributed/not individually exploitable), centrally controlled reproductive machinery (fair meiosis), (mutual) policing norms (e.g., nuclear control of mitochondrial reproduction). Here the property that controls particle fitness is, one way or another, "being in that collective" rather than anything specific to the particle. Fitness/reproduction of particles is not determined by their intrinsic properties (phenotypes or genotypes) but by properties of the whole.

There are advantages and disadvantages of each solution to individuality. The fraternal solution admits reproductive specialization: when particles have the same genotypes, reproducing any of them increases the fitness of that genotype. So if one particle has a phenotype that reproduces more than the others, or is otherwise biased in its ability to pass through a reproductive bottleneck, this does not undermine individuality because it is every particle's genotype (i.e., intrinsic properties of the particle that are common to all particles in the collective) that is getting reproduced (i.e., although their direct fitnesses may be different, they all have the same inclusive fitness). Examples include germ-soma separation in multicellular organisms and reproductive castes in eusocial insects. However, the fraternal solution creates a new problem—how can all the particles in the collective be intrinsically the same with respect to fitness but different with respect to their functional roles? This requires some way of separating what a particle is from what it does, that is, separating what the heritable material is from what the heritable material has the potential to create (i.e., the distinction between genotype and phenotype). For example, in multicellular organisms, cells are genetically homogeneous but cell phenotypes differentiate from one another during development. This requires particles that have plastic phenotypes such that the same particle can take one of several functional roles.[10]

Conversely, the egalitarian transition does not need particles that exhibit this phenotypic plasticity because each of the particles is intrinsically different (e.g., different genotypes and different phenotypes). For example, each gene in a chromosome produces different gene products, or each organelle of a eukaryote cell carries out different functions. However, egalitarian collectives cannot therefore exploit reproductive division of labor (they must all be reproduced in order to reproduce the collective), and the particles of egalitarian transitions now require a different sort of plasticity. The reproduction of egalitarian particles is somehow enslaved to the collective; their reproductive fitness is sensitive to (controlled by) the local context (i.e., the collective), and they thus share the same reproductive fate with all other particles in the collective. For example, the genes within a chromosome are all reproduced exactly once when the single molecule they reside on is copied. Thus, for the particles of egalitarian transitions, reproduction is strongly environmentally sensitive. This can be viewed as a special type of phenotypic plasticity if we view the fitness of a particle as a phenotype. To make the distinction clearer, we will refer to this type of plasticity as *reproductive plasticity*—that is, reproducing, or not, is a behavior of the particle that is dependent on extrinsic conditions, cues, or mechanisms.

In short, the fraternal solution utilizes particles that are phenotypically plastic, and the egalitarian transition utilizes particles that are reproductively plastic. That some form of plasticity (environmental sensitivity) is required makes sense given the nature of the "heterogeneous function with homogeneous fitness" problem. Function and fitness cannot have a one-to-one relationship; either the reproduction of a particle or the phenotypic function of a particle must be malleable in a context-specific way. Accordingly, control over the function or reproduction of a particle is ceded to the collective—as we might intuitively expect for a transition in individuality.

We conclude that particle plasticity (phenotypic or reproductive) is not just a contingent fact of evolutionary transitions but a requirement for a collective to exhibit heritable collective fitness. The argument for the requirement of plasticity is developed in (Tudge, Watson, and Brede 2016) and this work also shows that natural selection acts on plasticity traits in a manner that solves division of labor games given that particles have sensitivity to either the phenotype of the particles they interact with or the phenotype of the parent (parental effects).

3.3 Collectives Require Niche Construction

Above we argued for mechanisms of particle plasticity that cede control (over particle function or reproduction) to the collective in order to attain the benefits of complementary particle functions (i.e., to solve a division of labor game). Naturally, control over function or reproduction, and the distribution of the benefits that that confers, must be specific to the collective in which the particles reside. If there are to be fitness differences at the collective level, it cannot be the case that a particle's function influences, or is influenced by, the function of all other particles in the population. In other words, there must actually be collectives, rather than a well-mixed population of particles (this assumption is implicit in adopting a game-theoretic approach above). This may be provided by spatial localization, compartmentalization, or some other form of population structure[11] controlling which particles interact with one another functionally or which particles benefit from public goods or co-constructed reproductive machinery, for example, to restrict access (compartmentalization) and/or to restrict exit (vertical transmission).

However, a structure that is unchanging (exogenously imposed) cannot explain how individuality changes over time (Okasha 2006). In order for individuality to evolve, such structure must be created endogenously by evolvable properties of the particles. For example, individual traits that modify dispersal behavior (e.g., compartmentalization in membranes [Gánti 1975], seed radius [Pepper and Smuts 2002], or a stickiness trait that determines aggregation [Ispolatov et al. 2012]), the assortment of interactions (e.g., trait groups such as habitat choice [Wilson 1975]), the parameters of aggregation and dispersal reproduction (Powers, Penn, and Watson 2011), or the severity of a population bottleneck (Ryan, Powers, and Watson 2016).

The evolution of individual traits that modify fitness-affecting properties of the environment is niche construction (Odling-Smee, Laland, and Feldman 2003). And this particular type of niche construction where individually evolvable traits modify social interaction structure (who interacts with whom) is termed *social niche construction* (Powers 2010; Ryan et al. 2016). Previous work has shown that natural selection acting on such individual traits will modify interaction structures in the direction that is necessary to solve the social dilemma (Powers et al. 2011; Ryan et al. 2016; Tudge, Watson, and Brede 2016; Jackson 2016). For example, Powers et al. model individual traits that alter the initial group size preference of particles undergoing an aggregation and dispersal reproductive process. When initial group size is large, between-group variance in the proportion of co-operators is small and group selection is too weak for co-operators to increase in the population. Conversely, when initial group size is sufficiently small, between-group variance is large enough for selection to favor co-operators (Wilson 1975). Powers et al. show that individuals with traits that affect initial group size, although there is no intrinsic benefit to reduced size, are favored by selection.

Thus, while particle plasticity is necessary to create heritable collective fitness, it is also necessary for particles to exhibit niche construction in order for collectives to exhibit heritable fitness *differences*, that is, heritable collective fitness that is different from other collectives.

3.4 The "Development" of Collectives Requires Niche Construction "Signals" and Plasticity "Responses"

Niche construction and plasticity also have a more complex role in the individuality of the collective; the *development* of the collective phenotype.

The need for multiple functional roles within a collective can be partially satisfied by simple compartmentalization (i.e., well-mixed within each collective) and stochastic differentiation (i.e., each particle adopts a phenotype probabilistically). But the specific proportions of types cannot be fully controlled if differentiation is stochastic (Tudge, Watson, and Brede 2016). This may be particularly problematic when collectives have many different particle phenotypes (i.e., moving beyond the two-player games analyzed above). Put differently, large collectives may require internal *organization*; that is, structures that control how many particles of each phenotype are involved, which of them interact with which others, their relative locations, and so on.

Again, the requirements for such organization lead to different considerations for fraternal and egalitarian transitions. In fraternal transitions, a systematic symmetry-breaking mechanism is required: if genotypes are homogeneous but phenotypes heterogeneous then the factors that determine which particle takes which phenotype can neither be integral to the particle nor to environmental factors that are common to all particles. Particles that differentiate in a context-sensitive way can use cues from the environment to make their differentiation more predictable (e.g., cells on the outside of a cell cluster differentiate into surface cells by detecting that they are exposed to the outside; Schlichting 2003). When the environmental effects are homogeneous across the collective (differentiated external cues are not available), further differentiation can be facilitated by some form of communication among the particles. In this case, the factors that cue the differentiation of functional roles are created by the particles themselves in interaction with each other; the niche construction and phenotypic plasticity of the particles facilitate signals and responses, respectively.

The co-construction of localized conditions (e.g., axial gradients, tissue differentiation, organogenesis, cell-cell signals) that induce the right kind of plastic response (phenotypic form or function, abundance, timing, or movement, etc.) at the particle level is what we mean by *development* (of the collective phenotype). The ability of particles to signal and respond to one another, used during development, need not have been evolved de novo after the transition (Newman et al. 2003; Schlichting 2003; Newman and Bhat 2008; Solé and Valverde 2013; Duran-Nebreda and Solé 2015). The particles that organize themselves during development were, before the transition, evolutionary entities in their own right. The signals were locally differentiated environmental conditions, niche-constructed by other particles, and the responses were the plastic differentiation of the particles cued by those "environmental" conditions. In the transition, it can be these same relationships that are converted into developmental processes. For example, prior to the individualization of the collective, relative cell abundance may exhibit a reliable "climax community" composition, or may have multiple predictable attractors, through frequency dependent selection (Power et al. 2015). After the transition, these same ecological relationships are now internalized and can become an example of "internal selection" controlling the relative abundance of cells in the multicellular phenotype.

In egalitarian transitions, symmetry breaking is not required because the particles are intrinsically different. But removing fitness differences among intrinsically heterogeneous particles requires reproductive control—the creation and timing of reproductive machinery that controls the abundance of types (particle fecundity). The reproductive plasticity discussed in the previous section requires a radical form of niche construction—not just cuing or signaling the phenotypic plasticity of particles (as in the fraternal case), but cuing or signaling (or carrying-out) the reproduction of particles. This niche construction is not individual—a particle cannot construct the conditions for its own reproduction (because that would allow heterogeneous particle phenotypes to create heterogeneous particle fitnesses)—but collective; constructing environmental conditions that affect the selective environment/reproductive conditions of all particles in the same collective. Interestingly, from the particle's point of view this is *selective* niche construction, but from the collective's point of view this is *developmental* niche construction, or simply development (Stotz 2017; Uller and Helanterä 2017).

In sum, the phenotype of the collective thus depends on the organization of particles it contains and not individual particle characters. This organization requires some sort of communication; signal and response behaviors among the particles of the collective. The signal is provided by reproductive or functional niche-construction that cues the response provided by reproductive or functional particle plasticity. We conclude that niche construction of particles (as well as plasticity) is necessary for the development of the collective (controlling abundances of particles or types of particles). Particles that cannot niche construct in one of these ways cannot create individuality at the collective level. Intuitively, the niche construction and plasticity behaviors of the particles provide the "glue" that makes the collective more than the sum of the parts.

3.5 Reproductive Specialization Requires Extended Inheritance

Reproductive specialization is considered an important aspect of Darwinian individuality in fraternal transitions (Godfrey-Smith 2013). Although not required to define individuality, early determination of the germline means that particles can only increase their fitness by cooperating to increase the fitness of the collective (through the germline). The benefit of traits that implement reproductive specialization can be motivated by enabling the accumulation of reliably transmitted epigenetic information (Lachmann and Libby 2016), or by considering the fitness consequences of reversible particle differentiation. Specifically, whenever reversing particle differentiation is energetically costly or inefficient,[12] a collective that keeps some particles at a pluripotent state (at an energetic saddle point), and lets others differentiate (into energetic wells) is a more efficient solution. Examples include germ and soma, or reproductive and nonreproductive castes.

When particles are genetically identical, the differences between particles, such as the reproductive and the nonreproductive particle state, cannot be due to genetic differences (or more generally, to features intrinsic to particles). Nonetheless, the reproductive state must be maintained over multiple reproduction events for the continuation of the germ line. Accordingly, the reproductive state must be epigenetically heritable. Some form of extended inheritance is thus logically essential for reproductive specialization. Additionally, in cases where somatic particles reproduce within the collective (e.g., tissue growth in multicellular organisms), the inheritance of somatic types must likewise be epigenetic.

4. Discussion

We have argued that plasticity and niche construction of the particles (and extended inheritance when reproductive specialization is present) are necessary for a putative unit to be a Darwinian individual. A collective is nothing more than the sum of its parts unless its constituent parts have active phenotypes. Table 10.2 summarizes these points.

4.1 Active Particles "Standardize" the Collective as a Darwinian Unit

We started from the premise that a unit must exhibit heritable variation in fitness to be a bona fide evolutionary unit, and this led ultimately to the need for active phenotypes. Yet, the standard model takes the form it does, excluding the activity of phenotypes, because

Table 10.2.
Particle characteristics required for a putative unit to be a Darwinian individual.

• Particles **niche-construct** a social structure that creates nonrandom interactions among particles, that is, they create collectives.

• Within collectives, the particles must have (i) complementary functions (in order to produce collective fitness not attributable to individual particle phenotypes), and (ii) homogeneous fitness (so as not to compromise the heritability of the collective phenotype). They must therefore solve the "heterogeneous function with homogeneous fitness" problem.

• Accordingly, either their function or their reproduction must be context sensitive, or **plastic**; that is, determined by the characteristics of the collective (not their individual characteristics):

Fraternal	Egalitarian
• Particles are **functionally plastic**, and particularly in large collectives, particles **niche-construct** locally differentiated internal conditions for one another that signal functional particle differentiation.	• Particles are **reproductively plastic**, and particles **niche-construct** environmental conditions (internal to the collective) that affect, effect, or cue the reproduction of other particles.
• Reproductive specialization is beneficial whenever functional differentiation has energetic costs or is otherwise nonreversible. Maintenance of a pluripotent state cannot be genetic (if particles are genetically identical), thus the inheritance of at least this particle state requires **extended inheritance**.	

---

it describes a well-defined evolutionary unit in exactly the sense of supporting heritable variation in fitness. That is, the reason that the standard model excludes active phenotypes is not merely because it is simpler to do so, but because active phenotypes compromise the capability of a unit to exhibit heritable variation in fitness. To the extent that phenotypes are determined by environmental factors (plasticity), the ability of heritable material to control phenotypes and hence fitness differences is reduced. Likewise, to the extent that phenotype and environment are heritable factors (extended inheritance), the capability of random genetic variation to produce heritable fitness differences is reduced. And, to the extent that the salient properties of the selective environment experienced by other individuals or descendants are created by individuals (niche construction), this also reduces fitness differences. This suggests that an evolutionary unit can exhibit heritable fitness differences only to the extent that it is *not* active in these ways. Yet, it is exactly this necessity to create heritable fitness differences that led us to conclude above that the active roles of phenotypes were necessary to create an evolutionary unit.

This is not a contradiction because there are different levels involved. Specifically, in order for a *collective* to be a bona fide evolutionary unit, with standardized mechanisms of variation, selection, and inheritance it must be composed of component *particles* that are nonstandard, with active phenotypes. It is the nonstandard nature of the component particles that Darwinizes (standardizes) the collective. Moreover, as variation, selection, and inheritance of the particles are increasingly controlled by interactions with other particles in the collective and by properties of the collective as a whole (via plasticity, extended inheritance, and niche construction) the particles necessarily become de-Darwinized.

Table 10.3 gives an overview of how, before a transition, standard particles create a nonstandard collective, and after a transition, nonstandard particles create a "standardized" collective. In this context it is conceptually useful to distinguish between[13]:

a) **Accidental active phenotypes:** A sensitivity of variation, selection, and inheritance to the environment (or one's own phenotype) that "cannot be helped," that is, not adapted at this level of unit.[14]

b) **Adaptive active phenotypes:** Adaptive phenotypic activities (such as, plasticity, parental effects, and niche construction) that are adapted by natural selection to exhibit this specific sensitivity.

In order for the collective to become a proper Darwinian individual, accidental phenotypic activities are adapted (by natural selection acting on the lower-level particles [table 10.3, left], and in the context of one another) into organized plasticity, niche construction, and extended inheritance that reliably create heritable collective phenotypes that confer differences in collective fitness (table 10.3, right).

4.2   Is the Standard Model Still the "Prime Mover"?

If a bone fide evolutionary unit must be composed of nonstandard constituent parts, one might conclude that the standard model is an evolutionarily derived state, not causally prior. However, our depiction of the process of transition presupposes that the units at that lower level of organization are evolutionary units. That is, to explain why entities at one level organize into standardized replicators at a higher level of organization, we have appealed to the capabilities of those lower-level units to evolve their active phenotypes. This suggests that everything that happens here, including the transitions, is ultimately driven by heritable variation in reproductive success (acting on the lower-level units, until a standard model at the higher level takes over) as per the standard model. Fundamentally, the standard model remains as the "prime mover" because no other mechanism of adaptation is identified.

A multilevel view moves the issue downward in scale, but it does not in itself provide an alternative. That is, evolutionary subunits may be composed of sub-subunits, which must have nonstandard active phenotypes, and so on[15] (figure 10.2). But, is the original prime mover in this hierarchy of biological organization a standard unit, the first replicator? Or, is it a nonstandard unit, a loose, not-yet-standardized collection of active molecules, the first autocatalytic set? These two possibilities correspond to the replicator-first and metabolism-first hypotheses for the origin of life (Fry 2000; Shapiro 2007). The replicator-first view has the advantage that it begins at a level that already implements an adaptive mechanism, heritable variation in reproductive success. The metabolism-first view has the

Table 10.3.
Before a transition, standard particles create a nonstandard collective—and after a transition, nonstandard particles create a "standardized" collective.

Before a transition in individuality
Nonstandard collective. Particles are inactive or do not have active phenotypes adapted to access collective fitness.

Undifferentiated collectives
When particles are passive, the "collective" is nothing more than a subpopulation of self-reproducing particles and is not well differentiated from remaining population/ecosystem

The variability of the collective phenotype is accidentally plastic.
The phenotype of the collective is not internally controlled. The combination and arrangement of particles it contains is modified by environmental conditions and/or the heritable differences of *individual* particles. The collective phenotype has no meaningful development (it is nothing more than the population dynamics or "ecological" dynamics of the particles).

The reproduction of the collective is accidentally extended.
The collective has no reproductive specialization nor any reproductive policing—no germ/soma or genotype/phenotype distinction. Reproducing this subpopulation entails the reproduction of all the particles it contains (and this is not unified/holistic, see "selection" below).

The selection of the collective is undefined.
The effect of selection that the collective experiences is simply the sum of the selection that each of its component particles experiences (selective fate is not shared). The collective inevitably modifies its own selective environment (i.e., ecology happens) and this can be interpreted as accidental niche modification.

After a transition in individuality
Standardized collective. Particles have active phenotypes adapted to access collective fitness.

Differentiated collectives
Through the activity of the particle phenotypes, the collective is a well-defined evolutionary unit with localized interactions. The collective is well differentiated from the rest of the population/ecosystem.

The variability of the collective phenotype is internally controlled.
The "heterogeneous phenotypes with homogeneous fitness" problem is solved by either:

a) plastic particle phenotypes (fraternal)—The combination of particle phenotypes in the collective phenotype is due to the interaction of these particles in internally controlled conditions (niche construction + functional plasticity). Or

b) plastic particle reproduction (egalitarian)—The combination of particle *genotypes* in the collective is due to the interaction of these particles in internally controlled reproductive conditions (niche construction + reproductive plasticity).

The *ecological* dynamics of particles, internalized by the new collective, is the *development* of the collective.

The reproduction of the collective is by germ-particle only or gene-transmission only.*
a) (fraternal) Reproductive specialization requires particles with extended-inheritance of the reproductive state (the homogeneous genotypes of the individual particles control their inclusive fitness).

b) (egalitarian) The reproduction of the particles is controlled by the reproductive machinery of the collective (the heterogeneous genotypes of the individual particles do not control their fitnesses).

Selection of the collective is standard.
The selective fate of the particles in a collective is bound together (and separated from that of particles in other collectives). Either:

a) (fraternal) With reproductive specialization, selection is determined by the reproductive opportunity of the genotype (only) or germ particle (only).

b) (egalitarian) Without reproductive specialization, reproductive opportunity is equalized among the particles of the collective via, for example, shared public goods, internalized resource trading, and/or mutual policing.

*Reproductive specialization is optional in the fraternal case and not applicable in the egalitarian case.

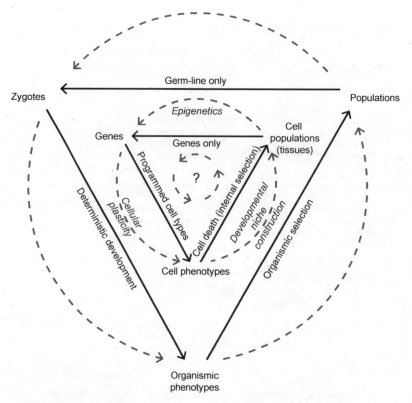

Figure 10.2
Active phenotypes "all the way down"? The standardization of the multicellular organism, for example, is enabled by the active phenotypes of the cells it contains. Looking upward to higher levels of organization, the multicellular organism may participate as a particle in super-organism units via the organized complementarity of functional roles enabled by its own active phenotypes. Looking downward, the active phenotypes of the cells are the evolutionary products of a standard model at a lower level of organization. Does this nested hierarchy of active and standard systems bottom out at particles that are active or passive?

advantage that it presupposes less, but to explain why entities that are not evolutionary units should organize into the first replicators (or organize at all), it requires that there is some mechanism of adaptation, or at least organization, that exists prior to the first evolutionary units.

Are active particles, without natural selection, enough to get things started? Can interaction with the environment cause a system to be modified in a way that makes it "well-fitted" to its environment without a Darwinian process? Complementary to conventional evolutionary theory, there are a number of proto, or non-Darwinian, adaptation mechanisms that have been proposed to provide adaptive change without natural selection. These include homeostasis and ultrastability (Ashby 1952), self-organization in dissipative structures

(Prigogine and Nicolis 1977), self-organized criticality (Bak 2013), adaptive improvisation (Soen, Knafo, and Elgart 2015), sequential selection (Betts and Lenton 2008; Worden 2010), associative induction (Watson, Mills, and Buckley 2011; Watson et al. 2016), and in a general sense, Le Chatelier's principle (Gall 2002). We note that each of these mechanisms is necessarily more akin to a transformational account of adaptation than the conventional variational account (Lewontin 1983).

Even if a replicator-first account were to turn out to be correct for the origin of life, it does not seem to us sufficient to assume that a variational account, at this or any other level, is sufficient to explain all the levels above it. Certainly, we cannot use a variational account at any given level to explain the existence of a variational process at that level (Williams 1992). And the observation that a system includes evolution by natural selection does not in itself allow the conclusion that the algorithm that is instantiated is merely evolution by natural selection (what may appear to be "proximal" details in the mechanisms or organization of evolution by natural selection, including a recursive structure, can instantiate quite different algorithmic processes; Watson 2012). Perhaps the more general position is to consider how each level of biological organization exhibits both variational (standard) and transformational (nonstandard) adaptive processes, or how variational processes at the particle level may describe transformative processes at the collective level (Power et al. 2016). More specifically, we need to understand how the adaptation that occurs at one level of organization interacts with adaptation at another, as developed below.

4.3   Why Might Active Phenotypes That De-Darwinize Particles Be Favored by Particle-Level Selection?

If the active phenotypes of particles create standardized collectives, are those phenotypes adapted for the benefit of the particle or for the benefit of the collective? A conventional answer, focusing on the fraternal type of transition, might say that this is a non-issue since all the particles of the collective are genetically identical to the germline. But this level of explanation overlooks the adaptation that occurs during the transition that led to this situation. The establishment of collectives, with functional integration and reproductive specialization, required particles to evolve plasticity, niche construction, and extended inheritance properties. We cannot pre-suppose a higher-level evolutionary unit to explain how these adaptations created such a unit. Since an evolutionary unit that does not yet exist cannot be the cause of anything, it cannot be the cause of adaptations that created it. So, the active phenotypes that created the collective were adaptations of the particles not the collective. It could be merely a fortuitous coincidence that the active phenotypes that were beneficial to the particles were those that happened to create individuality at the collective level. In some sense, we might see the disorganized ecology of active particles as an "excitable media" with the right generic properties for forming higher-level complexes

(Newman et al. 2003). For example, reaction-diffusion dynamics require particles that signal and respond to one another in a localized manner, and oscillation and multistability of dynamical states requires interactions with, what we would call, sign synergy.

But active phenotypes that are beneficial to particles may directly oppose the Darwinization of the collective, or perhaps more to the point, oppose the de-Darwinization of the particles (as is all the more obvious in nonclonal cases such as insect societies or eukaryote organelles). Why do active phenotypes that facilitate effective selection at the collective level, and disable selection at the particle level, evolve under particle-level selection?

In part, active phenotypes that standardize the collective are favored by particle-level selection because they enable the particles to access synergistic components of fitness interactions that are unavailable to inactive particles. In some cases, this may be sufficient for natural selection at the particle level to favor phenotypic activities that create higher-level evolutionary units and hence high-fitness collectives. For example, particle-level selection can favor phenotypic plasticity that solves a division of labor game (Tudge, Watson, and Brede 2016), or favor positive assortment (Snowdon, Powers, and Watson 2009; Powers et al. 2011) or compartmentalization and vertical transmission (Watson et al. 2009) that solves a coordination game. In some cases, particles form these associations because this enables them to *decrease* interactions with noncomplementary particles, not because they increase competition between collectives (Snowdon et al. 2009; Watson et al. 2009). This is sufficient because the associations that facilitate effective competition at the collective level are necessarily those that were *most reliably nondeleterious* at the particle level (Snowdon et al. 2009; Watson et al. 2009).

An analogy with the relationship between robustness and evolvability is also instructive, that is, the relationship between the short-term benefits of robust genotype-phenotype mappings and the longer-term benefits for evolvability that such mappings confer (Wagner and Altenberg 1996; Gerhart and Kirschner 2007). Correlations among components (e.g., by gene regulation) reduce the effective degrees of freedom of the phenotype at the system level (gene-expression profile), and selection ensures that deleterious correlations are removed. The immediate selective impetus derives from removing combinations of components that are unfit, but the systematic consequence of such dimensional reduction is an increase in *fitness variance* at the system level. This increase in fitness variance (occurring without significant decrease in mean fitness) enhances the efficacy of selection at the system level. One example is facilitated variation created by lines of least resistance aligned with the direction of selection (Schluter 1996; Gerhart and Kirschner 2007; Pavlicev, Cheverud, and Wagner 2010; Draghi and Whitlock 2012; Watson et al. 2014; Kounios et al. 2016; Kouvaris et al. 2017).

This dimensionality reduction perspective suggests that the alignment between active phenotypes that confer particle-level benefits and those that facilitate adaptation at the collective level is not mere coincidence—not just a fortuitous repurposing (exaptation) of individual-level adaptations by the later application of higher-level selection. For particles that initially interact in an uncontrolled way, active phenotypes will reduce the degrees of freedom in those interactions, but will be removed if they depress mean fitness. This increases fitness variance at the collective level. This can operate just by the evolution of associations that canalize the combinations of particles that are found at ecological equilibria—"species that occur together wire together" (Watson et al. 2009; Power et al. 2015; Watson and Szathmáry 2016). Specifically, selection on the lower-level particles (i.e., population dynamics, table 10.3, left) creates combinations of particles (particle communities) that are at least locally favored under particle-level selection. Subsequent canalization of these combinations allows competition between communities at the higher level of selection (Watson et al. 2009). Perhaps this can be understood as a special sort of phenotype-first evolution occurring across evolutionary levels, bottom-up. That is, selection acting on particles creates certain combinations of particles (local ecological attractors) more than others. These combinations of particles provide a context that favors evolved associations that canalize such combinations (or, more exactly, do not disrupt them) (Doolittle and Inkpen 2018). These particle combinations thus become the phenotypes of a new evolutionary unit, enabling competition between collectives at a higher level of organization. Deep associative learning, discussed below, may provide a suitable theoretical framework to formalize this dimensional reduction perspective on multilevel evolution.

4.4 Learning Theory Provides a Transformational Theory of Adaptation

Learning systems are systems that improve with experience, and learning theory is a rigorous theory of transformational adaptation. The relationship between learning and evolution is not a casual analogy but a functional isomorphism—evolution and learning are different ways of implementing the same algorithm (Frank 2009; Harper 2009; Chastain et al. 2014; Watson et al. 2014; Watson et al. 2016; Watson and Szathmáry 2016; Kouvaris et al. 2017). However, the type of learning that is possible depends on the model space that is assumed (e.g., additive models or correlation models) (Watson and Szathmáry 2016).

A strictly variational account of evolution constitutes a highly limited form of learning (random trial and error, and/or rote learning without generalization). In contrast, when natural selection can act on correlations between features, the kind of learning that is possible is *associative* or *correlation learning*—a basic but much more general kind of learning. Correlation learning is a theory of how the relationships between things change over time, rather than a theory of how the frequencies of things change over time (Watson et al. 2014). The equivalence between evolution and correlation learning has been developed in some detail, both for looking inward (at the organization of relationships inside the evolutionary unit) at the evolution of gene-regulation networks controlling development (Watson et al. 2014; Kounios et al. 2016; Watson and Szathmáry 2016; Kouvaris et al. 2017) and for looking outward (at the organization of relationships in which the evolutionary unit participates) at the evolution of ecological networks determining the selection that is experienced (Power et al. 2015). In the current context, the components of the collective pretransition constitute an untrained correlation model, and after the transition they are a trained model. Intuitively, the evolved relationships between particles control their coordination and thereby represent a model of the combinations of particles that work well together.

The relationship with learning becomes even more interesting when correlational learning is applied at multiple levels of organization. In learning systems, multiple layers of pairwise correlations are sufficient to represent arbitrary complex patterns, also known as *deep learning*. Moreover, training such networks can be carried out layer-by-layer, and the training of a lower-level layer can develop a representation that facilitates successful learning at the next layer without explicitly rewarding it for that purpose (i.e., before that higher-level layer is created) (Hinton, Osindero, and Teh 2006). This *unsupervised* learning technique is successful because it reduces the effective degrees of freedom in the data, even though it is not yet known how the data will be used at the next level of the network (Watson and Szathmáry 2016; Watson et al. 2016). In evolving systems, as we move through multiple transitions, a deep model of what works well together is constructed layer-by-layer. Each layer reduces the effective degrees of freedom in the system, and in so doing facilitates adaptation of the next layer, even though that level of organization is not yet instantiated. Deep evolution (i.e., incorporating evolutionary transitions in individuality) thus constitutes a quite sophisticated deep associative learning machine (Watson et al. 2016). Deep learning thus has potential to provide a formal framework to understand deep evolution, and specifically, to develop a more general account of evolutionary adaptation that incorporates variational and transformational processes, to help us understand how adaptation at one level of organization facilitates adaptation at the next level of organization, and to incorporate proximal mechanisms into ultimate explanations in a general way.

4.5   Is This Just about the Transitions?

With a sufficiently liberal mindset it is always possible to add detail to a model to make it more accurate and to cover more cases. Is this all that we are doing by adding active phenotypes to the standard model (Futuyma 2017)? Certainly, there are some things that the standard model cannot do, no matter how many additions are made. Specifically, a variational account, appealing to *ultimate* causes (differential survival and reproduction of units), relies on or asserts the abstraction of well-defined units. To maintain this abstraction, it must *assume* suitable variation, selection, and inheritance. Like any formalism, the standard model cannot explain its own axioms and therefore such an account cannot "look inside" the units to explain why variation or inheritance is like it is or "look outside" the units to explain why selection is like it is. Likewise, it cannot explain how variation, selection, and inheritance *change*; how they are affected by the products of evolution (Watson and Szathmáry 2016). Such issues thus become the remit of other disciplines such as evo-lutionary developmental biology and evolutionary ecology. In contrast, in an account includ-ing active phenotypes, evolutionary units are not presupposed but defined by context and behavior. Proximal mechanisms are necessary to define the relevant unit and are not merely mechanistic details for an evolutionary unit that is taken as an axiom from the outset.

In a general sense, to the extent that the standard model depends on a variational account, with presupposed and immutable units, it cannot be a theory of parts and wholes and their organization. The standard model is fundamentally *a theory of things* (fixed objects, and how they change in frequency in a population). What is needed is a transfor-mational theory of adaptation that explains how the *relationships between things* change resulting in the origin and maintenance of organizations.[16] This is what the standard model cannot provide.

None of this invalidates the standard model, of course, when the standard assumptions hold. So for the "normal" evolution that occurs in between transitions—at one level of organization with well-defined evolutionary units—do the considerations of this chapter have any implications that matter? We think they do. Mechanistic models of development, inheritance, and/or reproduction are necessary to address questions about how variation, selection, and inheritance change as a result of selection. In the absence of a mechanistic model of, for example, how developmental organization affects fitness variance we can still measure fitness variance and continue from there. However, it is increasingly apparent that such measurements have short-lived predictive power. Selection can change fitness variance or heritability or the salient components of selection in a few generations. This is why evolutionary developmental biology and evolutionary ecology are rapidly expanding—i.e., because we need theory that incorporates proximal mechanisms (in general ways) rather than abstracting-away such detail. Although our discussion has not been attached to any particular mechanistic details, we view an evolutionary unit as the result of adapted interactions between its component subunits and as a subpart within a higher-level unit. This means units are not inherently indivisible and context-independent, but systems that have internal organization and a position within a larger organization. This places constraints on the mechanisms of development and the organization of ecological relationships that can help us understand why variation, selection, and inheritance take the form they do and how they change over time.

5.   Conclusion

In order to be a meaningful Darwinian unit, a biological unit must exhibit heritable varia-tion in fitness that is not attributable to its component subunits (i.e., not merely due to the heritable variation in fitness at a lower level of organization). For the whole to be greater than the sum of the parts in this formal sense is not a mysterious, emergent phenomenon;

the difference between the sum of the parts and the whole is simply the organization of those parts. Since the adaptation of an evolutionary unit that does not yet exist cannot be a reason for such organization, this organization must be created bottom-up—by the adaptation of the parts (in interaction with one another), not by the adaptation of the whole. Bottom-up organization requires that parts communicate with one another—that they have mechanisms of signaling and responding. The "unfolding" of the organization of the collective, from the active interaction of its components, is the developmental process of the collective. We argue that this is not just a contingent observation about multicellularity, for example, but that, in some form or another, plastic responses (environmentally sensitive phenotypes or reproduction) and niche construction signals (localized changes to environmental and/or selective conditions) at the component level are essential to originate and maintain any evolutionary unit. Where reproductive specialization occurs, this further necessitates extended inheritance at the particle level. In short, the extensions are the "glue" that make the whole more than the sum of the parts.

Some assumptions remain in reaching this conclusion. Notably, the need for plasticity and niche construction depends on the need for functional heterogeneity, but the necessity for sign synergy does not logically exclude homogeneous groups. Furthermore, the need for extended inheritance depends on reproductive specialization, but reproductive specialization is not required for individuality. Nonetheless, it seems safe to conclude that the active roles of phenotypes cannot be treated merely as "optional extras" or extensions to evolutionary theory, that a much more complicated and equitable relationship exists between the standard and active mechanisms in evolutionary processes, and that new theoretical approaches are needed to integrate such processes with conventional evolutionary thinking.

Acknowledgments

We are grateful to the event organizers of The Royal Society meeting "New Trends in Evolutionary Biology: Biological, Philosophical and Social Science Perspectives" (2017) for the venue that motivated these works, and especially to Sonia Sultan for prompting the question "What would a theory of evolution look like if the extensions were at the center not the periphery?" We thank the attendees and organizers Tobias Uller and Kevin Laland of the workshop *Cause and Process in Evolution* (2017), hosted by the Konrad Lorenz Institute in Vienna, where many formative discussions shaped the work. Thanks also to Tobias Uller and Heikki Helanterä for their detailed feedback on the manuscript, and Kevin Laland, Michael Lachmann, Miguel Brun Usan, David Prosser, Frederick Nash, James Caldwell, and Eors Szathmáry for discussion and feedback on earlier drafts and ideas. This work was partly funded by the John Templeton Foundation (60501).

Notes

1. Here we take "economies of scale" to refer to efficiencies derived purely from increased size (e.g., susceptibility to predation) and not from any differentiation of functional roles that may (or may not) accompany an increase in size.

2. Note that the fitness of a collective is a function of particle *character* not particle *fitness* (Okasha 2006)—if particles attain their fitness from collectives, collective fitness cannot be separated from particle fitness.

3. This is the social evolution analogue of the genetic term "sign epistasis" (Weinreich, Watson, and Chao 2005).

4. Coordination and anti-coordination are relevant here for the same reason that the *nonlinearly separable* logical functions, IFF and XOR, respectively, feature prominently in machine learning theory. That is, there is no linear weighting of A and B such that A XOR B is a monotonic function of their weighted sum.

5. "Homogeneity" and "heterogeneity" (or, equality, inequality) are only well defined in games where component particles take strategies from the same domains (i.e., particles are fungible; Lachmann et al. 2003). When the domains of the particles are different (e.g., P1∈{big host,small host}, P2∈{motile symbiont, immotile symbiont}), it is nonetheless still possible to exhibit sign synergy (e.g., when <small host, motile symbiont> and <big host, immotile symbiont> are fit symbiotic pairs and <small host, immotile symbiont> and <big host, motile symbiont> are unfit). Fungible and nonfungible particles correspond to fraternal and egalitarian collectives, respectively, discussed later.

6. Queller may be right, however, that they form part of the story—for example, if increased scale enables, or is enabled by, division of labor.

7. **Selection:** In coordination games, competition between fit collectives (AA vs. BB) could be mistaken for competition between particles (A vs. B), for example, if $w(x)$ is the fitness of x, and $w(AA)>w(BB)$ then A will increase and B will decrease in frequency in the population, and this could have been a consequence of $w(A)>w(B)$. In anti-coordination games, there is no single winning particle type, and competition between fit collectives (AB vs. BA) cannot have the same effect on particle frequencies as competition between particles (A vs. B). **Variation:** In coordination games, differences between fit collectives are indistinguishable from differences between particles (e.g., AA≠BB because A≠B). In anti-coordination games, variation at the collective level arises from the interaction, arrangement, and complementarity of the particles in the collective and not from differences among the particles per se. **Inheritance:** If the homogeneity of the collective is maintained by a single-particle population bottleneck or propagule, then the heritability of particle characters is indistinguishable from the heritability of collective characters (AA→AA because A→A). Whereas, in a heterogeneous collective, even though the collective may grow from a single progenitor particle, the character of that particle that is relevant to the collective is the propensity of that particle, when suitably cued, to differentiate predictably and form a collective with heterogeneous functional roles (e.g., A→AB).

8. This intrinsic property may be a property that enables them to express functionally diverse phenotypes in an interactive, context-sensitive manner—but it is this intrinsic property, shared by all particles, and not the specific phenotype expressed by a particle that determines its fitness.

9. If there is also reproductive specialization, then technically particles have the same *inclusive* fitness. Note that the property that determines which particles become germ and which become soma (which have high direct fitness and which do not) cannot be determined by genetic differences since the particles are genetically homogeneous.

10. While it is conventional to think of cellular differentiation (i.e., pluripotent cells acquiring different final states as a result of the developmental dynamics) as a different kind of process from phenotypic plasticity (i.e., phenotypic sensitivity to environmental cues), we emphasize that this distinction is only one of levels of description. That is, the collective is the environment of the particle, and the phenotypic plasticity of the particles, in interaction with one another, is the development of the collective phenotype (see section 4.1).

11. In the most-studied social dilemmas, positive assortment, for example via limited dispersal, is sufficient but division of labor games require slightly more careful consideration. When solved fraternally, positive assortment of genotypes can be beneficial, but it is the anti-assortment of phenotypes that is important (Tudge, Watson, and Brede 2016). When solved in egalitarian fashion, positive assortment is not at all the right way to think about population structure (because both genotypes and phenotypes are heterogeneous).

12. For example, if particle differentiation is implemented by descending an epigenetic energy landscape a la Waddington (1957), or "falling" into an attractor state (Kauffman 1969).

13. Badyaev, Foresman, and Young (2005) distinguish between passive and active, respectively—but it seems to us that a non-adapted response to the environment can nonetheless be active.

14. Accidental active phenotypes at the collective level may be the result of adaptive active phenotypes at the particle level.

15. See also *combogenesis*, "the building-up from combination and integration to produce new things with innovative relations" (Volk 2017) and *compositional evolution* (Watson 2006).

16. In principle, a theory of how relationships change over time could be developed within a variational account if all possible relationships preexist and then change in frequency. But since the relationships at one level of organization become the things at the next level of organization (Volk 2017), this approach seems untenable.

References

Ashby, W. R. 1952. *Design for a Brain: The Origin of Adaptive Behaviour*. London: Chapman & Hall.

Badyaev, A. V., K. R. Foresman, and R. L. Young. 2005. "Evolution of Morphological Integration: Developmental Accommodation of Stress-Induced Variation." *The American Naturalist* 166 (3): 382–395.

Bak, P. 2013. *How Nature Works: The Science of Self-Organized Criticality*. New York: Springer Science and Business Media.

Betts, R. A., and T. M. Lenton. 2008. "Second Chances for Lucky Gaia: A Hypothesis of Sequential Selection." *Gaia Circular* 1 (1): 4–6.

Bourke, A. F. 2011. *Principles of Social Evolution*. Oxford: Oxford University Press.

Buss, L. W. 2014. *The Evolution of Individuality*. Princeton, NJ: Princeton University Press.

Cavalli-Sforza, L. L., and M. W. Feldman. 1981. *Cultural Transmission and Evolution: A Quantitative Approach* (No. 16). Princeton, NJ: Princeton University Press.

Charlesworth, D., N. H. Barton, and B. Charlesworth. 2017. "The Sources of Adaptive Variation." In *Proceedings of the Royal Society of London B* 284 (1855): 20162864. The Royal Society.

Chastain, E., A. Livnat, C. Papadimitriou, and U. Vazirani. 2014. "Algorithms, Games, and Evolution." *Proceedings of the National Academy of Sciences USA* 111 (29): 10620–10623.

Clarke, E. (2014). "Origins of Evolutionary Transitions." *Journal of Biosciences* 39 (2): 303–317.

Corning, P. A., and E. Szathmáry. 2015. "'Synergistic Selection': A Darwinian Frame for the Evolution of Complexity." *Journal of Theoretical Biology* 371:45–58.

Day, T., and R. Bonduriansky. 2011. "A Unified Approach to the Evolutionary Consequences of Genetic and Nongenetic Inheritance." *The American Naturalist* 178 (2): E18–E36.

Dickins, T. E., and Q. Rahman. 2012. "The Extended Evolutionary Synthesis and the Role of Soft Inheritance in Evolution." *Proceedings of the Royal Society of London B* 279 (1740): 2913–2921.

Doolittle, W. F., and S. A. Inkpen. 2018. "Processes and Patterns of Interaction as Units of Selection: An Introduction to ITSNTS Thinking." *Proceedings of the National Academy of Sciences USA* 201722232.

Draghi, J. A., and M. C. Whitlock. 2012. "Phenotypic Plasticity Facilitates Mutational Variance, Genetic Variance, and Evolvability along the Major Axis of Environmental Variation." *Evolution* 66 (9): 2891–2902.

Duran-Nebreda, S., and R. Solé. 2015. "Emergence of Multicellularity in a Model of Cell Growth, Death and Aggregation under Size-dependent Selection." *Journal of the Royal Society Interface* 12 (102): 20140982.

Frank, S. A. 2009. "Natural Selection Maximizes Fisher Information." *Journal of Evolutionary Biology* 22 (2): 231–244.

Fry, I. 2000. *The Emergence of Life on Earth: A Historical and Scientific Overview*. New Brunswick, NJ: Rutgers University Press.

Futuyma, D. J. 2017. "Evolutionary Biology Today and the Call for an Extended Synthesis." *Interface Focus* 7 (5): 20160145.

Gall, J. 2002. *The Systems Bible: The Beginner's Guide to Systems Large and Small*. Walker, MN: General Systemantics Press.

Gánti, T. 1975. "Organization of Chemical Reactions into Dividing and Metabolizing Units: The Chemotons." *BioSystems* 7 (1): 15–21.

Gerhart, J., and M. Kirschner. 2007. "The Theory of Facilitated Variation." *Proceedings of the National Academy of Sciences USA* 104 (suppl 1): 8582–8589.

Gilbert, S. F., and D. Epel. 2009. *Ecological Developmental Biology: Integrating Epigenetics, Medicine, and Evolution.* Sunderland, MA: Sinauer.

Godfrey-Smith, P. 2007. "Conditions for Evolution by Natural Selection." *Journal of Philosophy* 104 (10): 489–516.

Godfrey-Smith, P. 2011. "Darwinian Populations and Transitions in Individuality." In *The Major Transitions in Evolution Revisited*, edited by B. Calcott and K. Sterelny, 65–81. Cambridge, MA: MIT Press.

Godfrey-Smith, P. 2013. "Darwinian Individuals." In *From Groups to Individuals: Evolution and Emerging Individuality*, edited by F. Bouchard and P. Hunema, 17–36. Cambridge, MA: MIT Press.

Gomez-Mestre, I., and R. Jovani. 2013. "A Heuristic Model on the Role of Plasticity in Adaptive Evolution: Plasticity Increases Adaptation, Population Viability and Genetic Variation." *Proceedings of the Royal Society of London B* 280 (1771): 20131869.

Gomulkiewicz, R., and M. Kirkpatrick. 1992. "Quantitative Genetics and the Evolution of Reaction Norms." *Evolution* 46 (2): 390–411.

Hammerschmidt, K., C. J. Rose, B. Kerr, and P. B. Rainey. 2014. "Life Cycles, Fitness Decoupling and the Evolution of Multicellularity." *Nature* 515 (7525): 75.

Harper, M. 2009. "The Replicator Equation as an Inference Dynamic." *arXiv* preprint arXiv:0911.1763.

Hinton, G. E., S. Osindero, and Y. W. Teh. 2006. "A Fast Learning Algorithm for Deep Belief Nets." *Neural Computation* 18 (7): 1527–1554.

Ispolatov, I., M. Ackermann, and M. Doebeli. 2012. "Division of Labour and the Evolution of Multicellularity." *Proceedings of the Royal Society of London B*: Biological Sciences 279 (1734): 1768–1776.

Jablonka, E. 2006. "Genes as Followers in Evolution–A Post-synthesis Synthesis?" *Biology and Philosophy* 21 (1): 143–154.

Jackson, A. 2016. "Metagames: The Evolution of Game-Changing Traits" (Doctoral dissertation, University of Southampton).

Kauffman, S. A. 1969. "Metabolic Stability and Epigenesis in Randomly Constructed Genetic Nets." *Journal of Theoretical Biology* 22 (3): 437–467.

Kounios, L., J. Clune, K. Kouvaris, G. P. Wagner, M. Pavlicev, D. M. Weinreich, and R. A. Watson. 2016. "Resolving the Paradox of Evolvability with Learning Theory: How Evolution Learns to Improve Evolvability on Rugged Fitness Landscapes." *arXiv* preprint arXiv:1612.05955.

Kouvaris, K., J. Clune, L. Kounios, M. Brede, and R. A. Watson. 2017. "How Evolution Learns to Generalise: Using the Principles of Learning Theory to Understand the Evolution of Developmental Organisation." *PLoS Computational Biology* 13 (4): e1005358.

Lachmann, M., N. W. Blackstone, D. Haig, A. Kowald, R. E. Michod, E. Szathmáry, L. Wolpert, et al. 2003. "Conflict in the Evolution of Genomes, Cells, and Multicellular Organisms." In *Genetic and Cultural Evolution of Cooperation*, edited by P. Hammerstein, 327–356. Cambridge, MA: MIT Press.

Lachmann, M., and E. Libby. 2016. "Epigenetic Inheritance Systems Contribute to the Evolution of a Germline." *Philosophical Transactions of the Royal Society of London B* 371 (1701): 20150445.

Laland, K. N., F. J. Odling-Smee, and M. W. Feldman. 1999. "Evolutionary Consequences of Niche Construction and Their Implications for Ecology." *Proceedings of the National Academy of Sciences USA* 96 (18): 10242–10247.

Laland, K. N., J. Odling-Smee, W. Hoppitt, and T. Uller. 2013. "More on How and Why: Cause and Effect in Biology Revisited." *Biology and Philosophy* 28 (5): 719–745.

Laland, K. N., K. Sterelny, J. Odling-Smee, W. Hoppitt, and T. Uller. 2011. "Cause and Effect in Biology Revisited: Is Mayr's Proximate-Ultimate Dichotomy Still Useful?" *Science* 334 (6062): 1512–1516.

Laland, K., T. Uller, M. Feldman, K. Sterelny, G. B. Müller, A. Moczek, J. Odling-Smee, et al. 2014. "Does Evolutionary Theory Need a Rethink?" *Nature* 514 (7521): 161.

Laland, K. N., T. Uller, M. W. Feldman, K. Sterelny, G. B. Müller, A. Moczek, J. Odling-Smee, et al. 2015, August. "The Extended Evolutionary Synthesis: Its Structure, Assumptions and Predictions." In *Proceedings of the Royal Society of London B* 282 (1813): 20151019. The Royal Society.

Lande, R. 2009. "Adaptation to an Extraordinary Environment by Evolution of Phenotypic Plasticity and Genetic Assimilation." *Journal of Evolutionary Biology* 22 (7): 1435–1446.

Lewontin, R. C. 1970. "The Units of Selection." *Annual Review of Ecology and Systematics* 1 (1): 1–18.

Lewontin, R. C. 1983. "The Organism as the Subject and Object of Evolution." *Scientia* 118 (1–8): 65–95.

Maynard Smith, J., and E. Szathmáry. 1995. *The Major Transitions in Evolution*. Oxford: W. H. Freeman.

Mayr, E. 1989. "Speciational Evolution or Punctuated Equilibria." *Journal of Social and Biological Structures* 12 (2–3): 137–158.

Michod, R. E. 2000. *Darwinian Dynamics: Evolutionary Transitions in Fitness and Individuality*. Princeton University Press.

Michod, R. E. 2006. "The Group Covariance Effect and Fitness Trade-Offs during Evolutionary Transitions in Individuality." *Proceedings of the National Academy of Sciences USA* 103 (24): 9113–9117.

Müller, G. B., and S. A. Newman (Eds.). 2003. *Origination of Organismal Form: Beyond the Gene in Developmental and Evolutionary Biology*. Cambridge, MA: MIT Press.

Newman, S. A., and R. Bhat. 2008. "Dynamical Patterning Modules: Physico-Genetic Determinants of Morphological Development and Evolution." *Physical Biology* 5 (1): 015008.

Newman, S. A., G. Forgacs, and G. B. Muller. 2003. "Before Programs: The Physical Origination of Multicellular Forms." *International Journal of Developmental Biology* 50 (2–3): 289–299.

Odling-Smee, F. J., K. N. Laland, and M. W. Feldman. 2003. *Niche Construction: The Neglected Process in Evolution* (No. 37). Princeton, NJ: Princeton University Press.

Okasha, S. 2006. *Evolution and the Levels of Selection*. New York: Oxford University Press.

Pavlicev, M., J. M. Cheverud, and G. P. Wagner. 2010. "Evolution of Adaptive Phenotypic Variation Patterns by Direct Selection for Evolvability." *Proceedings of the Royal Society of London B*: Biological Sciences, rspb20102113.

Pepper, J. W., and B. B. Smuts. 2002. "A Mechanism for the Evolution of Altruism among Nonkin: Positive Assortment through Environmental Feedback." *The American Naturalist* 160 (2): 205–213.

Pichugin, Y. 2015. "Theoretical Investigation into the Origins of Multicellularity" (Doctoral dissertation, Massey University).

Pigliucci, M. 2009. "An Extended Synthesis for Evolutionary Biology." *Annals of the New York Academy of Sciences* 1168 (1): 218–228.

Power, D. A., Watson, R. A., Szathmáry, E., Mills, R., Powers, S. T., Doncaster, C. P., and Czapp, B. 2015. "What Can Ecosystems Learn? Expanding Evolutionary Ecology with Learning Theory." *Biology Direct* 10 (1): 69.

Powers, S. T. 2010. "Social Niche Construction: Evolutionary Explanations for Cooperative Group Formation" (Doctoral dissertation, University of Southampton).

Powers, S. T., Penn, A. S., and Watson, R. A. (2011). "The Concurrent Evolution of Cooperation and the Population Structures That Support It." *Evolution* 65 (6): 1527–1543.

Prigogine, I., and G. Nicolis. 1977. *Self-Organization in Non-Equilibrium Systems*. New York: John Wiley & Sons.

Queller, D. C. 1997. "Cooperators since Life Began." *Quarterly Review of Biology* 72 (2): 184–188.

Rueffler, C., J. Hermisson, and G. P. Wagner. 2012. "Evolution of Functional Specialization and Division of Labor." *Proceedings of the National Academy of Sciences USA* 109 (6): E326–E335.

Ryan, P. A., S. T. Powers, and R. A. Watson. 2016. "Social Niche Construction and Evolutionary Transitions in Individuality." *Biology and Philosophy* 31 (1): 59–79.

Schlichting, C. D. 2003. "Origins of Differentiation via Phenotypic Plasticity." *Evolution and Development* 5 (1): 98–105.

Schluter, D. 1996. "Adaptive Radiation along Genetic Lines of Least Resistance." *Evolution* 50 (5): 1766–1774.

Schwander, T., and O. Leimar. 2011. "Genes as Leaders and Followers in Evolution." *Trends in Ecology and Evolution* 26 (3): 143–151.

Scott-Phillips, T. C., K. N. Laland, D. M. Shuker, T. E. Dickins, and S. A. West. 2014. "The Niche Construction Perspective: A Critical Appraisal." *Evolution* 68 (5): 1231–1243.

Shapiro, R. 2007. "A Simpler Origin for Life." *Scientific American* 296 (6): 46–53.

Snowdon, J. R., S. T. Powers, and R. A. Watson. 2009, September. "Moderate Contact between Sub-populations Promotes Evolved Assortativity Enabling Group Selection." In *Proceedings of European Conference on Artificial Life*, 45–52. Berlin: Springer.

Soen, Y., M. Knafo, and M. Elgart. 2015. "A Principle of Organization which Facilitates Broad Lamarckian-like Adaptations by Improvisation." *Biology Direct* 10 (1): 68.

Solé, R. V., and S. Valverde. 2013. "Before the Endless Forms: Embodied Model of Transition from Single Cells to Aggregates to Ecosystem Engineering." *PLoS ONE* 8 (4): e59664.

Stotz, K. 2017. "Why Developmental Niche Construction Is Not Selective Niche Construction: and Why It Matters." *Interface Focus* 7 (5): 20160157.

Szathmáry, E. 2015. "Toward Major Evolutionary Transitions Theory 2.0." *Proceedings of the National Academy of Sciences* 112 (33): 10104–10111.

Tal, O., E. Kisdi, and E. Jablonka. 2010. "Epigenetic Contribution to Covariance between Relatives." *Genetics* 184 (4): 1037–1050.

Tudge, S. J., R. A. Watson, and M. Brede. 2016. "Game Theoretic Treatments for the Differentiation of Functional Roles in the Transition to Multicellularity." *Journal of Theoretical Biology* 395:161–173.

Uller, T., S. English, and I. Pen. 2015. "When Is Incomplete Epigenetic Resetting in Germ Cells Favoured by Natural Selection?" *Proceedings of the Royal Society B* 282 (1811): 20150682.

Uller, T., and H. Helanterä. 2011. "When Are Genes 'Leaders' or 'Followers' in Evolution?" *Trends in Ecology and Evolution* 26 (9): 435–436.

Uller, T., and H. Helanterä. 2017. "Niche Construction and Conceptual Change in Evolutionary Biology." *British Journal for the Philosophy of Science*, axx050.

Via, S., and R. Lande. 1985. "Genotype-Environment Interaction and the Evolution of Phenotypic Plasticity." *Evolution* 39 (3): 505–522.

Volk, T. 2017. *Quarks to Culture: How We Came to Be*. New York: Columbia University Press.

Waddington, C. H. 1957. *The Strategy of the Genes*. London: Allen.

Wagner, G. P., and L. Altenberg. 1996. "Perspective: Complex Adaptations and the Evolution of Evolvability." *Evolution* 50 (3): 967–976.

Walsh, D. M. 2015. *Organisms, Agency, and Evolution*. Cambridge: Cambridge University Press.

Watson, R. 2012. "Is Evolution by Natural Selection the Algorithm of Biological Evolution?" *Procs. Artificial Life* 13:121–128.

Watson, R. A. 2006. *Compositional Evolution: The Impact of Sex, Symbiosis and Modularity on the Gradualist Framework of Evolution*. Cambridge, MA: MIT Press.

Watson, R. A., R. Mills, and C. L. Buckley. 2011. "Global Adaptation in Networks of Selfish Components: Emergent Associative Memory at the System Scale." *Artificial Life* 17 (3): 147–166.

Watson, R. A., R. Mills, C. L. Buckley, K. Kouvaris, A. Jackson, S. T. Powers, D. Power, et al. 2016. "Evolutionary Connectionism: Algorithmic Principles Underlying the Evolution of Biological Organisation in Evo-Devo, Evo-Eco and Evolutionary Transitions." *Evolutionary Biology* 43 (4): 553–581.

Watson, R. A., N. Palmius, N., Mills, R., Powers, S. T., and Penn, A. 2009, September. "Can Selfish Symbioses Effect Higher-Level Selection?" In *European Conference on Artificial Life*, 27–36. Berlin/Heidelberg: Springer.

Watson, R. A., and E. Szathmáry. 2016. "How Can Evolution Learn?" *Trends in Ecology and Evolution* 31 (2): 147–157.

Watson, R. A., G. P. Wagner, M. Pavlicev, D. M. Weinreich, and R. Mills. 2014. "The Evolution of Phenotypic Correlations and 'Developmental Memory.'" *Evolution* 68 (4): 1124–1138.

Weinreich, D. M., R. A. Watson, and L. Chao. 2005. "Perspective: Sign Epistasis and Genetic Constraint on Evolutionary Trajectories." *Evolution* 59 (6): 1165–1174.

West, S. A., C. El Mouden, and A. Gardner. 2011. "Sixteen Common Misconceptions about the Evolution of Cooperation in Humans." *Evolution and Human Behavior* 32 (4): 231–262.

West, S. A., R. M. Fisher, A. Gardner, and E. T. Kiers. 2015. "Major Evolutionary Transitions in Individuality." *Proceedings of the National Academy of Sciences USA* 112 (33): 10112–10119.

West-Eberhard, M. J. 2003. *Developmental Plasticity and Evolution*. New York: Oxford University Press.

Williams, G. W. 1992. *Natural Selection: Domains, Levels, and Challenges*. New York: Oxford University Press.

Wilson, D. S. 1975. "A Theory of Group Selection." *Proceedings of the National Academy of Sciences USA* 72 (1): 143–146.

Worden, L. 2010. "Notes from the Greenhouse World: A Study in Coevolution, Planetary Sustainability, and Community Structure." *Ecological Economics* 69 (4): 762–769.

Wray, G. A., H. E. Hoekstra, D. J. Futuyma, R. E. Lenski, T. F. Mackay, D. Schluter, D., and J. E. Strassmann. 2014. "Does Evolutionary Theory Need a Rethink?-COUNTERPOINT No, All Is Well." *Nature* 514 (7521): 161.

11 The Paradox of Population Thinking: First Order Causes and Higher Order Effects

Denis M. Walsh

Darwin's discovery of descent with modification was inaugurated by a shift in perspective that Ernst Mayr (1975) has dubbed "population thinking." Darwin realized that the explanation of the fit and diversity of organic form should be approached as a question about the constitution of populations. Rather than ask how individual organisms come to acquire their remarkable features—their complexity, their functional integration, their exquisite adaptedness to their conditions of existence—we should ask how populations come to comprise such individuals. Reframing the question in this way, and deploying a rudimentary grasp of the principles of population change, Darwin was able to demonstrate that the daily activities of organisms' lives (plus some heritable variation) suffice to account for the array of "endless forms most beautiful and most wonderful." There is no need to invoke nonsubstantival forms, entelechies, vital forces, or providential design, as prior theories of fit and diversity of form were wont to do (Ariew 2008). Though this is a simple conceptual maneuver, it ushered a seismic change in the project of explaining biological form that Mayr (1982) is quite right to acclaim as both revelatory and revolutionary.

The Modern Synthesis version of population thinking is an advance on Darwin's. It involves a radical reconceptualization of the object of evolutionary study. In place of assemblages of individual organisms, the Modern Synthesis casts biological populations as ensembles of abstract types, commonly gene types (Fisher 1930 [2000]; Morrison 2002). The study of evolution is the study of the kinematics of these ensembles. The principal virtue of this version of population thinking lies in its capacity to account for evolutionary change *without* having to advert to the complex, multitudinous properties of individuals. Evolutionary explanations in this mode cite only the ensemble-level properties. By way of illustration, Fisher draws an analogy between the study of evolution and the "Theory of Gases" in which:

... it is possible to make the most varied assumptions as to the accidental circumstances, even the essential nature of the individual molecules, and yet develop the general laws as to the behaviour of gases. (Fisher 1930 [2000], 36)

This rarefied form of population thinking permits biologists to explain, predict, and quantify evolutionary change in a way that would have been impossible without the ascent to this ontology of abstract types.

The orthodox interpretation of Modern Synthesis population thinking is that it articulates the causes of evolution that operate exclusively at the level of populations.[1] Natural selection, for example, is that processes that occurs to a population when and only when there is predictable variation in the fitnesses of the trait types (Sober 2013). It is taken to be a discrete population-level cause, and the only process that introduces a systematic adaptive bias to evolution: "allele frequency change caused by natural selection is the only credible process underlying the evolution of adaptive organismal traits" (Charlesworth, Charlesworth, and Barton 2017, 10). The orthodox interpretation has the implication that if the causes of evolutionary change can be exhaustively described without adverting to the properties and activities of individual organisms, then organisms would appear to be expendable as parts of the apparatus of evolutionary theory. Modern Synthesis population thinking appears to imply that the lives and deaths of individual organisms play no (or at best a peripheral) role in the explanation of evolutionary population change.[2]

The population is an entity, subject to its own forces, and obeying its own laws. The details concerning the individuals who are parts of this whole are pretty much irrelevant. Describing the single individual is as theoretically peripheral to a populationist as describing the motion of a single molecule is to the kinetic theory of gases. In this important sense population thinking involves *ignoring individuals*. (Sober 1980, 344 [emphasis added])

And yet, the cardinal lesson to be drawn from the startling empirical advances of twenty-first-century biology is exactly the opposite. Evo devo, Eco-evo devo, extended inheritance, niche construction, epigenetics—these fields unanimously point to the significance of individual organisms as difference makers in evolution. The organismal-level processes of development, ecology, behavior, learning, immunity, epigenetic imprinting, alternative splicing, genome engineering, and ecosystem engineering (to name a mere few) influence the tempo, mode, and direction of evolutionary change. These are among the processes whose marginalization from Modern Synthesis evolutionary thinking has evidently been licensed by population thinking.[3]

So, it seems that our best current biology is embroiled in something resembling a paradox: the causes of individual living, dying, and reproducing both are and are not indispensable parts of the explanation of evolutionary change. More simply, current evolutionary thinking is lumbered with a "Paradox of Population Thinking" (PPT):

PPT: (i) the processes occurring within and between organisms in their day-to-day lives are (among) the causes of evolution.
(ii) these processes are precluded from the explanation of evolution.

The tension highlighted by the PPT is clearly evident in the most fervid dispute in current evolutionary biology, the so-called Extended Evolutionary Synthesis (EES) debate.

In essence, the debate concerns whether the Modern Synthesis, as standardly interpreted, can assimilate the insights of twenty-first-century biology. Those who call for a wholesale revision (Shapiro 2011; Laland et al. 2015; Noble 2017) argue that ignoring the various activities of individual organisms renders an impoverished understanding of evolution. Those who defend the Modern Synthesis against these challenges (Wray et al. 2014; Charlesworth et al. 2017) maintain that natural selection operating on genetic variation is wholly adequate to the explanation of evolutionary change. Proponents of the EES press considerations in support of (i), while supporters of the status quo argue from (ii). There seems to be no resolution in sight.

But, if the PPT really is at the heart of this debate, then it is no real dispute at all. It is possible to reconcile Modern Synthesis population thinking with a substantive causal/explanatory role for organisms in evolution. Properly understood, population thinking is in no way inimical to according a substantive explanatory role to individual organismal processes in evolution, *even if* in some important sense it "involves ignoring individuals": (ii) is false (at least on one reasonable reading). The fault lies not so much with population thinking itself but in the orthodox interpretation. The objective of this chapter is to take a step toward the reconciliation between Modern Synthesis population thinking with a substantive causal/explanatory role for organismal processes as the causes of evolution. The place to start is back at the *Origin*.

1.   Darwin's Answer

Early in chapter 3 of the *Origin*, Darwin poses the question around which the entire work is structured: "How have all those exquisite adaptations of one part of the organization to another part, and to the conditions of life, and of one distinct organic being to another being, been perfected?" (1859 [1968], 114). We might be excused for supposing that his answer is "natural selection." Natural selection, after all, is the process that (alongside Wallace) Darwin discovered, the one most intimately associated with his name. But that is not the answer that Darwin gives. The very same paragraph ends: "All these results ... follow inevitably from the struggle for life" (Darwin 1859 [1968], 115). Struggling for life is not natural selection. Natural selection happens to populations, whereas struggling is something that organisms do. Moreover, Darwin says these population changes "follow inevitably" from what organisms do. Two important features of Darwin's answer—*following inevitably* and the *struggle for life*—point to the reconciliation between Modern Synthesis population thinking and a substantive role of organisms in evolutionary explanation.

When a population undergoes evolutionary change, typically three things happen to it: (1) It changes in its lineage structure. That is to say, as organisms are born, survive, reproduce and die, some ancestor/descendent lineages become more prevalent than others. (2) It changes in its adaptedness; the population comes to comprise individuals generally better suited to their conditions of existence. (3) It changes in its trait structure; some trait types increase in their relative frequency at the expense of others.[4] Darwin's theory is obviously primarily an account of the causes of the first kind of population change. Yet, he is explicit that, in certain circumstances (specifically, that the trait types that confer greater survival and reproduction are heritable), changes of the second and third kind happen as consequences of the first. Richard Lewontin, for example, has repeatedly argued that Darwin's theory of evolution requires heritable variations in individual fitness. The general idea is that when individual fitnesses and the traits that confer them are reasonably heritable, then as a population changes in its lineage structure, two further changes will happen: the fitness conferring trait types will increase in their relative frequency, and the population will tend to comprise better adapted individuals. So, we may see Darwin's version of population thinking as offering two things. First, it is an account of changes in the adaptedness of form (type 2 change), and changes in population trait structure (type 3), as *concomitants* of changes in lineage structure. Second, it is a detailed account of the individual-level causes of changes in lineage structure. "All these results … follow … from the struggle for life."

The Modern Synthesis theory of evolution has a different principal objective. Like Darwin's theory, it explains changes of the third kind—change in the trait structure—in a population.[5] It does so by appeal to population-level processes, selection, and drift in a way that dispenses with the causes of type 1 change—those that impinge on the lives of individual organisms (Walsh, Lewens, and Ariew 2002; Walsh, Ariew, and Matthen 2017). It plots changes in trait structure as a function of variation in the growth rates (fitnesses) of the trait *types*. The varying growth rates of the trait types at one time predict and explain the trait frequencies at a later time.

As mentioned, the orthodox interpretation supposes that in identifying population-level processes that explain change in relative trait frequencies, the Modern Synthesis version of evolution has articulated the causal structure of evolution; selection and drift explain population change because they are population-level causes of evolution.[6] Maybe so; but a naïve interpretation of this metaphysical picture threatens to obscure Darwin's great insight. "Following inevitably" suggests that change in trait structure is an *analytic consequence* of the struggle. Moreover, it is a *higher order effect*. These two concepts, *analytic consequence* and a *higher order effect*, are crucial to an understanding of the relation between individual-level organismal processes and population-level change. They will need some explication.

1.1   Analytic Consequence

Imagine two particles in a container, p_1 and p_2, moving away from one another at constant velocity, such that their center of mass, c, remains fixed.[7] Suppose now that p_1 contacts the wall of the container and rebounds back, while p_2's motion continues as it was. Immediately, the location of c changes. It is tempting to say that the change in c is caused by the change in p_1, or by the change in the relative motions of p_1 and p_2. I have no desire to legislate the use of these terms, but if this is causation, it is no run-of-the-mill causal relation. It couldn't be the case, for example, that some causal signal is propagated from p_1 to c.[8] The change in c is instantaneous. Such a signal would have to travel faster than the speed of light, and we're told that no causal signal can. Rather, the change in c is *entailed* by the change in p_1. The location of c, at any time, is a function of the straightline distance between p_1 and p_2 and the product of their masses. The motion of c is simply *constituted by* the relative motions of p_1 and p_2. The change in c, I shall say, is an *analytic consequence* of the locations and masses of p_1 and p_2.

Similar considerations apply to changes in trait structure. Given that a biological population—even an ensemble of trait types—is constituted of individual organisms, there are basically two ways that it can change in its constitution: arrival and departure. Organisms arrive by immigrating or being born (hatched, budded, germinated, etc.). They depart by emigrating or dying.[9] As they enter or leave the population they bring or take their trait tokens with them. As a consequence of all this arrival and departure, the trait types that those tokens belong to may become more or less numerous.

The arrival and departure of individual organisms doesn't *cause* the frequency of the trait types in a population to decrease or increase (in any conventional sense), any more than the movement of the particles, p_1 and p_2, causes the change in the center of mass, c. The change in trait structure *just is*, or is *entailed* by, the aggregate of individual arrivals and departures. Equivalently, the disproportionate deaths of white morph (*Biston betularia*) individuals, for example, doesn't *cause* the increase in relative frequency of the black morph (type), it just *is*, or *entails*, the increase in relative frequency of the black morph. If the objective of Modern Synthesis population dynamics is to quantify, predict, and explain these changes in trait structure, then it is a theory of the analytic consequences of arrival and departure.

1.2 Higher Order Effect

Change in trait structure is, moreover, a special kind of analytic consequence; it is a *higher order effect*. A higher order effect is the effect on an ensemble of the aggregated effects of the causes affecting the components severally. That's a bit of a mouthful, but the idea is simple. Consider the changing pattern of colored balls in a billiard game. The first player breaks. If the cue ball contacts a red ball, which scatters the other reds, one of which in turn hits the brown, and then ricochets off to bump the green, which nudges the yellow, the forces acting on the balls (severally) cause each one to move. As a result, the pattern of balls on the table also changes: perhaps the average distance between balls has increased, perhaps the reds are more evenly distributed across the table. The change in pattern is a higher order effect of the forces acting on each the billiard balls severally.

Higher order effects are familiar to us and highly significant. Passive diffusion is a helpful example. If we put a drop of concentrated potassium permanganate into a beaker of water, it diffuses. The highly localized deep-purple droplet becomes an evenly distributed light pink. Erwin Schrödinger (1944) explains diffusion in the following way. Suppose we were to place a membrane into the solution some time before it has reached equilibrium. Even if the motions of the permanganate molecules are random (i.e., none has any greater propensity to move in one direction over another), we would find that there are more collisions with the membrane on the high-density side than on the low-density side. That is to say that there are more molecules traveling from high density to low than the other way around. And this will continue until the solution has reached an equilibrium distribution of permanganate molecules.

Diffusion is thus a—highly significant, non-trivial—higher order effect: the effect on an ensemble of the aggregate of causes acting on the individual components. It is a regular, projectable occurrence that is discernible only at the level of ensembles; it is a population-level phenomenon. We must ascend to the level of population structure to explain it, or even to see it. But Schrödinger's account demonstrates that we do not need to posit a population-level force or cause in order to explain it. All the causes in this system are those that move the individual molecules around. To posit a further, population-level, *diffusive* cause or force would be to overpopulate the world with causes.

So too with Darwin's account of evolution. According to Darwin, the changes in a population's structure—both trait structure and adaptedness—are also higher order effects. Each is the effect on the structure of a population of the complete suite of causes of individual birth, survival, death, and migration, in the "struggle for life." The answer that Darwin gives to his central question, then, is that—to paraphrase—evolutionary change in the trait structure of a population is both a higher order effect and an analytic consequence, of the individual-level causes of the arrival and departure of organisms.

2. Higher Order Effect Explanations

Where there are individual-level causes and higher order effects, we can distinguish two kinds of explanations of ensemble change. In one, we cite the way the causes acting on individual components of an ensemble affect those components. I'll call these "first order cause" explanations. In the other, we cite the way the distribution of these effects affects the structure of the ensemble over time. I'll call these "higher order effect" explanations. These two forms of explanation have different characters and tell us different things about the change in the structure of an ensemble.

The first order cause explanation of diffusion demonstrates that the final arrangement of the molecules (their respective locations and momenta) is a result of whatever caused each molecule to move in the way it did. The final arrangement is sensitively dependent upon the properties of individual-level molecules, where they started out, and what external forces acted on them. This explanation (causally) entails that were we to have this exact combination of initial conditions and individual-level causes again, the final arrangement of locations and momenta would be exactly the same. The explanation is highly specific and detailed. But there is a lot about the dynamics of this system that it doesn't explain. For example, it tells us that diffusion, a macroproperty of the system, is dependent upon its microproperties—the arrangement of individual causes—but it says nothing about how sensitive diffusion is in general to variations in these microproperties. It fails to answer a version of the question that Woodward (2003) takes to be the mark of a good explanation: "What if things had been different (in relevant ways)?" As such, it doesn't tell us much about what this particular instance of diffusion has in common with any other. The first order cause explanation lacks the generality and projectability that we usually want from an explanation of diffusion.

The higher order effect explanation fills in these lacunae. It demonstrates that while diffusion is causally dependent upon the arrangement of molecules and their velocities, it is largely insensitive to the differences, and rather more sensitive to the *distribution* of velocities. In doing so, the higher order effect explanation represents diffusion as a robust and general phenomenon, realizable in a huge array of systems, given virtually any of a range of arrangements. The higher order effect explanation of diffusion has particular virtues; while it does not have much causal detail, it is robust and generalizble.

The first order cause and the higher order effect explanations demonstrate that the phenomenon to be explained—change in population structure—exhibits a modal dependence on two different kinds of properties. First order cause explanations demonstrate that the phenomenon depends upon the behavior of individuals. Higher order effect explanations demonstrate that the phenomenon to be explained counterfactually depends upon a property of ensembles.

The relation between higher order effect explanations and first order cause explanations has a significant pedigree. It is implicit, for example, in Maxwell and Boltzmann's development of statistical mechanics. Maxwell and Boltzmann demonstrated that the macroscopic behavior of gases are higher order effects of the aggregated motions of individual molecules. Boltzmann discerned that a gas of a given kind at a given temperature would have an equilibrium distribution of molecular velocities. This distribution explained the gases' typical behavior.

The molecules are likewise just so many individuals having the most varied states of motion, and it is only because the number of them that have, on the average, a particular state of motion that is constant, that the properties of the gas remain unchanged. (Boltzmann 1872; cited in Uffink 2017, translation by Uffink)

Moreover, each microstate of the molecules (i.e., the arrangement of individual momenta) realizes a particular macrostate (i.e., a distribution of energies). Those microstates that instantiate high entropy macrostates occupy a much greater volume of the state space than those that instantiate low entropy macrostates (Frigg 2011). It is thus overwhelmingly more likely that a volume of gas will move from lower to higher entropy macrostates than the other way around. Classical mechanics delivers a first order cause explanation of the change or stasis in the structure of a gas. Statistical mechanics offers a higher order effect explanation of the same phenomena. In implicitly exploiting the distinction between first order cause explanations (of classical mechanics) and the higher order effect explanations (of statistical mechanics) Maxwell and Boltzmann were able to ground the thermodynamics of ensembles in the Newtonian physics of forces impinging on individuals. They were able to demonstrate that thermodynamic phenomena are simply the consequences (higher order effects) of first order causes.

The distinction between first order cause explanations and higher order effect explanations is equally applicable to evolutionary biology. Suppose we wish to explain the change in trait structure (i.e., the relative frequencies of trait types) of a particular population. We can do so *either* by citing the first order causes *or* the higher order effects. The first order cause explanation describes the ways in which biological, ecological processes impinged on the lives, deaths, and reproductions of individual organisms, and how the aggregate of these in turn was realized changes in trait structure. We could say, for example, that the visibility of white morph moths on soot-blackened trees caused disproportionately heavy predation on black moths. The aggregate effect of these predations is registered at the population level as a decrease in relative frequency of the white morph individuals.

This explanation has its limitations.[10] It does not tell us, for example, how sensitive the change in trait structure is to variations in individual causes of living and dying. For that we need to offer the sort of higher order effect explanation we find in Modern Synthesis population thinking. In order to do so, we first reconfigure the population as an ensemble abstract entities (trait types) as Fisher did. Then we assign a parameter (trait fitness) to the abstract trait types that measures their relative growth rates. Assigning differential fitnesses to the trait types will allow us to explain and predict their change in relative frequency (trait structure), irrespective of the details of the individual-level causes in operation. These explanations are highly robust, abstracted, projectable, and largely substrate neutral (Matthen and Ariew 2002; Walsh et al. 2017).

The first order cause explanations and the higher order effect explanations explain the *same phenomena*, change in population trait structure. But they do so in different ways. First order cause explanations represent population change as a consequence of those processes that impinge on individual organisms. Higher order effect explanations represent the change in trait structure as the consequence of differential rates of change in the relative frequencies of abstract trait types. First order cause and higher order effect explanations are each complete in the sense that neither augments the other. Once we know that rate of change of the various trait types, we can give a higher order effect explanation. Any further information about which causes impinge on which individuals is strictly redundant. Furthermore, once we know the individual-level causes of evolution, and which traits each individual has, the growth rates of the abstract trait types are explanatorily redundant (though they can be deduced). Nevertheless, the explanations are complementary and non-competing. Each tells us something different about the dynamics of the population; neither replaces the other (Walsh et al. 2017).

3.   The Two-Force Model

It has not been customary to represent the relation between the individual-level causes of evolution and the population-level phenomena of change in trait frequency in this way. Under the sway of the orthodox interpretation of Modern Synthesis population thinking, population-level processes (selection and drift), and individual-level processes are conceived as distinct causes of evolutionary change. I call this the "two-force model" of evolution (Walsh 2003, 2015). It is the two-force model, I contend, rather than population thinking itself, that has perpetuated the apparent conflict between the Modern Synthesis approach to explaining population change, and the explanatory role of individual-level biological processes in evolution, of the sort promoted by advocates of the Extended Evolutionary Synthesis. The two-force picture makes two fundamental kinds of error. It is wrong on empirical grounds, and it is wrong on conceptual grounds.

The two-force model treats populations as concrete particulars and selection and drift as forces acting on them. This metaphysical maneuver promoted the marginalization of organisms that progressively infiltrated twentieth-century biology.[11] Organismal development provides a striking example. Viktor Hamburger (1980) notes that development was "blackboxed" in Modern Synthesis evolutionary biology. There didn't seem to be much that its details could offer one way or another to the account of how populations change. Development is simply treated as a conduit that delivers traits to the arena of selection; whereupon the population-level forces have at them. Bruce Wallace's suggestion that "problems concerned with the orderly development of the individual are unrelated to those of the evolution of organisms through time" (Wallace 1986, 149) seems to have been reasonably representative of the attitude to the relevance of organisms to evolution that prevailed at the time. As John Maynard Smith declared, "it is possible to understand genetics, and hence evolution, without understanding development" (Maynard Smith 1982, 6).

One traditional way to implicate development in evolution was to cast it as a constraint, a particular kind of "bias in form" (Maynard Smith et al. 1985). If selection is the sole force that imparts an adaptive bias to evolution, then development will show up only as an impediment to that force. Even among those who advocated the importance of development to evolution, it was generally seen as a constraint on the adaptation promoting power of selection.

The nature of the existing developmental system somehow constrains or channels acceptable change [of form in evolution], so that selection is limited in what it can achieve given some starting anatomy. (Raff 1996, 294–295)

For instance developmental constraints frustrate selection by restricting the phenotypic variation selection has to act upon. Adaptations would be able to evolve only to optima within the constrained space of variability. (Wagner and Altenberg 1996, 973)

The processes occurring in the lives of individual organisms are explanatorily relevant, then, only when selectional explanations cannot fully account for the distribution of biological form.

This has led to a 'dichotomous approach' in which constraint is conceptually divorced from natural selection and pitted against it in a kind of evolutionary battle for dominance over the phenotype ... much of the constraint literature over the last 25 years has explicitly sought to explain evolutionary outcomes as either the result of selection or constraint. (Schwenk and Wagner 2004, 392)

The picture generalizes to all individual-level processes, inheritance, mutation, migration, niche construction, plasticity, epigenetic marking. The list of such processing is long and expanding. Selection and the individual-level causes of organismal form are both represented as "forces" that move a population around its state space (or impede its motion). Moreover, individual-level processes and population-level processes are distinct, and autonomous. One force promotes adaptation, while the others either oppose adaptive change, or are adaptively neutral.[12] As they typically have different effects on population structure, the supposition goes, they must be different kinds of causes.

The empirical error of the two-force model has been comprehensively exposed over recent years. Evo devo, eco-evo devo, Developmental Systems Theory, multiple inheritance systems, epigenetics, ecosystem engineering, genome engineering, niche construction: these fields of study have all brought to light the various ways in which the processes that occur within and to individuals make a substantive positive contribution to evolution. The responses of organisms to their circumstances introduce stable evolutionary novelties into a population (West-Eberhard 2003). They structure the inheritance of characters (Uller and Helanterä 2017). They affect the kind and degree of variation in a population (Sultan 2003). They introduce heritable, adaptive biases into evolution (Moczek et al. 2011; Laland et al. 2013; Herman et al. 2016).

The volume and variety of empirical corroboration is mounting daily. Organisms are no longer "irrelevant." Nor are organismal processes a mere check on those processes that promote adaptation. They are *causes* of adaptive evolutionary change. One cannot understand evolution unless one understands the ways that individual-level processes systematically drive bias and constitute evolution. I take it to be the single-most significant achievement of twenty-first-century evolutionary biology that it has established the contribution made to the adaptiveness of evolution by the processes that occur within and between individual organisms.

The traditional two-force picture could be amended to accommodate the great empirical advances of recent evolutionary biology, without being dismantled altogether. We could acknowledge, for instance, that individual-level forces work alongside, or in conjunction with, the population-level processes of selection and drift. They both promote adaptive evolution; they are both reflected as changes in trait structure.[13] The complete account of the causes of evolution, then, must incorporate *both* individual-level and population-level processes. We might even find ourselves seeking a division of explanatory labor, apportioning a certain amount of causal responsibility to individual-level processes and a distinct portion to the population-level processes. Indeed, many of those who advocate for an expansion or extension of the Modern Synthesis work with this modified two-force model (implicitly or explicitly) in mind. Strategy might provide a welcome corrective to the marginalization of individual-level processes from evolutionary explanations. It might further offer a partial solution to the empirical error of the two-force model.[14]

But reconfiguring the relation between individual-level causes of evolution and population-level processes in this way would perpetuate the conceptual error of the two-force model. It is the conceptual error that generates the Paradox of Population Thinking. It is also responsible for the marginalization of organisms that progressively took hold throughout twentieth-century Modern Synthesis biology. The revised two-force picture encourages us to ask: "how much of evolutionary change is due to individual-level processes and how much to selection?" But one cannot partition causal responsibility between the first order causes of population change and their higher order effects in this way. Consider the analogue in the case of diffusion; we might ask, "How much of the change in distribution of the molecules is due to individual-level forces and how much is due to diffusion?" Similarly, in the case of the behavior of a gas we might ask, "How much of the change in his gas is due to the forces acting on individual molecules, and how much to entropy?" These questions make no sense. The incoherence results treating the causes of molecular motion on the one hand and (say) entropy on the other as discrete, interacting causes, as one might treat, say, gravitational and electromagnetic forces. But they are not; they are effects of a suite of common causes. Both the change in entropy and motions of molecules are effects of the same cause, of the forces acting on individual molecules. Change in entropy is the higher order effect; the motion of molecules is the first order effect. Analogously, if change in population trait structure—selection and drift—is a higher order effect of individual-level processes, then it is strictly illegitimate to ask how much of the change is due to selection and drift (on one hand) and how much is due to individual-level processes (on the other). Selection and drift, are higher order effects, and organismal living, dying, reproducing, are first order effects *of the same causes*. All these causes operate at the level of individual organisms.

The error of the two-force model lies in construing individual-level causes and ensemble-level processes as somehow on an ontological par, as interacting causes of ensemble change. This generates the competition for explanatory relevance that is the mark of the two-force picture. But this is a category error. A new picture is needed.

4. The Two-Level Model

My suggestion is that we should replace the "two-force model" with the "two-level model." According to the two-level model, there is one level of causation; *all the causes of evolution are the causes of arrival and departure* (the "struggle for life"). Yet, there are two discernible levels of effect. There are effects on individual organisms (first order effects), and there are effects on the distribution of abstract trait types in a population (higher order effects). Consequently, there are two wholly distinct kinds of evolutionary explanation. First order cause explanations tell us how the processes that impinge on individuals *cause* evolutionary change in population structure. Higher order effect explanations tell us how changes in trait structure at one time *depend* upon the distribution of growth rates of the abstract trait types at an earlier time. So, one and the same phenomenon—evolutionary change in trait structure—can be explained in two different ways: by citing first order causes or higher order effects. They each have their uses. These two kinds of explanation tell us very different things about the evolution of a population. First order cause explanations identify the causes of population change but are typically less good at quantifying and predicting rates of population change. Higher order effect (natural selection) explanations measure and project rates of change in trait structure. But they are silent on which individual-level causes impinge on individuals, and on how changes in trait structure depend upon individual-level causes.

First order cause explanations and higher order effect explanations can both either be specific, applying to a single population at a time, or highly general, applying widely across biological populations and times. But they generalize in different ways.[15] First order cause explanations can be generalized to show, for example, how *in general*, individual-level processes—the plasticity of development, epigenetic inheritance, the construction of one's niche, social learning, the interaction with one's microbiome, the engineering of one's own genome—can impact the dynamics of evolution. These processes can cause the origin and maintenance of novelties, or the resemblance of offspring to parent. They may account for the resistance to perturbing effects of mutations or environmental disturbances. They explain the origin of adaptively biased novelties, the capacity of gene regulatory networks to search adaptive space. These are exactly the sorts of contributions to evolution that those who advocate the extension or revision of the Modern Synthesis highlight. Higher order effect (selection and drift) explanations, for their part, generalize in another way. The principles they appeal to (distribution of trait fitness, population size) apply to *any* population in which the abstract trait types can be assigned growth rates, no matter what the causes are or who they affect. These two modes of biological explanation are autonomous, complementary, and wholly noncompeting kinds of explanations. Evolutionary biology needs both kinds of explanations.

The take-home message of the two-level model, in a slogan, is *one level of causation; two levels of effect*. All the causes of evolution are the individual causes of arrival and

departure. The effects of these causes can be registered as the differential growth of lineages (first order causes), which has as an analytic consequence change in trait structure.
Alternatively, the effects of individual-level causes can be measured as the differential
growth of abstract trait types (higher order effects). Alongside some conceptual clarity,
the two-level model offers a resolution of the paradox of population thinking. Recall that
the PPT says:

(i): the processes occurring within and between organisms in their day-to-day lives are
(among) the causes evolution.

(ii): these processes are precluded from the explanation of evolution.

The two-level model endorses (i) but shows (ii) to be ambiguous. The problem with
(ii) is that "the explanation of evolution" can mean either of two things: "first order cause
explanation" or "higher order effect explanation." Only one of these kinds of explanation
ignores individual processes. These are the higher order effect explanations (those that
invoke selection and drift). They are indispensable to evolution. But they are not the only
sorts of evolutionary explanations we might want. First order cause explanations, however,
do not ignore individuals; they cite the causes of individual arrival and departure. These
too are indispensable to evolutionary biology. On one reading of (ii), the PPT is true but
wholly unparadoxical, on the other, the PPT is paradoxical, perhaps, but nevertheless false.

5.   Proximate and Ultimate

The misunderstanding of population thinking encapsulated in the PPT pervades much of
the discourse in current evolutionary biology and its philosophy.[16] A case in point is the
recently renewed debate over the propriety, or otherwise, of Mayr's (1961, 1982) proxi-
mate/ultimate distinction.

   Mayr (1961) distinguished between two kinds of questions we might pose in evolutionary biology, the "how" and the "why." We may ask how an individual came to acquire
a particular trait, or we may ask why a particular trait is prevalent in a population. Mayr
correctly perceives that these questions call for different kinds of explanation. The former
is an individual-level explanation, the latter is a population-level explanation. He
proposes that these different kinds of explanations correspond to different kinds of
causes—proximate and ultimate causes respectively (Scholl and Pigliucci 2015). The
proximate causes are "biological" processes—processes that occur within individual's
lifetimes. The ultimate causes are "evolutionary processes," those that happen to populations over evolutionary time (specifically selection and drift). The upshot of the proximate/
ultimate distinction seems to be that "proximate causes," those individual-level processes
that cause organisms severally to have the traits they have, do not appear in evolutionary
explanations.

Understandably, those who seek to extend or expand the Modern Synthesis, by incorporating these very processes into the explanations of evolutionary change, resist the proximate/ultimate distinction: "The proximate ultimate distinction has given rise to a new confusion, namely, a belief that proximate causes of phenotypic variation have nothing to do with ultimate, evolutionary explanation" (West-Eberhard 2003, 6). They do so on two eminently reasonable grounds. First, they point to the fact that (as per thesis (i) of the PPT) individual-level processes most certainly do contribute to evolution (see Odling-Smee, Laland, and Feldman 2000; Laland et al. 2011; Laland et al. 2015).

Proximate mechanisms both shape and respond to selection, allowing developmental processes to feature in both proximate and ultimate explanations. (Laland et al. 2011, 1512)

One prime example of this contribution is "developmental bias," the capacity of organismal development to introduce adaptive bias.

Developmental bias is potentially widespread in nature and can contribute to evolutionary stasis... or promote evolutionary adaptation.... If the proximate biology of a lineage makes some variants more likely to arise than others, these proximate mechanisms help construct evolutionary pathways. (Laland et al. 2011, 1513)

As a consequence, the contributive role of organismal processes to evolution, like development, has been undersold.

... too much causal significance is afforded to genes and selection, and not enough to the developmental processes that create novel variants, contribute to heredity, generate adaptive fit, and thereby direct the course of evolution. (Laland et al. 2015, 6)

Opponents of the proximate/ultimate distinction further point out that, given the nature of causation in evolution, we cannot partition the causes neatly into the "evolutionary" and the "biological":

... causation is reciprocal. This means that "ultimate explanations" must include an account of the sources of selection (what caused the selective environment?) when they are modified by the evolutionary process itself. (Laland et al. 2013, 725)

There should be little wonder, then, that there is widespread resistance to the proximate/ultimate distinction.

Those who wish to preserve the distinction argue that there is explanatory value to be had by distinguishing the population-level explanations of selection and drift from individual-level biological explanations of the causes of living and dying (Wray et al. 2014; Dickins and Rahman 2013; Dickins and Barton 2012).

To answer how an individual operates within a particular environment is to give a proximate account. To answer why that individual operates in that way is to give an ultimate explanation. In this way ultimate accounts address the reasons why a trait evolved, typically in terms of selection, but also potentially including traits that are by-products of selection or selectively neutral traits spread by genetic drift. (Dickins and Barton 2012, 749)

These authors argue that the prevalence of proximate mechanisms that contribute to evolution are themselves the consequences of evolutionary processes. They should be explained in the way that the prevalence of any adaptive trait is explained. Take, for example, developmental bias. "Developmental bias, if real, is the result of proximate mechanisms that themselves have ultimate explanations" (Dickins and Barton 2012, 753). The quite reasonable point is that the individual-level processes that contribute to evolution have evolved. To cast them merely as causes of evolutionary change, and not as consequences of evolution, is to risk overlooking the fact that they too have evolutionary "why" explanations. The Dickins and Barton point is a platitude of course. It is agreed on all sides that prominent traits have evolutionary explanations, and that biologists quite legitimately distinguish between "how" and "why" questions.[17] As I understand it, their point is that this platitude alone speaks strongly in favor of the proximate/ultimate distinction.

We seem to have reached an impasse. On the one hand, preserving the proximate/ultimate distinction seems to debar us from allowing merely "biological" processes from participating in evolution. This is not only implausible, it threatens to undo the great empirical advances of twenty-first-century biology. On the other hand, abandoning the distinction seems to debar us from distinguishing genuine evolutionary explanations of the distribution of trait types in populations ("why" explanations) from "biological" explanations of the occurrence of trait tokens in individuals ("how" explanations). But this distinction too is undeniable. It looks like evolutionary biology needs but cannot have a proximate/ultimate distinction.

The dispute is spurious, of course. In fact, the proximate/ultimate debate is just a rehash of the PPT. Clearly, those who argue against the proximate/ultimate distinction, and those who defend it, construe "evolutionary explanation" differently. The former (implicitly) interpret it as "first order cause" explanations, and the latter as "higher order effect" explanation. Those who oppose the distinction argue (i) that individual-level processes are among the causes of evolution, and should figure in evolutionary explanations. Those who wish to preserve the distinction do so from the perspective of a particular reading of clause (ii). They argue, correctly, that (a certain kind of) evolutionary explanation ignores individual-level causes. But once the equivocation on "evolutionary explanation" becomes clear, we see that these positions are consistent.

Consider the arguments on each side of the debate. Those who propose to abandon the distinction do so on causal grounds.[18] Those who wish to preserve the proximate/ultimate distinction cite explanatory considerations. They appeal to the different kinds of "accounts" we give of individual-level and population-level phenomena. So, revisionists argue that there is no defensible *causal* proximate/ultimate distinction, while retentionists argue that there is a genuine *explanatory* distinction. They are both right.

Just as the two-level model resolves the PPT, it defuses the proximate/ultimate debate. The revisionists are right in insisting that there is no such thing as an ultimate cause as distinct from a proximate cause of evolution. All the causes of evolution are causes of individual living and dying. It is "proximate" causes all the way down. Nevertheless, there is a workable distinction between proximate and ultimate *explanations*.[19] The proximate explanations are the first order cause explanations that trace the way that causal processes affecting individual organisms bring about change in population trait structure. The ultimate explanations are the higher order effect explanations that trace the way that the distribution of growth rates of the abstract trait types in a population results in change in trait structure.

Those who insist that the ultimate explanations of selection and drift somehow preclude, or compete with, the proximate explanations have simply misunderstood the relation between individual-level (first order) causes and population-level (higher order) effects. I detect the influence of the two-force model at work here. The two-level model demonstrates that evolution needs both levels of explanation, but it has no use for two levels of causation.

6.   Conclusion

For far too long evolutionary biology has been gudgeoned by an unparadoxical "paradox":

PPT: (i): the processes occurring within and between organisms in their day-to-day lives are (among) the causes evolution.
   (ii): these processes are precluded from the explanation of evolution.

The apparent paradox is the result of an equivocation on "evolutionary explanation," which is aided and abetted by the two-force model of evolution. That model makes the erroneous supposition that the processes that occur within individual lives constitute one set of causes, and the processes that occur to populations (e.g., selection, drift) constitute another, wholly distinct set of causes.

The resolution of the paradox lies in giving due weight to Darwin's discovery that all the causes of evolution are to be found in the "the struggle for life"—*inter alia* in the living, dying, and reproducing of individuals—and that evolutionary changes "follow inevitably." The population change we are interested in as evolutionists is change in trait structure. It has two kinds of explanation. One kind adverts to the activities of individuals: organisms live, they die, they migrate, they mate, they give birth, they produce adaptive novelties, they change their environments, they regulate their genome structure and function. These in turn bring about changes in trait structure *as an analytic consequence*. The explanation of these changes invokes the causes that impinge on individual organisms; they are first order cause explanations. The other kind of explanation adverts to the rates of change of the abstract trait types. Some trait types increase in relative frequency with respect to others. A selection (higher order effect) explanation accounts for change in trait structure exclusively in terms of the variation in the growth rates of abstract trait types. The first order cause explanations are "proximate"; the "higher order effect explanations are ultimate."

There is no Paradox of Population Thinking. Thesis (i) is true; thesis (ii) is ambiguous. There are two kinds of explanation of population change, only one of which involves "ignoring individuals." Population thinking is in no way inimical to according individual organisms an indispensable place in evolutionary explanation. This is a lesson we should have learned from the answer that Darwin gave to the central question that structures the *Origin of Species*: "All these things follow inevitably from the struggle for life."

Acknowledgments

I thank the audience at the *Cause and Process in Evolution* workshop at the Konrad Lorenz Institute in Vienna and André Ariew for comments. I am also grateful for comments provided by two anonymous referees. This chapter has benefited greatly from extensive suggestions made by Kevin Laland and Tobias Uller. I thank the IEA, Paris, for the opportunity to apply the final touches.

Notes

1. The *locus classicus* is Sober (1984), but there are many advocates of this view, for example Millstein (2006), Brandon (2006).

2. It was not always supposed to be that way. Many of those who forged the Modern Synthesis, like Ernst Mayr and (early) Theodosius Dobzhansky, stressed the significance of the activities of individual organisms to evolution. Gould (1983) calls the progressive exclusion of organisms from evolutionary theory the "hardening of the synthesis."

3. There are many such appeals. See West-Eberhard (2003), Pigliucci (2009), Pigliucci and Müller (2010), Shapiro (2011), Laland et al. (2013), Laland et al. (2015), Noble (2017). See Huneman and Walsh (2017) and essays therein.

4. By "trait structure" here I simply mean the pattern of relative frequencies of the abstract trait types. If the black morph of *Biston betularia* becomes more frequent than the white morph, this is a change in trait structure.

5. It sometimes further assumes (or attempts to demonstrate; see Grafen 2014) that the other two kinds of population change are concomitants of change in trait structure.

6. See, for example, Sober (1984), Okasha (2006), Shapiro and Sober (2007), and Pence and Ramsey (2015).

7. This example originates with Matthen and Ariew (2009).

8. On one prominent approach to understanding causation (Salmon 1984), a causal process is one that transmits a signal (or alternatively, a conserved quantity; Dowe 1992). According to Salmon, the motion of c is a pseudoprocess.

9. I am leaving aside for now the increasingly important process of organisms changing their evolutionarily significant characters. I'm grateful to Earnshaw-Whyte (2012) for this discussion.

10. I discuss below the ways that first order cause explanations can be made general.

11. Hence the significance of the quote from Sober (above) that begins: "The population is an entity, subject to its own forces, and obeying its own laws," and ends: "... population thinking is about ignoring individual" (1984, 344).

12. Inheritance and mutation are neutral in the sense that the former is strictly conservative and the latter is random.

13. As Pocheville (this volume) argues, they may be "entangled" or "intertwined" to varying degrees.

14. I thank Kevin Laland for pressing this point.

15. I thank Tobias Uller for pressing this issue.

16. I suggest it is the principal motivation behind the "levels of selection debate," though that debate could, in some etiolated form, survive the demise of the two-force model (see Walsh 2003).

17. I thank Kevin Laland and Tobias Uller for help here.

18. In effect, they are arguing from thesis (i) of the PPT.

19. Others including Ariew (2003), Calcott (2013), Scholl and Pigliucci (2015), and Gardner (2013) have argued that the distinction should be seen as primarily an explanatory one. My own view is an extension of Ariew's.

References

Ariew, A. 2003. "Ernst Mayr's 'ultimate/proximate' distinction reconsidered and reconstructed." *Biology and Philosophy* 18:553–565.

Ariew, A. 2008. "Population Thinking." In *Oxford Handbook of Philosophy of Biology*, edited by M. Ruse, 64–86. Oxford: Oxford University Press.

Boltzmann, L. 1872. "Weitere Studien über das Wärmegleichgewicht unter Gasmolekülen." *Wiener Berichte* 66:275–370; in WA I, paper 23.

Brandon, R. N. 2006. "The Principle of Drift: Biology's First Law." *Journal of Philosophy* 103 (7): 319–335.

Calcott, B. 2013. "Why How and Why Aren't Enough: More Problems with Mayr's Proximate–Ultimate Distinction." *Biology and Philosophy* 28:767–780.

Charlesworth, D., B. Charlesworth, and N. Barton. 2017. "The Sources of Adaptive Variation." *Proceedings of the Royal Society B; Biological Sciences,* doi: 10.1098/rspb.2016.2864.

Darwin, C. 1859 [1968]. *Origin of the Species*. New York: Penguin

Dickins, T. E., and R. A. Barton. 2012. "Reciprocal Causation and the Proximate–Ultimate Distinction." *Biology and Philosophy* 28:747–756.

Dickins, T. E., and Q. Rahman. 2013. "The Extended Evolutionary Synthesis and the Role of Soft Inheritance in Evolution." *Proceedings of the Royal Society B.* doi: 10.1098/rspb.2012.0273.

Dowe, P. 1992. "Wesley Salmon's Process Theory of Causality and the Conserved Quantity Theory." *Philosophy of Science* 59:195–216.

Earnshaw-Whyte, E. 2012. "Increasingly Radical Claims about Heredity and Fitness." *Philosophy of Science* 79:396–412.

Fisher, R. A. (1930 [2000]). *The Genetical Theory of Natural Selection: A Complete Variorum Edition*. Oxford: Oxford University Press.

Frigg, R. 2011. "What Is Statistical Mechanics?" In *History and Philosophy of Science and Technology, Encyclopedia of Life Support Systems*, Volume 4, edited by Carlos Galles, Pablo Lorenzano, Eduardo Ortiz, and Hans-Jörg Rheinberger. Isle of Man: Eolss.

Gardner, A. 2013. "Ultimate Explanations Concern the Adaptive Rationale for Organism Design." *Biology and Philosophy* 28:787–791.

Gould, S. J. 1983. "The Hardening of the Modern Synthesis." In *Dimensions of Darwinism: Themes and Counterthemes in Twentieth Century Evolutionary Theory*, edited by M. Grene, 71–93. Cambridge: Cambridge University Press.

Grafen, A. 2014. "The Formal Darwinism Project in Outline." *Biology and Philosophy* 29:155–174.

Hamburger, V. 1980. "Embryology and the Modern Synthesis in Evolutionary Biology." In *The Evolutionary Synthesis*, edited by E. Mayr and W. Provine, 97–112. Cambridge, MA: Harvard University Press.

Herman, J. J., S. Sultan, T. Horgan-Kybelski, and C. Riggs. 2016 "Adaptive Transgenerational Plasticity in an Annual Plant: Grandparental and Parental Drought Stress Enhance Performance of Seedlings in Dry Soil." *Integrative and Comparative Biology* 52:77–88.

Huneman, P., and D. M. Walsh, eds. 2017. *Challenging the Modern Synthesis: Adaptation Development, Inheritance*. Oxford: Oxford University Press.

Laland, K. N., J. Odling-Smee, E. W. Hoppit, and T. Uller. 2013. "More on How and Why: Cause and Effect in Biology Revisited." *Biology and Philosophy* 28:719–745.

Laland, K. N., K. Sterelny, J. Odling-Smee, W. Hoppitt, and T. Uller. 2011. "Cause and Effect in Biology Revisited: Is Mayr's Proximate-Ultimate Distinction Still Useful?" *Science* 334:1512–1516.

Laland, K., T. Uller, M. W. Feldman, G. B. Müller, A. Moczek, E. Jablonka, and J. Odling-Smee. 2015. "The Extended Evolutionary Synthesis: Its Structure, Assumptions and Predictions." *Proceedings of the Royal Society B* 282:20151019. http://dx.doi.org/10.1098/rspb.2015.1019.

Matthen, M., and A. Ariew. 2002. "Two Ways of Thinking about Fitness and Selection." *Journal of Philosophy* 99:58–83.

Matthen, M., and A. Ariew 2009. "Selection and Causation." *Philosophy of Science* 76:201–223.

Maynard Smith, J. 1982. *Evolution and the Theory of Games*. Cambridge: Cambridge University Press.

Maynard Smith, J., R. Burian, S. Kauffman, P. Alberch, J. Campbell, B. Goodwin, R. Lande, D. Raup, and L. Wolpert 1985. "Developmental Constraints and Evolution." *Quarterly Review of Biology* 60:265–287.

Mayr, E. 1961. "Cause and Effect in Biology." *Science* 134:1501–1506.

Mayr, E. 1975. *Evolution and the Diversity of Life*. Cambridge, MA: Harvard University Press.

Mayr, E. 1982. *The Growth of Biological Thought*. Cambridge, MA: Harvard University Press.

Millstein, R. L. 2006. "Natural Selection as a Population-Level Causal Process." *British Journal for the Philosophy of Science* 57 (4): 627–653.

Moczek, A., S. Sultan, S. Foster, C. Ledón-Rettig, I. Dworkin, H. F. Nijhout, E. Abouheif, et al. 2011. "The Role of Developmental Plasticity in Evolutionary Innovation." *Proceedings of the Royal Society B* 278:2705–2713.

Morrison, M. 2002. "Modelling Populations: Pearson and Fisher on Mendelism and Biometry." *British Journal for the Philosophy of Science* 53 (1): 39–68.

Noble, D. 2017. *Dance to the Tune of Life*. Cambridge: Cambridge University Press.

Odling-Smee, F. J., K. Laland, and M. Feldman. 2000. *Niche Construction: The Neglected Process in Evolution*. Princeton, NJ: Princeton University Press.

Okasha, S. 2006. *Evolution and the Levels of Selection*. Oxford: Oxford University Press.

Pence, C. H., and G. Ramsey. 2015. "Is Organic Fitness at the Basis of Evolutionary Fitness?" *Philosophy of Science* 82:1081–1091.

Pigliucci, M. 2009. "An Extended Synthesis for Evolutionary Biology." *Annals of the New York Academy of Sciences* 1168:218–228.

Pigliucci, M., and G. Müller. 2010. *Evolution: The Extended Synthesis*. Cambridge, MA: MIT Press.

Raff, R. 1996. *The Shape of Life: Genes, Development and the Evolution of Animal Form*. Chicago: Chicago University Press.

Salmon, W. 1984. *Scientific Explanation and the Causal Structure of the World*. Princeton, NJ: Princeton University Press.

Scholl, R., and M. Pigliucci. 2015. "Ultimate Explanations Concern the Adaptive Rationale for Organism Design." *Biology and Philosophy* 28:787–791.

Schrödinger, E. 1944. *What Is Life?* New York: Dover.

Schwenk, K., and G. Wagner. 2004. "The Relativism of Constraints on Phenotypic Evolution." In *The Evolution of Complex Phenotypes*, edited by M. Pigliucci and K. Preston, 390–408. Oxford: Oxford University Press.

Shapiro, J. A. 2011. *Evolution: A View from the 21st Century Perspective*. Upper Saddle River, NJ: FT Press Science.

Shapiro, L., and E. Sober. 2007. "Epiphenomenalism—the Do's and Don'ts of Epihenomenalism." In *Studies in Causality: Historical and Contemporary*, edited by G. Wolters and P. Machamer, 235–264. Pittsburgh, PA: University of Pittsburgh Press.

Sober, E. 1984. *The Nature of Selection*. Cambridge, MA: MIT Press.

Sober, E. [1980] 2006. "Evolution, Population Thinking and Essentialism." In *Conceptual Issues in Evolutionary Biology*, edited by E. Sober, 329–359. Cambridge, MA: MIT Press

Sober, E. 2013. "Trait Fitness Is Not a Propensity, But Fitness Variation Is." *Studies in History and Philosophy of Biological and Biomedical Sciences* 44 (3): 336–341.

Sultan, S. E. 2003. "Commentary: The Promise of Ecological Developmental Biology." *Journal of Experimental Zoology (Mol Dev Evol)* 296B:1–7.

Uffink, Jos. 2017. "Boltzmann's Work in Statistical Physics." In *The Stanford Encyclopedia of Philosophy* (Spring 2017 Edition), edited by Edward N. Zalta. https://plato.stanford.edu/archives/spr2017/entries/statphys-Boltzmann/.

Uller, T., and H. Helanterä. 2017. "Heredity and Evolutionary Theory." In *Challenging the Modern Synthesis: Adaptation, Development, Inheritance*, edited by P. Huneman and D. M. Walsh, 280–316. Oxford: Oxford University Press.

Wagner, G., and L. Altenberg. 1996. "Complex Adaptations and the Evolution of Evolvability." *Evolution* 50:967–976.

Wallace, B. 1986. "Can Embryologists Contribute to an Understanding of Evolutionary Mechanisms?" In *Integrating Scientific Disciplines*, edited by W. Bechtel, 149–163. Dordrecht: M. Nijhoff.

Walsh, D. M. 2003. "Fit and Diversity: Explaining Adaptive Evolution." *Philosophy of Science* 70:280–301.

Walsh, D. M. 2015. *Organism, Agency and Evolution*. Cambridge: Cambridge University Press.

Walsh, D. M., A. Ariew, and M. Matthen. 2017. "Four Pillars of Statisticalism." *Philosophy, Theory and Practice in Biology*, 9:1.

Walsh, D. M., T. Lewens, and A. Ariew. 2002. "The Trials of Life." *Philosophy of Science* 69:452–473.

West-Eberhard, M. J. 2003. *Developmental Plasticity and Evolution*. Oxford: Oxford University Press.

Woodward, J. 2003. *Making Things Happen*. Oxford: Oxford University Press.

Wray, G. A., H. E. Hoekster, D. J. Futuyma, R. E. Lenski, T. F. C. Mackay, D. Schluter, and J. E. Strassman. 2014. "Does Evolutionary Theory Need a Rethink? No, All Is Well." *Nature* 514:161–164.

12 Ontology, Causality, and Methodology of Evolutionary Research Programs

Jun Otsuka

Introduction

Disagreements and conflicts are part and parcel of scientific practice. Mutual criticisms among opposing hypotheses, interpretations, and approaches are crucial ingredients for a healthy and productive science that prevent it from lapsing into dogmatism. Some of these problems are empirical and settled by direct observations or experiments. Others are less so and concern meta-scientific beliefs about how to interpret data or conduct experiments. The recent debate (e.g., Laland et al. 2014) between the Modern Synthesis (MS) and the so-called Extended Evolutionary Synthesis (EES) falls into the latter category. As I argue in this chapter, what are at stake in the debate are less particular facts and processes of evolution than the ontological and methodological frameworks of the theory itself. That is, the controversy largely stems from dissident views on both the ontological units of evolution and the proper methodology for studying their dynamics.

By ontology I mean a set of beliefs shared (mostly implicitly) by a group of scientists about basic entities and properties thereof that populate the domain of their investigation. Any scientific theory has its ontological assumptions as to what constitute the bricks and mortar of the world-picture according to that theory. The world of classical mechanics is filled with particles having definite location and momentum. Freud postulated an unconscious *id*, libidos, and so on, as basic psychological mechanisms. These ontological postulates are deeply linked to the notion of causality. To be is to be causally effective—no empirical theory postulates the existence of a thing that would never affect or be affected by anything else, and if it posits something it must specify how and in what way that thing is causally related to the rest of the world. The goal of a scientific theory is to elucidate this causal structure of the world. This implies that the ontology of a theory, along with its concept of causality, prescribes scientific methodology and practice. Psychoanalysis thus tries to bring unconscious conflicts into the conscious mind through therapeutic intervention. Dalton's atomic theory led chemists to study chemical reactions in terms of combinations, separations, and rearrangements of atoms. Conversely, methodological postulates and abilities (i.e., what we can do and observe) may influence our idea of what there is in the first place ("electrons exist if we can spray them," says Hacking 1983). Thus the ontology, the concept of causality, and the methodology of a theory together determine a conceptual framework through which we see and investigate the world.

Shared ontology and methodology facilitate communication among scientists working in the same field and enable them to evaluate each other's work. The lack thereof, in contrast, often times leads to miscommunication or fierce discussions, or what Kuhn (1962) notoriously called the incommensurability between different paradigms. Even to a lesser extent, discordance in meta-scientific assumptions prevents mutual understanding and collaborations in a scientific community. This is especially so with biology, which comprises a vast variety of disciplines with different traditions, agendas, and methodologies. The spirit of the Modern Synthesis, epitomized by Dobzhansky's "nothing in biology makes sense except in the light of evolution," was to bring a unity to the heterogeneous biological sciences under the same conceptual framework of evolutionary theory. But what is evolutionary theory? Since Darwin, the principle of descent with modification (due to various factors, including natural selection) has been conceptualized in different, sometimes incompatible, ways, and that has given rise to debate over what *the* theory of evolution really is, even among the constructors of the Modern Synthesis (Smocovitis 1996).

As I see it, what are often at issue in these debates are the ontology and methodology of evolutionary theory. That is, the dissonance comes from different views on what count as units and causes of evolution, and how we should study their dynamics. This chapter develops this thesis by analyzing meta-scientific assumptions underlying three contemporary evolutionary thoughts: Ernst Mayr's population thinking, the gene-centered view of the Modern Synthesis, and the recently proposed Extended Evolutionary Synthesis. They view evolution differently. For Mayr, it is a change in a population of heterogeneous organisms triggered by environmental factors; for the gene-centered view, it is a shift in gene frequencies resulting from the variation in their phenotypic effects; and the EES takes the whole organism-environmental complex as both units and causes of evolutionary changes. Even in cases where they refer to the same "phenomenon," its interpretation and significance depend on the conceptual framework of each tradition. Whence arises the importance of clarifying these ontological and methodological assumptions in order to understand the source of their disagreements and to facilitate further communication and discussion among different schools of evolutionary thought.

The chapter unfolds as follows. "Population Thinking" begins the aforementioned meta-scientific analysis by taking up Ernst Mayr's population thinking as the first explicit attempt to build a unified ontology and methodology for evolutionary theory. At variance with the conventional interpretation, however, I argue that it falls short of the conceptual framework for the Modern Synthesis because of its inherent inconsistency with the Mendelian, gene-centered ontology. As will be discussed in "The Gene-Centered View," the ontological framework of the MS was completed when genes were recognized as exclusive units underlying both organismic and evolutionary phenomena. This ontological shift came to feature population genetics as the default methodology to study evolutionary change. "The Extended Evolutionary Synthesis" then characterizes the EES as an alternative to this gene-centric picture that tries to treat whole complex causal mechanisms over genotype, phenotype, and environment as extended units of evolution. I illustrate a part of this attempt using the causal graph theory, and describe some research questions suggested from this perspective. In conclusion, "Quo Vadis?" summarizes the MS and the EES frameworks as competing *research programs* (Lakatos 1980), and sketches challenges and tasks for both camps with a view toward generating constructive discussion and theory development in future investigations.

Population Thinking

Great scientific works do not just reveal novel phenomena or regularities in the world but they also change how we see the world itself. That was the case for Darwin's *Origin* (Darwin [1859] 2003), in which he not only presented mechanisms and an overwhelming mass of evidence of evolutionary changes but also provided a new way of looking at life. Darwin's conceptual revolution was eloquently epitomized by Ernst Mayr as the replacement of typological thinking by population thinking. Typological thinking, according to Mayr, has been the canonical view of living things since Aristotle, and defines an organism in terms of its essence. An essence is what specifies the nature of an individual organism as a member of a certain species. It is an unchanging standard ontologically prior to any within-specific difference or variation, which is conceived only as deviation from the norm. In contrast, population thinking puts the variation as primitive; biological populations are inherently heterogeneous and there is no fixed property shared by all members of a given species. Evolution is a necessary consequence of this variability when it is heritable and related to fitness. This made the assumption of fixed type unnecessary for, or even orthogonal to, evolutionary theory. What is crucial is rather that organisms vary from each other, and that some of their variations are heritable and have fitness effects.

Population thinking is an ontological thesis. It takes organisms as ontological primitives in the sense that they are all idiosyncratic, with any apparent commonality traditionally conceived as essence or type being merely a statistical abstraction that has no real existence.

The populationist stresses the uniqueness of everything in the organic world.... All organisms and organic phenomena are composed of unique features and can be described collectively only in statistical terms.... Averages are merely statistical abstractions; only the individuals of which the populations are composed have reality. (Mayr 1975, 28)

For Mayr, this is the only ontological doctrine compatible with the Darwinian theory of evolution. Natural selection could not occur if populations were homogeneous, and evolution would be impossible if there were fixed types. To deny population thinking, therefore, is to deny selection and evolution.

The replacement of the typological ontology by the populationist ontology is accompanied by a conceptual shift in the notion of causality and scientific methodology. For typological thinkers, the essence of a thing is also the locus of its causal power, so that behavior, properties, and regularities exhibited by an object are determined and accounted for by its internal nature or structure. Typically, such inherent regularities are revealed in experiments that control all external variables and carefully isolate the causative agent under study so that it exhibits its "own course of nature." This, according to Mayr (1961), is the canonical concept of causation and methodology in physics, chemistry, and also what he calls "functional" biology that studies *proximate causes* or mechanisms of organic phenomena. Population thinking, however, implies that this understanding of causality is utterly inadequate and inapplicable for the study of evolution, simply because there is no fixed essence. Evolutionary biology thus requires a different concept of causality that does not rely on essences, and that is what Mayr calls *ultimate cause*. Ultimate causes are historical factors; what explains a particular evolutionary episode, say the radiation of mammals during the Paleocene and Eocene, is a past environmental condition such as the extinction of dinosaurs triggered by an asteroid impact. Mayr (1982) claims that such causal relationships cannot be confirmed by laboratory experiments but are rather reconstructed by *historical narratives* that connect points of past singular events into a consistent sequence of causal story.

The typological/populationist dichotomy is thus closely tied to the methodological distinction of proximate/ultimate causation. The onto-methodological dichotomy served for Mayr as a criterion to distinguish what does and does not count as an appropriate study of evolution. With this standard he condemned ahistorical approaches such as the phenetic species concept or mathematical genetics as typological (Mayr 1982, 41), and dismissed the accumulating knowledge in molecular, developmental, and physiological biology as irrelevant to the study of evolution because they are concerned with only proximate but not ultimate causes. In contrast, natural history, which studies natural habitats of populations and historical change thereof, is expected to play the central role in identifying ultimate causes of evolutionary trajectories. Although Mayr's appraisal here might have reflected his territorial ambition to some extent (Beatty 1994), what should be noted is that his criticism was made on a conceptual, rather than empirical, ground. Mayr did not, for example, question theoretical results in mathematical genetics or empirical findings in molecular biology; but for him they contribute little, if anything, to our understanding of evolution, as it requires a different notion of causality and methodology.

In sum, population thinking for Mayr was not just a metaphysical corollary of Darwin's theory but rather a conceptual framework that helped him define the very discipline of evolutionary theory both ontologically (what it is about) and methodologically (how it works). The flip side of this demarcation was the ostracism or marginalization of molecular and developmental studies on proximate mechanisms as they were deemed irrelevant or peripheral to the study of evolution. Another important omission is Mendelian genetics. Population thinking is just Darwinian—Mayr presents it as Darwin's exclusive legacy with no mention of Mendel. This raises serious doubts as to whether population thinking serves as an adequate ontological basis for the Modern Synthesis, which by definition was the synthesis of Darwin's theory of evolution *and* Mendel's theory of inheritance.[1] In fact, Mendelian genetics brings an inconvenient tension into population thinking by reintroducing the notion of types (allelo- or genotypes), so that studying evolution as a change in gene frequencies requires a different ontology. This is the core claim of Dawkins's gene-centered view, to which we now turn.

The Gene-Centered View

Although variation is certainly essential for natural selection as population thinkers emphasize, it is not sufficient to produce significant adaptive change. Since most variation has an insignificant effect, selection must be *cumulative* and operate for several generations to yield any notable difference. This posed a serious problem for Darwin, who postulated a sort of blending inheritance; if parents' characters "blended" in their offspring, the variation in a population would quickly fade away before selection produces any significant change, making it a negligible factor of evolution. As is well known, the problem came to be settled by the integration of Mendelian genetics into Darwin's theory (Provine 1971), but this has also reintroduced the notion of type into evolutionary thinking by reducing organismal features into exchangeable and infusible genes. The genetic reduction was completed by Fisher's (1918) seminal work, which showed that even apparently continuous variation can be expressed as differential combinations of discrete alleles. The Darwinian notion of a population of heterogeneous individual organisms was thus replaced by the concept of a *gene pool*, which consists of relatively few kinds of genotypes or allelotypes.

Through the twentieth century the gene-centered view has been buttressed by the works of T. H. Morgan, W. D. Hamilton, and G. C. Williams, to name a few, but it was Richard Dawkins who most emphasized the conceptual significance of the discrete inheritance for adaptive evolution and erected it into *the* ontological basis of evolutionary theory. Although he does not mention it explicitly, Dawkins's target was the populationist-like ontology. Population thinking stressed the uniqueness of individual organisms as its fundamental axiom. Dawkins accepts this premise, but he concludes that *because of this uniqueness* adaptive evolution cannot occur at the individual level. "Each individual is unique. You cannot get evolution by selecting between entities when there is only one copy of each entity!" (Dawkins 1976, 34). Most selective pressure is weak, so if it is to produce a significant adaptation a population must undergo multiple rounds of selection and the result of each round must accumulate. This requires units of adaptation to be persistent and to exert a constant fitness effect over generations. Whereas individual organisms, being idiosyncratic and ephemeral, cannot do this job, genes responsible for phenotypic differences are "virtually immortal" and can serve as bearers of cumulative adaptation. Hence it is genes that evolve in the first sense, while changes in individual phenotypes are epiphenomena of the underlying genetic evolution.

Dawkins describes himself as offering "a particular way of looking at animals and plants" (Dawkins 1982, 1). I would call it an ontology, for the claim really concerns what kinds of entities make up the living world. The fundamental building blocks of evolution in his view are called *active germ-line replicators*. They play two important roles in adaptive evolution; namely, (i) to form a possibly infinite lineage of identical copies (replicas), making themselves a bearer of cumulative adaptation, and (ii) to exert constant causal effects on phenotype and fitness, thereby influencing their probability of being copied. Note that these are both important characteristics of what philosophers call "type." A type is characterized by its identical nature and causal power; for example, every instance of gold is assumed to have the same atomic structure and chemical activity, and such an in-class homogeneity is what makes gold a chemical type. Likewise, every instance of an active germ-line replicator is identical by descent, and assumedly has the same fitness effect if put in the same environmental condition. Adaptive evolution is understood as a race among such *types*, each of which increases or decreases its share as a consequence of its type-specific fitness effect.[2]

The gene-centered ontology does not leave the notion of evolutionary cause intact. The focus on the gradual and cumulative nature of adaptive evolution pushes historical and/or environmental conditions to the background and instead features genes as a major driving force of evolutionary changes. An active replicator "exerts phenotypic power over its world" (Dawkins 1982, 91)—*ADH1B* allele enables rapid alcohol breakdown and the *SD* gene has a capacity to kill flies' sperms that do not carry its copy to favor its own transmission—and by doing so they affect their frequency in future generations. The study of evolutionary dynamics, then, boils down to an analysis of the "phenotypic power" of the underlying gene.[3] Each allele has its own capacity—presumably thanks to its nucleotide sequence and place within the genome—to produce a phenotype, to segregate, to mutate, to recombine, and so on. Population genetics quantifies manifestations of such capacities in a given environmental condition and deduces equations that describe evolutionary dynamics under that environment. In this sense, the practice of population genetics is quite Aristotelian *sensu* Cartwright (1999): it studies the natural capacity of genes and derives laws of evolution as deployments of their nature under a specified condition.

The gene-centered view thus presents quite a different ontology and methodology than those of population thinking. A biological population now consists of combinations of discrete types rather than heterogeneous individuals and changes its composition as a function of the phenotypic power of such types. The redefinition of evolution as a change in gene frequencies features population genetics as the primary methodology to study evolutionary dynamics. In order for it to be regarded as a bona fide evolutionary process, any hypothesized evolutionary mechanism must be expressed in terms either of the phenotypic effect of genes or of environmental conditions, and its evolutionary significance must be evaluated with an appropriate population-genetic model. That was how natural selection and drift became accepted as major forces of evolutionary changes around a century and a half-century ago, respectively, so surely shouldn't any other hypothetical mechanism go through the same procedure? The onto-methodological framework of the gene-centered view has thus come to serve as the touchstone to distinguish what is and what is not a proper study of evolution in the Modern Synthesis framework—even to the extent that "nothing in evolution makes sense except in light of population genetics" (Lynch 2007).

The Extended Evolutionary Synthesis

The Modern Synthesis and its underlying gene-centered ontology has proven to be a very productive framework that has promoted a host of empirical findings as well as theoretical developments including, to name a few, neutral theory and inclusive fitness theory. Its hegemony, however, has recently been challenged by those who work on "non-standard" evolutionary mechanisms such as developmental bias, plasticity, niche construction, or epigenetic inheritance. Growing evidence from these studies suggests the mechanisms underlying evolution are too complex and diverse to be adequately captured by the MS framework—phenotypic variations are not random but canalized through developmental processes; organisms are not just passive "vehicles" of adaptation but actively interact with their environment to influence their evolutionary fate; and inheritance is mediated not just by genes but also by other epigenetic, environmental, and cultural means. Some believe that these findings go beyond the scope of the traditional, gene-centered approach and necessitate a conceptual update, which they call the Extended Evolutionary Synthesis or EES (Müller and Pigliucci 2010; Laland et al. 2015).

The plea for reform is countered by those who work on the forefront of the Modern Synthesis framework (Laland et al. 2014; Welch 2017). They doubt the alleged "novel" evolutionary mechanisms are anything more than just "add-ons" or other fancy ways genes exert their causal power. Developmental bias, for example, is just another name for a particular way genes get expressed, plasticity an environmental-dependent expression of genotype, niche construction a kind of "extended phenotype," and epigenetic factors are merely other gene-like entities with a peculiar transmission pattern. Construed as such, these mechanisms can be well handled by the existing population-genetic models with modest adjustments and proper parameterization. Of course these modifications should affect evolutionary trajectories, but by themselves they are not essential, or at least not anything that require a significant extension or overhaul of the current evolutionary theory, or so the skeptics argue.

The contention here is not whether these phenomena exist—both camps agree that developmental bias, niche construction, and so on, are real, and maybe common, phenomena—but rather it is about their significance to evolutionary theory. The populationist thinker, for example, would dismiss these factors as mere variations in proximate mechanisms with no direct implication for evolutionary studies of ultimate causes. From the gene-centered perspective, they are background parameters regulating genes' actions whose evolutionary significance is to be evaluated only in terms of genetic changes. Proponents of the EES resist such interpretations on two fronts, first by denying the sharp distinction between proximate/ultimate causation in favor of "reciprocal" relationships between development and evolution (Laland et al. 2011), and second by arguing for a holistic view of evolution where epigenome, developmental processes, and niche all count as their own evolutionary units that together constitute population dynamics irreducible to changes in gene frequencies (Laland et al. 2015). The new empirical findings thus call for an alternative perspective or "conceptual framework" that has different research questions and focus than those of the Modern Synthesis, or so it is argued.

At the same time, Laland and colleagues stress that the EES is not a refutation or replacement but an *extension* of the traditional view. Their claim, therefore, must be understood and assessed in relationship to the Modern Synthesis framework—which part of it is to be extended, and in what way? In what follows, I try to answer these questions from the perspective developed so far, by characterizing the proposed extension as concerning the ontology and methodology of evolutionary theory.

Extended Ontology

One of the central tenets of the EES is that mechanisms underlying evolution are more complex and diverse than the Modern Synthesis often presupposes, and that makes a difference in evolutionary trajectories. But how do proximate causal mechanisms influence ultimate evolutionary dynamics? Despite Mayr's strict segregation, the fact that the two sorts of causes are systematically related can be shown by the *causal graph theory*, which deals with formal relationships between a causal structure represented by a directed graph and a probability distribution generated from it (see box 12.1). The graphical representation has been used to illustrate causal assumptions underlying the standard quantitative genetic models (Frank 1997; Rice 2004; Otsuka 2016a), but it also allows us to characterize "non-standard" evolutionary mechanisms as modifications or extensions of the basic causal model (Otsuka 2015). In the extended causal models, epigenetic inheritance introduces an extra pathway between the parental and offspring phenotypes, while a niche is represented as a "contextual variable" constructed by individuals in a local population and inherited to the next generation (figure 12.2). Upon parameterization, these models induce

Box 12.1

The causal reconstruction of evolutionary models stands on two theoretical bases. The first is the Price equation (Price 1970), which gives the evolutionary change of any phenotypic mean between two generations by

$$\Delta \bar{Z} = cov(W, Z')/\bar{W} + \bar{Z}' - \bar{Z} \tag{1}$$

where W, Z, and Z', respectively, denote fitness, phenotype, and the average offspring phenotypic value of each individual. The second is the premise of the *causal graph theory* that the probability distribution of the population is generated from a *causal model* (Wright 1920; Spirtes, Glymour, and Scheines 1993; Pearl 2000). A causal model $M = \langle G, F, P \rangle$ has three components: the causal or path *graph G* that qualitatively represents cause–effect relationships with directed edges, the set of *structural equations F* that specify the quantitative nature of each causal link, and the probability distribution P over the exogenous variables (those that have no causal input in the model). Figure 12.1 is an example of a simple causal model of selection and reproduction in an asexually reproducing population. Given a genetic distribution $P(X)$ as an input, the model yields a joint distribution out of which the right-hand side of the Price equation (1) can be calculated. If all the causal relationships are linear, the model in figure 12.1 yields the *breeder's equation* $\Delta \bar{Z} = Sh^2$ (Otsuka 2016a). By modifying the causal model (graph and functions) one may obtain evolutionary dynamics for different mechanisms.

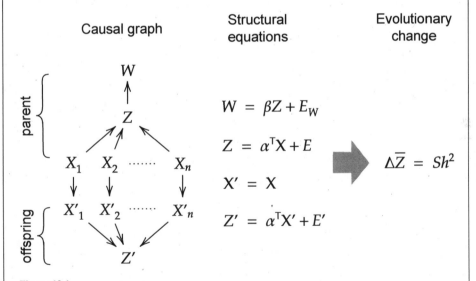

Figure 12.1
The causal model of the breeder's equation for an asexually reproducing population. X, Z, and W, respectively, represent allele counts, phenotype, and fitness of a parent, whereas X' and Z' denote averages over offspring. Environmental influences (E) are assumed to be independent and omitted from the graph. Parameters α and β represent additive genetic effects and the selection gradient, respectively. The graph and the structural equations (in the middle) together induce a probability distribution, from which evolutionary change is calculated.

probability distributions, from which evolutionary dynamics under each causal hypothesis can be calculated. The resulting equations unequivocally and quantitatively show that difference in the underlying mechanisms is reflected in the evolutionary change of the population.

In the traditional model (figure 12.1), genes take full responsibility for both selection (as the eventual cause of fitness) and reproduction (as the sole mediator of parent–offspring resemblance). The streamlined picture justifies taking genes as the primary causal agent of evolution, while pulling together development and other environmental factors as background conditions that regulate the genetic action (figure 12.3, left). As a natural consequence, organismic features or behaviors come to be understood as properties or capacities of a genome, e.g., niche construction as an "extended phenotype" and plasticity as a conditional gene expression in response to a changing environment. The gene-centric picture, however, does not fit well with the extended models where extra-genetic and environmental factors form additional causal pathways underlying selection and inheritance in parallel with the genome (figure 12.2). Here, these factors serve no longer as background parameters but rather as variables that change their distribution—that is, evolve—in response to their fitness effect. Hence, in order for a model to be dynamically sufficient and to correctly predict evolutionary change, an evolving population must be characterized not by its genetic composition alone but also by epigenetic, phenotypic, and even environmental variables related to each other via a certain causal network. This suggests an inclusive ontology that takes the entire causal structure as the generator of evolutionary dynamics and the basic unit of its analysis (figure 12.3, right). Accordingly, various evolutionary mechanisms such as niche construction or developmental plasticity are re-construed as structural properties (e.g., as a particular configuration of a causal graph or a form of structural equations).

The shift in the ontological focus from genes to structures has implications for the notion of evolutionary cause. Traditional evolutionary thinking has assumed a dualistic, figure-ground concept of causality, where the notion of causes is divided into two categories, only one of which is granted primary importance in evolutionary studies. Population thinkers set a strict boundary between ultimate causes of evolution and proximate mechanisms. Neo-Darwinians draw genes as the central "figure" that drives evolutionary change on the "ground" of environmental or developmental conditions. In contrast, the EES upholds a univocal and democratic view of causality, which denies the sharp distinction between proximate and ultimate causation, or between a primary causal agent and background conditions. The systematic derivation of evolutionary equations from causal models of proximate mechanisms demonstrates that proximate causes *are* ultimate causes of evolution (Otsuka 2015). It also shows that evolutionary dynamics is determined not solely by a gene's capacity, but from the entire causal structure including genotype, phenotype, and environment as its parts. Of course this does not mean all these factors make an equal contribution; the genome will remain one of or perhaps the most important factor

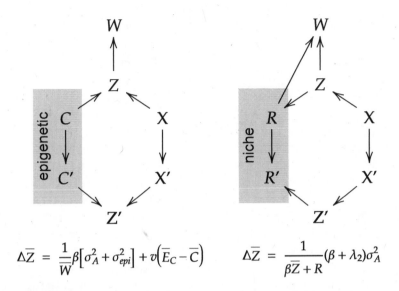

$$\Delta \overline{Z} = \frac{1}{\overline{W}}\beta\left[\sigma_A^2 + \sigma_{epi}^2\right] + v\left(\overline{E}_C - \overline{C}\right) \qquad \Delta \overline{Z} = \frac{1}{\beta \overline{Z} + R}(\beta + \lambda_2)\sigma_A^2$$

Figure 12.2
Causal graphs and resulting evolutionary equations for epigenetic inheritance (left) and niche construction (right). The Price equation implies that adaptive responses of a population, summarized by the covariance $cov(W, Z')$ of the parental fitness and the offspring phenotype, is influenced by the causal connection between these two variables. In addition to the genetic pathway through X, the epigenetic factor C and niche R both create alternative routes that influence this covariance. The evolutionary response under each mechanism is obtained under the assumption of linear causal relationships. See Otsuka (2015) for details.

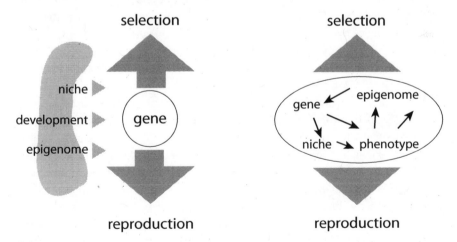

Figure 12.3
Schematic representations of adaptive evolution according to the MS and the EES. **Left**: In the gene-centered view, genes are the principal factor of both selection and reproduction, with other phenotypic and environmental factors regulating the genetic action. **Right**: In the extended picture, it is the whole causal structure involving genetic as well as nongenetic variables that generates evolutionary dynamics. Various developmental and environmental processes are parts or properties of this causal structure, rather than the genome alone.

and for that reason it is privileged in studies of evolution, but its causal and evolutionary significance is to be evaluated only within the network of organismal and environmental factors. Evolutionary change is a function not of genes alone but of the entire causal structure, so that there is no ontological justification to distinguish figure and ground in the kinds of evolutionary causes.

Extended Methodology

In the Modern Synthesis framework, genes serve as the primary unit of evolutionary analysis. Models of population genetics study how evolutionary dynamics, defined in terms of genetic frequencies, arises from genes' capacities or properties under a given condition. In contrast, evolution from the EES perspective is a change in the entire organismal structure over genotypic, phenotypic, and environmental factors and is generated from their causal connections. Does this justify treating causal structures as an additional unit of evolution? The argument for genes as the primary unit derives its force from their ability to influence evolutionary processes, their stability across generations, and their changeability via mutation—these three features together qualified genes as *the* inductive basis of evolutionary analyses. If causal structures are to play a similar role, we must likewise ask how they affect evolution, are conserved across generations or even among species, and "mutate" and evolve through time. This prompts three research questions as described below.

Recent studies in evo devo have suggested that developmental processes may bias or constrain phenotypic variability and thereby influence the rate and direction of evolutionary change (see Laland et al. 2015 and references therein). Analyzing this relationship will constitute the first line of research to evaluate the role of causal structures on evolutionary dynamics. Although short-term constraints on adaptive processes have often been studied through the (eigen) structure of the additive genetic variance-covariance or **G** matrix (e.g., Hansen 2006; Blows 2007), it only provides a temporary measure because genetic variation keeps changing due to selection, drift, mutation, and so on. To understand a longer-term *evolvability* one needs to turn to the source of these genetic variations, which is in turn determined by structural features of the underlying developmental/causal pathways. It is expected, for example, that two sets of traits will evolve independently from each other if they do not share developmental resources and form distinct *modules* (Wagner and Altenberg 1996). On the other hand, *generative entrenchment* contributes to the evolutionary conservatism of early developmental stages through a strong stabilizing selection on their broad downstream effect (Wimsatt and Schank 1988). Qualitative features of developmental/causal structures thus determine both properties of the mutation matrix and selective regime, thereby influencing evolutionary dynamics. The importance of causal, and not just statistical, knowledge in the study of evolution invites applications of causal search algorithms that aim to infer causal relationships just from observational data (Spirtes et al. 1993; Pearl 2000; see also figure 12.4). Applied to phenotypic records, these

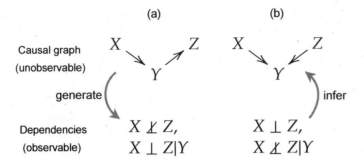

Figure 12.4
Although causal structures are usually unobservable, they leave some trace in the probability distribution they generate. In the above figure (a), Z is expected to depend on its remote cause X (i.e., $X \not\perp Z$) but become independent if conditioned on the intermediate variable Y ($X \perp Z \mid Y$). The pattern reverses in the case of (b), where X and Z are unconditionally independent but become dependent when conditioned on Y ($X \perp Z$ but $X \not\perp Z \mid Y$). The causal search algorithms (Spirtes et al. 1993; Pearl 2000) exploit these and other relationships between causality and probability to infer causal relationships among variables.

algorithms can be used to generate possible path diagrams over measured traits or to find a modular structure (Shipley 2000; Valente et al. 2010; Otsuka 2016b).

The second type of research concerns extrapolations of evolutionary models from one context to another. Many models of population/quantitative genetics are *local*: they are built and parameterized for particular populations at a given time to track their evolutionary trajectory for a few generations. In contrast, some of our questions on evolutionary processes have more *global* scopes: do different populations or species of the same genus respond to selection in a similar fashion? How far can one extrapolate the result of a particular population-genetic model? Answers to these questions hinge on the stability of the underlying causal structures, with one hypothesis being that different populations (or the same population at different times) exhibit qualitatively similar evolutionary dynamics as far as they share the same causal structure. The reported similarity in the patterns of genetic and/or phenotypic correlations in related species of mammals (Young and Hallgrímsson 2005) and of flowers (Bissell and Diggle 2010) indicates that the conserved developmental structures generate a similar evolutionary dynamics in these populations. If this is a general phenomenon, causal models may serve as units of induction of a broader scope, unifying evolutionary dynamics of different populations.

The final, but not least important, question is how causal structures themselves evolve. Population-genetic models study changes in genetic and phenotypic frequencies under the assumption of a fixed causal structure, but from a broader perspective these causal structures are themselves products of evolution. Past studies have revealed cases of both wide-range homologies *and* radical changes in developmental mechanisms (e.g., True and Haag 2001; Shubin, Tabin, and Carroll 2009). These empirical findings prompt a theoretical inquiry on laws that govern change and stasis of causal/developmental structures. Are these macro-evolutionary changes still in reach of the traditional population genetics apparatus, or do they require a new approach? In the causal modeling framework, qualitative changes in causal structure (graph G and/or structural equations F) and quantitative changes in genetic or phenotypic distribution (P) are conceptually distinguished. Whether they represent distinct evolutionary *processes*, however, is a question that awaits further theoretical as well as empirical investigations. Given that developmental mechanisms are underpinned by some genes, changes in causal parameters could in principle be modeled as the epistatic evolution of regulative genes that control the strength of causal relationships (Hansen and Wagner 2001). Although the complexity of nonlinear interactions and the sensitivity to local conditions make predictions and analytical studies along this line difficult, some general insights may still be obtained through simulation studies (e.g., Jones, Bürger, and Arnold 2014). Another possibility is that structural changes are caused by different kinds of genes; Carroll (2005) suggests that major differences among species are attributed to variation in *cis*-regulatory regions, and that evolution in regulatory structure obeys different dynamics than that of the coding regions as modeled in population genetics (but see also Hoekstra and Coyne 2007).

With these three questions the EES introduces an additional layer of research methodology, putting forward a two-story view of evolutionary theory where each level deals with different timescales, ontological units, and epistemic goals. Traditional population genetics constitutes the first layer of this extended picture which focuses on micro-scale quantitative changes in genetic and phenotypic frequencies in a fixed environment. The second layer, on the other hand, examines conditions and assumptions of these quantitative analyses by attending to the underlying causal structures and evolution thereof. It asks, for instance, how developmental homology or various types of organism–environment interactions are established and conserved, and how they affect evolutionary dynamics of local populations. The focus here is more on breadth than depth, or on qualitative characterizations of large-scale evolutionary trends rather than qualitatively accurate predictions of particular populations.

The two-story methodology is a corollary of the extended ontological framework discussed above. Introducing causal/developmental structures as additional ontological units beside genes and treating developmental plasticity, inclusive inheritance, niche construction, and so on, as irreducible properties of such units requires a matching methodology to study the causal and evolutionary implications of those units. Inversely, if these phenomena turn out to be well handled by extrapolation of the existing methodology, no addition would be needed to the ontological inventory of evolutionary theory. The extensions of the ontology and methodology of evolutionary theory are, therefore, two sides of the same coin of the proposed new "conceptual framework." From this perspective, the success of the EES depends on whether it can provide a productive and unified research program that integrates various evolutionary phenomena in a way not captured by the traditional methodology. This section has sketched a few research directions suggested from the causal modeling framework, though they are admittedly partial and possibly biased interpretations of the EES project as a whole. Unifying all these approaches under the same methodological framework is a challenging but also essential task for the EES to become a new *synthesis* of evolutionary studies.

Quo Vadis?

This chapter has compared conceptual frameworks of evolutionary theory in terms of their ontological and methodological assumptions, with a special focus on the Modern Synthesis and the Extended Evolutionary Synthesis. From this perspective, the debate between the MS and the EES can be seen as a reductionist vs. anti-reductionist tug-of-war in evolutionary biology. The MS is characterized by its reductionist tendency, trying to represent and study organismal features and evolutionary phenomena in terms of underlying genes. Genes (and other gene-like entities) are the exclusive building blocks of evolution, so that all complex evolutionary mechanisms and phenomena can and must be captured through elaborations or extrapolations of the current population-genetic models. Proponents of the EES contest this view by pointing out that the phenomena they call constructive development cannot be reduced to just properties of the genome. Rather, they are properties of causal structures which, as higher-level units of evolution, prompt new lines of research questions that demand an alternative approach.

Where does the debate go from here? I for one doubt there is a single or a handful of empirical litmus tests that would settle the issue once and for all. When the contention concerns different conceptual frameworks, such a Popperian falsification is the last thing we could hope for. But that does not mean we should be Kuhnian fatalists and accept that debates between two different paradigms do not come to an end until one of them dies out. A more constructive picture would be to view the MS and the EES as two competing *research programs* in the sense of Lakatos (1980). A research program, according to Lakatos, consists of its core tenets or "hard core" and a "protective belt" that relates the core claims to empirical data or tests. When results from the tests do not turn out as expected, or unpredicted phenomena are observed, researchers rarely give up the hard core of their theory but try to accommodate anomalies by modifying the protective belt. In this way the same research program can change and "evolve" in the face of apparent difficulties while keeping its central claim intact.

The gene-centric ontology and methodology constitute the hard core of the Modern Synthesis. Historically this program faced a number of challenges, including neutral evolution, the problem of altruism, and punctuated equilibrium, to name a few, but each time it survived to evolve by modifying its protective belt (e.g., drift as an important evolutionary process, the concept of inclusive fitness, etc.). In this sense it was a *progressive* program, and this history of past successes underlies the expectation of many that the current challenge from the EES can also be handled by a similar maneuver—as new "add-ons" to the same hard core. In contrast, proponents of the EES are skeptical and complain that the handling of the MS program of complex developmental and environmental inter-actions has begun to show a sign of *degeneration*, issuing ad-hoc patches just to save its core doctrines but without leading to novel research questions or predictions. What we need is not a superficial treatment but a major surgery of the hard-core ontology and methodology, and only within a renewed research program can we appreciate the full evolutionary implications of these mechanisms, or so they argue.

The question we should ask, therefore, is not which research program is correct, but rather which is more progressive and offers novel predictions and insights on evolution. From this perspective, the challenge for the MS is not just to show the consistency of phenomena like developmental plasticity, inclusive inheritance, and so on, with its core framework, but to devise a suite of auxiliary theories that provide a novel insight into these phenomena from the gene-centered perspective, just as it did in the cases of neutral theory and inclusive fitness theory. In contrast, what is needed for the EES camp is to provide effective means to bring its core tenets to actual scientific investigations along the direction it envisages. Provided that constructive development and reciprocal causation are ubiquitous and important, how can we forge these concepts into an empirical hypoth-esis that makes testable predictions? In the dawn of the Modern Synthesis, population genetics has served as a formal platform both to quantitatively verify Darwin and Wallace's idea of natural selection (which was at its time quite dubitable), and to stimulate subse-quent investigations, say, on the amount and the source of genetic variation in natural populations. Can the EES provide a similar, unifying theoretical framework? Recent works on evo devo, epigenetic inheritance, and complex organism–environment interactions have laid seeds of hitherto neglected or marginalized research directions; whether these seeds grow into a new progressive research program or get incorporated as an auxiliary "protective belt" of the existing one hinges on future conceptual, theoretical, and empirical investigations.

Acknowledgments

I thank Tobias Uller, Kevin Laland, an anonymous referee, and the participants of the KLI Workshop on *Cause and Process in Evolution* for stimulating discussions and insightful comments on the draft of this chapter. This work is supported by JSPS grant-in-aid 16K16335.

Notes

1. Witteveen (2015, 2016) tracks down the historical process in which Mayr has appropriated, with his own twist, the type/population distinction from its earlier articulation by Simpson and Dobzhansky.

2. It is no wonder, therefore, that the controversy over the gene-centered view has focused on these two alleged features of gene-as-type. The notion of identical genes must be reconciled with mutation and recombination (Dawkins 1982, ch. 5), whereas the assumption of constant phenotypic effect has been criticized as being untenable on the face of gene-by-gene or gene-by-environment interactions (Wimsatt 1980; Lloyd 1988).

3. Note that the "phenotypic power" of a gene may be silent or neutral. In this sense, neutral evolution stands on the same, gene-centered ontology as neo-Darwinism. They disagree on the relative importance of various evolutionary forces (drift vs. selection), but share the same ontological framework.

References

Beatty, J. 1994. "The Proximate/Ultimate Distinction in the Multiple Careers of Ernst Mayr." *Biology and Philosophy* 9 (3): 333–356.

Bissell, E. K., and P. K. Diggle. 2010. "Modular Genetic Architecture of Floral Morphology in Nicotiana: Quantitative Genetic and Comparative Phenotypic Approaches to Floral Integration." *Journal of Evolutionary Biology* 23 (8): 1744–1758.

Blows, M. W. 2007. "A Tale of Two Matrices: Multivariate Approaches in Evolutionary Biology." *Journal of Evolutionary Biology* 20 (1): 1–8.

Carroll, S. B. 2005. "Evolution at Two Levels: On Genes and Form." *PLoS Biology* 3 (7): e245.

Cartwright, N. 1999. *The Dappled World: A Study of the Boundaries of Science*. Cambridge: Cambridge University Press.

Darwin, C. [1859] 2003. *On the Origin of Species: A Facsimile of the First Edition*. Cambridge, MA: Harvard University Press.

Dawkins, R. 1976. *The Selfish Gene*. Oxford: Oxford University Press.

Dawkins, R. 1982. *The Extended Phenotype: The Long Reach of the Gene*. Oxford: Oxford University Press.

Frank, S. A. 1997. "The Price Equation, Fisher's Fundamental Theorem, Kin Selection, and Causal Analysis." *Evolution* 51 (6): 1712–1729.

Fisher, R. A. 1918. "The Correlation between Relatives on the Supposition of Mendelian Inheritance." *Transactions of the Royal Society of Edinburgh* 51:399–433.

Hacking, I. 1983. *Representing and Intervening: Introductory Topics in the Philosophy of Natural Science*. Cambridge: Cambridge University Press.

Hansen, T. F. 2006. "The Evolution of Genetic Architecture." *Annual Review of Ecology, Evolution, and Systematics* 37 (1): 123–157.

Hansen, T. F., and G. P. Wagner. 2001. "Modeling Genetic Architecture: A Multilinear Theory of Gene Interaction." *Theoretical Population Biology* 59 (1): 61–86.

Hoekstra, H. E., and J. A. Coyne. 2007. "The Locus of Evolution: Evo Devo and the Genetics of Adaptation." *Evolution* 61 (5): 995–1016.

Jones, A. G., R. Bürger, and S. J. Arnold. 2014. "Epistasis and Natural Selection Shape the Mutational Architecture of Complex Traits." *Nature Communications* 5:3709.

Kuhn, T. S. 1962. *The Structure of Scientific Revolutions*. Chicago: University of Chicago Press.

Lakatos, I. 1980. *The Methodology of Scientific Research Programmes: Philosophical Papers*, Vol. 1. Edited by J. Worrall and G. Currie. Cambridge: Cambridge University Press.

Laland, K. N., K. Sterelny, J. Odling-Smee, W. Hoppitt, and T. Uller. 2011. "Cause and Effect in Biology Revisited: Is Mayr's Proximate-Ultimate Dichotomy Still Useful?" *Science* 334:1512–1516.

Laland, K. N., T. Uller, M. W. Feldman, K. Sterelny, G. B. Müller, A. Moczek, E. Jablonka, and J. Odling-Smee. 2014. "Does Evolutionary Theory Need a Rethink?" *Nature* 514 (7521): 161–164.

Laland, K. N., T. Uller, M. W. Feldman, K. Sterelny, G. B. Müller, A. Moczek, E. Jablonka, and J. Odling-Smee. 2015. "The Extended Evolutionary Synthesis: Its Structure, Assumptions and Predictions." *Proceedings of the Royal Society B: Biological Sciences* 282 (1813): 20151019.

Lloyd, E. A. 1988. *The Structure and Confirmation of Evolutionary Theory*. Princeton, NJ: Princeton University Press.

Lynch, M. 2007. "The Frailty of Adaptive Hypotheses for the Origins of Organismal Complexity." *Proceedings of the National Academy of Sciences USA* 104 (Suppl 1): 8597–8604.

Mayr, E. 1961. "Cause and Effect in Biology." *Science* 134:1501–1506.

Mayr, E. 1975. *Evolution and the Diversity of Life: Selected Essays*. Cambridge, MA: Harvard University Press.

Mayr, E. 1982. *The Growth of Biological Thought: Diversity, Evolution, and Inheritance*. Cambridge, MA: Belknap Press of Harvard University Press.

Müller, G. B., and M. Pigliucci. 2010. *Evolution: The Extended Synthesis*. Cambridge, MA: MIT Press.

Otsuka, J. 2015. "Using Causal Models to Integrate Proximate and Ultimate Causation." *Biology and Philosophy* 30 (1): 19–37.

Otsuka, J. 2016a. "Causal Foundations of Evolutionary Genetics." *British Journal for the Philosophy of Science* 67:247–269.

Otsuka, J. 2016b. "Discovering Phenotypic Causal Structure from Nonexperimental Data." *Journal of Evolutionary Biology* 29 (6): 1268–1277.

Pearl, J. 2000. *Causality: Models, Reasoning, and Inference*. New York: Cambridge University Press.

Price, G. R. 1970. "Selection and Covariance." *Nature* 227:520–521.

Provine, W. B. 1971. *The Origins of Theoretical Population Genetics*. Chicago: University of Chicago Press.

Rice, S. H. 2004. *Evolutionary Theory: Mathematical and Conceptual Foundations*. Sunderland, MA: Sinauer.

Shipley, B. 2000. *Cause and Correlation in Biology: A User's Guide to Path Analysis, Structural Equations and Causal Inference*. Cambridge: Cambridge University Press.

Shubin, N., C. Tabin, and S. Carroll. 2009. "Deep Homology and the Origins of Evolutionary Novelty." *Nature* 457 (7231): 818–823.

Smocovitis, V. B. 1996. *Unifying Biology: The Evolutionary Synthesis and Evolutionary Biology*. Princeton, NJ: Princeton University Press.

Spirtes, P., C. Glymour, and R. Scheines. 1993. *Causation, Prediction, and Search*. Cambridge, MA: MIT Press.

True, J. R., and E. S. Haag. 2001. "Developmental System Drift and Flexibility in Evolutionary Trajectories." *Evolution and Development* 3 (2): 109–119.

Valente, B. D., G. J. M Rosa, G. De los Campos, D. Gianola, and M. A. Silva. 2010. "Searching for Recursive Causal Structures in Multivariate Quantitative Genetics Mixed Models." *Genetics* 185 (2): 633–644.

Wagner, G. P., and L. Altenberg. 1996. "Complex Adaptations and the Evolution of Evolvability." *Evolution* 50 (3): 967–976.

Welch, J. J. 2017. "What's Wrong with Evolutionary Biology?" *Biology and Philosophy* 32 (2): 263–279.

Wimsatt, W. 1980. "Reductionistic Research Strategies and their Biases in the Units of Selection Controversy." In *Scientific Discovery: Historical and Scientific Case Studies*, edited by T. Nickles, 213–259. Dordrecht: Reidel.

Wimsatt, W., and J. C. Schank. 1988. "Two Constraints on the Evolution of Complex Adaptations and the Means for Their Avoidance." In *Evolutionary Progress*, edited by M. Nitecki, 231–273. Chicago: University of Chicago Press.

Witteveen, J. 2015. "'A Temporary Oversimplification': Mayr, Simpson, Dobzhansky, and the Origins of the Typology/Population Dichotomy (Part 1 of 2)." *Studies in History and Philosophy of Biological and Biomedical Sciences* 54 (C): 20–33.

Witteveen, J. 2016. "'A Temporary Oversimplification': Mayr, Simpson, Dobzhansky, and the Origins of the Typology/Population Dichotomy (Part 2 of 2)." *Studies in History and Philosophy of Biological and Biomedical Sciences* 57 (C): 96–105.

Wright, S. 1920. "The Relative Importance of Heredity and Environment in Determining the Piebald Pattern of Guinea-Pigs." *Proceedings of the National Academy of Sciences USA* 6 (6): 320–332.

Young, N. M, and B. Hallgrímsson. 2005. "Serial Homology and the Evolution of Mammalian Limb Covariation Structure." *Evolution* 59 (12): 2691–2704.

13 A Darwinian Dream: On Time, Levels, and Processes in Evolution

Arnaud Pocheville

Introduction

Darwin put to the fore the idea that natural selection explains adaptation. His project reflected preoccupations of his time, when adaptation was precisely that which British natural theology found so significant (Ruse 1992, 78). Darwin's mechanism of natural selection was embedded in a larger (and arguably somewhat ever changing) picture, where variation would also play a central role (Hoquet 2009). The laws of variation would famously retain aspects of the old Lamarckism, such as the effects of use and disuse, as a complementary path to adaptation (Darwin 1859, chap. V).

By the sole means of natural selection, however, Darwin had brought contingency as a means to teleology. Accidental variations, requiring apparently no further explanation, would feed natural selection and, when hereditary, lead to adaptation over generations (209). This contrasted with previous, evolutionary but orthogenetic views, such as Lamarck's, according to which lineages would tend to complexify following to-be-discovered physical laws (Lamarck 1809; Corsi 2011).

Contingency did not bring teleology for free. It was to be compensated by a shift in focus on populations, with geometrical powers of increase (Darwin 1859, chap. III), and on long biological and geological timescales (Darwin 1859, chap. IX, XIV; Burchfield 1990, chap. III). The latter required that living beings be able to transmit characteristics conserved over long timescales, a question that would fuel decades of controversy until—and even after—the advent of the Modern Synthesis in the 1930s (Mayr and Provine 1998).

For early neo-Darwinians such as Weismann, not only could variation be accidental and still permit adaptation—variation actually *had* to be accidental, "sealed off"[1] from challenges encountered during lifetime (e.g., Weismann 1891, 190). Reasons for this assumption ranged from cytological—no mechanisms of hereditary directed variation were known (Weismann 1893b, chap. XIII)—and theoretical—the characters and their determinants were supposed to be of different natures (Weismann 1904b, 107–108)[2]—to epistemological—

natural selection made the assumption of directed beneficial variation superfluous (Weismann 1893a). The explanatory virtue of natural selection was that it explained adaptation and that it was the only known naturalistic explanation for it.[3]

Nowadays, adaptation may be just one among many explanatory targets of natural selection—which also explains rock-paper-scissors games (Sinervo and Lively 1996), evolutionary suicides (Ferrière and Legendre 2013), and so on—but it seems to have kept all of its charm and appeal. Eminent challengers of a natural selection–centered view of evolution often seek to provide alternative explanations for adaption. Niche construction, for instance, has been proposed as a second route to adaptation, a route supplementary and irreducible to natural selection (Odling-Smee, Laland, and Feldman 2003, 290; see also Lewontin 1983, 282; Odling-Smee 1988, 74; Day, Laland, and Odling-Smee 2003, 81; Laland 2004, 321; Laland and Sterelny 2006, 1751, but see 1759). Directed, beneficial variation associated with nongenetic inheritance is also frequently proposed as a nonselective route to adaptation (reviewed in Pocheville and Danchin 2017).

I aim here to show that niche construction and nongenetic inheritance represent challenges to neo-Darwinian views of evolution in that they retain the same explanatory target (adaptation), but dismantle the beams of the explanation: the possibility of reasoning over long timescales on populations of individuals. Indeed, both niche construction and nongenetic inheritance enable interactions to spread across biological levels, space, and, more importantly, time. I argue that their sole effect on the neo-Darwinian scheme are sufficient for significant explanatory change. The retention of adaptation as an explanatory target could be seen as a vestige of history.

In the first section, I propose a formal interpretation of an explanation of adaptation by the means of natural selection, rooted in current modeling techniques. In the second section, I expose some threats to this explanatory scheme posed by niche construction and nongenetic inheritance. In the third section, I expose ways of rescuing the explanatory scheme in a number of cases, by the means of timescales and levels separation. In the fourth section, I discuss the implications of this account for current debates in evolutionary biology and offer some perspectives on whether we should expect to be able to rescale and separate processes, and on what this means for our conception of adaptation. A glossary at the end gathers the technical meaning of the keywords used in this contribution.

1. A Darwinian Dream

Darwin, in *The Origin*, did not provide empirical demonstrations of natural selection—only of artificial selection. Neither did he provide clear-cut principles enabling one to judge whether so-called adaptations, of which he offered a number of putative examples, were indeed adaptations. What Darwin provided instead was a verbal proof of concept that a naturalistic explanation of adaptation, when identified, could be given—that is, a how-possibly explanation of adaptation. Darwin's explanation, although familiar to us, is far from trivial. I trace below an idealized version of it, not meaning to give any realistic model of the living, but a flavor of what it means to predict or explain adaptation by the means of natural selection. I call it a Darwinian dream—a dream Darwin might have had, had he had a mathematical mind. Readers can think of it as an example of what is known, in the niche construction literature, as the "standard evolutionary theory."[4]

1.1   Verbal Narrative

Let us imagine a world where an almost infinite number of individuals compete over an almost infinite number of generations. Individuals survive, reproduce, and their offspring can vary with variations that are almost infinitely rare, almost indefinitely heritable, and sometimes beneficial (with a positive probability). The appearance of a given variation is accidental: it requires no further explanation. Say that variation can only occur in a gradual and cumulative fashion. Say that environmental challenges are set once and for all (in particular, they are frequency-independent).[5] Now, survival and reproduction coming with geometrical powers of increase (or decrease), episodes of competition lead the more beneficial variants (at a point in time) to invade. Since the environmental challenges are stable, the less beneficial variants are lost over a finite number of generations—they become rare below any infinitesimal threshold of rarity. After, so to speak, an infinite number of generations, an observer with a God's eye view would expect a priori to observe almost only indefinitely adapted individuals—in particular, they should be more adapted than expected if chance alone were to set their characters. As the number of individuals would be almost infinite, the observer could also a posteriori state that if a variation did invade, it was indeed beneficial.

In such a world, natural selection would explain the enrichment in adaptive traits over generations. It would not explain the existence of beneficial variations in the space of possible variations (this would be explained by, for example, calls to organization), only their accumulation through time as realized variations. Variation would only need to be blind, that is, explanations of adaptation would not need to refer to any laws of "beneficial" variation. Even if beneficial variations were almost infinitely rare in the space of possible variations, natural selection would explain why we should expect to see some of them realized in the living world. In this sense contingency, compensated with infinite time and an infinite number of individuals, would provide a means to teleology.

Of course, organic evolution seems to exist in a finite world, ruling out such a priori and a posteriori deductions on adaptation—if individuals are in finite numbers, beneficial variations can be undiscovered or lost by chance; if evolutionary time is finite, evolution might not be complete when one observes the result.[6] Still, even in a finite world, natural selection would skew evolution toward adaptation. The observer can only hope that the numbers be large enough—that is, that the idealizations reasonably hold—then, the explanatory and predictive powers of natural selection grow with population size and timescale of evolution. Population genetics and adaptive dynamics are two theoretical frameworks aiming at drawing this hope more precisely.[7]

1.2   Formal Narratives

Population genetics formalizes the dynamics of gene frequencies in populations featuring Mendelian heredity.[8] Population genetics does not generally aim at digging into the dynamics of individual development, it rather focuses on the succession of generations of genes. A core aspect (though by no means the only one) of population genetics is the study of evolutionary equilibria: assuming that the number of generations tends to infinity, the observer can draw expectations as regards polymorphism, average fitness, and the like—an aspect Haldane would call the "statics" of evolution (Haldane 1954; Crow and Kimura 1970, chap. 6; Gayon and Montévil 2017).

Adaptive dynamics is another framework, specifically designed to deal with cases where fitness is density or frequency-dependent (Metz et al. 1996; Geritz, Mesze, and Metz 1998; for a critical introduction, see Waxman and Gavrilets 2005). It is based on the assumption that mutations are rare, and have small phenotypic effects. In most applications, the population is assumed to be monomorphic and subject to the possible invasion of an initially rare mutant. If the mutant is deemed successful, a process of replacement occurs and the population is assumed to reach a new monomorphic state, where the former mutant has become the new resident (figure 13.1). Calculations in adaptive dynamics are simplified by the assumption that mutations are rare enough to enable any replacement to be complete before a new mutation occurs (thus, only two types interact at any moment). As with population genetics, individual development is classically assumed to be instantaneous with regards to the selective dynamics.

Equilibrium and more generally asymptotic thinking (i.e., assuming that a quantity tends to infinity), as exemplified by population genetics and adaptive dynamics, enables one to step back from historical contingencies—but for evolutionary branching toward one or another equilibrium (Sober 1983)—and to ignore supposedly transient and fast processes. In evolutionary theory, the rationale for asymptotic thinking is manifold. On the one hand, the explanatory power of geometrical dynamics (such as natural selection) geometrically grows with time. On the other hand, development is assumed to last only one generation, while genetic heredity is supposed to last for thousands, if not millions, of generations. This promotes explanations segregating individual, short-term developmental dynamics, from long-term, population-level dynamics of selection on heritable materials, and more generally evolution. Individual development would be the target of population-level selective processes operating over long timescales. Such a segregation would culminate in writings such as Mayr's, who would distinguish two kinds of causes in biology: the proximate causes, dealing with individual, short-term development, and the ultimate causes, dealing with long-term, population-level evolution (Mayr 1961). Thus construed, the

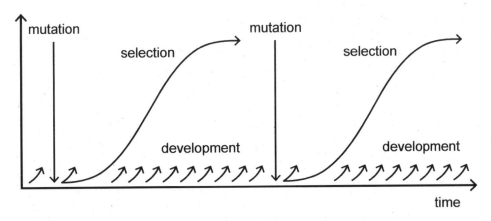

Figure 13.1
The separation of processes in adaptive dynamics. Rare mutations constraining fast developmental processes undergo slow selective processes. The vertical axis represents the values of the variables relevant for each process (e.g., size for development, gene frequencies for selection).

hypothesis of separation between development and selection is not just a mathematical simplification. It is a principled separation, that is, a principle of explanation: for development to be understood as being "screened," and possibly adapted, by selection, it *must be* supposed separated from, instantaneous with respect to selection.[9] This is the very separation many challengers to the Modern Synthesis want to tear apart (Laland et al. 2011).

2. Challenges to the Dream

For the past three decades, evolutionary theory has delivered a growing movement of contestation seeking at reconceptualizing adaptation, dismantling Mayr's distinction, and more generally re-examining key hypotheses about variation and selection (Lewontin 1983; Odling-Smee 1988; Jablonka and Lamb 1995; West-Eberhard 2003; Odling-Smee, Laland, and Feldman 2003, 18; Laland 2004, 316; Laland et al. 2011). Eminent challengers have called for a new, extended, evolutionary theory (Odling-Smee, Laland, and Feldman 2003, 370–385; Pigliucci and Müller 2010; Danchin et al. 2011; Laland et al. 2015). I now go through some of the challenges raised. For the sake of conceptual organization, I will sometimes borrow from different schools of thought.

2.1 Processes Are Causally Connected

Process entanglement is a first kind of challenge. Evolutionary processes which were previously thought of being separated (such as hereditary variation, development, and natural selection) now seem to be causally intertwined (Uller and Helanterä 2017). For instance, nongenetic inheritance introduces causal connections between development and evolution, between the challenges experienced by an organism and subsequent variation (Danchin and Pocheville 2014). Process entanglement is also one key aspect of niche construction theory. Niche construction, its proponents say, is in a situation of reciprocal causation with natural selection. It is an evolutionary process on its own, which cannot be subservient to natural selection:

Yet the standard view is that niche construction should not be regarded as a process in evolution because it is determined by prior natural selection. The unstated assumption is that the environmental source of the prior natural selection is independent of the organism.... However, in reality, the argument that niche construction can be disregarded because it is partly a product of natural selection makes no more sense than the proposition that natural selection can be disregarded because it is partly a product of niche construction. One cannot assume that the ultimate cause of niche construction is the environments that selected for niche constructing traits, if prior niche construction had partly caused the state of the selective environment.... Ultimately, such recursions would regress back to the beginning of life, and as niche construction is one of the defining features of life... there is no stage at which we could say natural selection preceded niche construction. (Odling-Smee, Laland, and Feldman 2003, 18–19, 375; see also Day, Laland, and Odling-Smee 2003, 83; Laland 2004, 319; Laland and Sterelny 2006, 200; Laland, Kendal, and Brown 2007, 54; Odling-Smee 2007, 282; Laland, Odling-Smee, and Gilbert 2008, 552; Laland et al. 2009, 199)

I will argue that the standard view listed in the quote above is not what it *should* be. There is a "standard" view according to which this kind of reciprocal causation between niche construction and natural selection does not forbid one to consider that one is subservient to the other. To make this point, I will distil process entanglement into two finer kinds of entanglement: that of levels (or, similarly, space-scales),[10] and, more importantly, timescales.

2.2   Levels Are Entangled

Level entanglement and space-scale entanglement relate to the identification of individuals constituting a population. The non-entanglement case is when individual-level processes and population-level processes can be studied independently (e.g., each individual makes a negligible contribution to a population). For individuals to be the only (or higher) relevant level of description in a population, they must not differentially affect one another. The physicist would say: they must be symmetrical with respect to the laws of variation and selection. This hypothesis of symmetry can be hindered in several ways.

The population may act as a functional constraint on individual variation.[11] Individual variation can be directed or bet-hedged at the level of the population (rather than accidental and independent in each individual), thus making dubious the idea that adaptation, if any, is occurring solely at the level of the individual. Population-level constraints might seem more plausible when the "population" itself is already identified as a biological level, such as an organism, or a social group. In cancer biology, for instance, the mainstream, so-called somatic mutation theory of cancer is that cancer is a Darwinian process among cells

(Hanahan and Weinberg 2000). By contrast, the opponents say, the tissue organization constrains cell fate and individual cell variation in a way which "de-Darwinizes"[12] cells, even in cancerous tissues (Sonnenschein and Soto 1999). Population-level phenomena could also underlie recent intriguing results obtained on cell cultures (Braun 2015). Populations of *Saccharomyces cerevisiae* were inflicted an "unforeseen challenge" by the means of experimentally rewiring an essential gene. On average, 50 percent of individuals survived and adapted (and the adaptation was partly inherited). Interestingly, the expression of the rewired gene showed slow collective modes (10–20 generations), and populations showed slow relaxation toward a stable state (ca. 100 generations), suggesting that population-level phenomena could constrain individual behaviors. Finally, culture and learning are other examples where individual variation can be non-accidental at the individual level, for population-level reasons (Claidière and André 2012).

Even without resorting to the definition of a higher level of causation (such as populations acting as constraints on individuals), there is a well-established tradition of thought according to which the individual scale might not be the only one to look at to tell a selection story. Individuals can have effects extending beyond individual fates in space, a phenomenon Dawkins named the "extended phenotype" (Dawkins 1982). As Laland and Sterelny (2006, 1759) recall, using the example of exudation of mucus by worms in the soil, these extended phenotypes can have no clear-cut boundaries. With spatialized interactions, the environmental challenge undergone by individuals may depend on the phenotype expressed by their neighbors, and not solely on their own phenotype (Van Baalen and Rand 1998).[13] This can result in spatial structures and heterogeneities in the population hindering the idea that evolution by natural selection, if it occurs, is to be understood solely at the scale of the individual.

2.3 Timescales Are Intertwined

Timescale entanglement relates, broadly construed, to the possibility of identifying generations (even overlapping ones) of individuals embedded in a larger, evolutionary process. Much like with space, individuals can have effects extending beyond individual fates *in time*, leading to "posthumous" extended phenotypes (Lehmann 2008). Examples are numerous in the niche construction and nongenetic inheritance literatures: beavers build dams that last for generations, mothers lay eggs imprinted by their own experience of the environment, worms produce mucus that durably alters the soil, photosynthetic organisms have increased dioxygen levels over billions of years, and so on (reviewed in Jablonka and Lamb 1995; Mousseau and Fox 1998a; Odling-Smee et al. 2003; Jablonka and Lamb 2005; Gissis and Jablonka 2011). Posthumous extended phenotypes extend developmental time beyond the boundaries of individual generations: the biological history of individuals continues after their own death. If posthumous effects extend far enough in time, their dynamics can overlap with the selective dynamics, thus questioning the distinction between the developmental time and the time of selection (Pocheville 2010).

Timescale entanglement can also come from the other side of the evolutionary coin: variation and evolution can be rapid, possibly commensurate with developmental time (Thompson 1998; Hairston et al. 2005; Richards, Bossdorf, and Pigliucci 2010; Shefferson and Salguero-Gomez 2015; Braun 2015; reviewed in Gingerich 2009). In addition, the timescale (and propensity to horizontal transmission) of hereditary materials can depend on their adaptive value, as exemplified by plasmids (Turner et al. 2014) and cultural variants (Claidière and André 2012).

Last, short-term, nongenetic variation can also lead to long-term effects, as exemplified by culture, gene-culture coevolution—the lactase gene is a paradigmatic example (Laland, Odling-Smee, and Myles 2010)—and mutational assimilation (Jablonka and Lamb 1995, 167–171). Mutational assimilation refers to a situation where a stimulus would durably alter the chromatin in a way which affects the mutability or reparability of the chromatin region, sometimes turning the region into a mutational "hot spot." For instance, dysfunctional genes can be up-regulated (which can be mutagenic) until they become functional or the individual dies (Wright 2000). This may lead to effects comparable to those of classical genetic assimilation (Jablonka and Lamb 1995, 167; Danchin et al. 2018). This kind of argument invites us to a view of evolution where fast processes "cause" slow ones—while, usually, slow ones "constrain" fast ones—leading to bidirectional relationships in timescales.

3.   Responses to the Challenges

I now wish to examine what, in my view, the "standard" response to these challenges should be—especially, when it comes to explaining adaptation. While I would not consider myself as a "standardist," I worry that, on a number of occasions, esteemed colleagues, proponents of an extended view of evolution, seem to embrace similar responses—thus, apparently undermining their own view. Hopefully, the following argument will help sharpening the disagreement between the diverse schools of thought. I will keep it simple and general—it could be further implemented in virtually any dynamical system. I will argue, in brief, the following:

Arguing for reciprocal causation in a dynamical system, or rather, for the nonseparation of two processes, requires showing that these processes significantly interact on the timescales of interest.

Characterizing timescales empirically is a fair challenge, the burden of which naturally falls on those who study the considered processes. My argument, here, remains theoretical. It is, however, not anecdotal, and will apply to any account of variation, development, or evolution as dynamical systems.

3.1 Level and Space Rescaling

Space provides a pedagogical entry to the question. Even if a phenotype is extended in space, it can have one or several defined space-scales, separable from the space-scales of the environmental challenge faced at the population level. For instance, say that an organism modifies its nearby environment—it builds a structure, such as a spider web, or releases a substance which diffuses in the environment. The rate of decay in space of the environmental modification defines a spatial scale[14]: some structures will extend farther, some substances will diffuse farther than others, and the influence of the organism may be farther reaching accordingly. This extended phenotype, with its spatial dimension, is what is "seen" by selection. A spider can be selected for building a larger or a smaller web, a predator can be selected for releasing less kairomones, etc. An interesting case is when individuals are so distributed in space that they share some parts of the modified environment—think, for instance, of social spiders sharing their web (Avilés and Tufino 1998). Such spatialized interactions can lead to the identification of clusters of neighbors constituting new, extended units of selection (as exemplified by the models of Van Baalen and Rand 1998; Lehmann 2008). These units need not have clear-cut boundaries to be targets of selection, and are thus better conceived as spatial scales of selection—these are the scales at which, for instance, adaptation for sharing goods between neighbors could occur. Considering these spatial scales extending beyond the individual boundaries amounts to rescaling (in space) evolution by natural selection.

3.2 Time Rescaling

Similarly, even if phenotypes are extended in time, development can have a defined timescale. Furthermore, this developmental timescale may be separable from the timescale of evolution by natural selection. The timescale of a phenotype is, roughly speaking, how long this phenotype will last, and, as far as selection matters, how long this phenotype will be "seen" by selection. Where a classical phenotype would have a timescale of one generation (or less), the timescale of a posthumous phenotype would typically span several generations (possibly affecting future individuals). The timescale of selection is characterized by the rate at which variants exclude each other (commensurate with the inverse of the selection coefficients). Assuming that heritable variation is replenished, the timescale of evolution (here by natural selection) is characterized by the rate at which cumulative change occurs.

Timescale separation means that one process is assumed to be much faster than the other, in a way which enables one to separate their respective dynamics (see box 13.1 and figure 13.2). Timescale separation is a way to "orthogonalize" two dynamics: the fast dynamics looks instantaneous with respect to the slow one, and the slow one looks immobile with respect to the fast one, making their time dimensions independent. When two processes are at play, the ratio between their respective timescales leads to four possible cases: either the first is much faster than the second, or the second is much faster than the first, or both have similar timescales, or at least one of them has no characteristic timescale.

Box 13.1

Principle of Timescale Separation

Let a dynamical system be described by two differential equations, where a small parameter $\tau \to 0$ appears in front of the highest derivative term of the putative fast variable.

$\tau \, dO/dt = f(O,E)$

$dE/dt = g(O,E)$

This system can be simplified by a timescale decoupling (of factor τ) of fast and slow variables. Variable E is first "frozen" at a fixed value, and the system is solved for O. This enables one to obtain an asymptotic value (if there is any) $O(E)$ such that $f(O(E), E) \equiv 0$. (Fast oscillating variables can similarly be accommodated by taking the average of their oscillations—as if the slow variable did not "see" them.) This asymptotic value can then be injected in the equation describing the dynamics of E. This leads to obtaining the equation for the slow variable:

$dE/dt = g(O(E), E) \equiv G(E)$

Generally speaking, if equations contain fixed parameters, these may be variables the dynamics of which are purposely ignored.

This box is adapted from Lesne (2013, 9–10), using Lewontin's famous metaphorical equations (Lewontin 1983, 282). The case presented here corresponds to the cyanobacteria example.

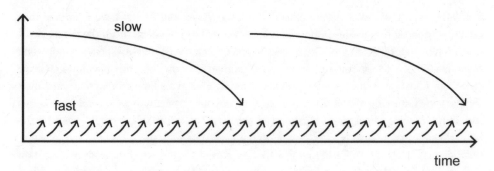

Figure 13.2
When timescales are well separated, slow processes constrain fast ones. There is no reciprocal causation.

To keep things simple, I now expound these four cases, focusing on the two processes of development (possibly posthumous development: the way present individuals affect future ones) and evolution by natural selection, assuming that the replenishment of heritable variation (e.g., by mutation) is slower than both development and selection. This last assumption is compatible with a gene-centered view of adaptation—which I will here stretch to the limit—but is independent from my general argument. This subsection is summarized in table 13.1.

3.2.1 Fast development vs. slow evolution by natural selection ("standard" view)

The paradigmatic view of evolution is one where organisms develop and die at the scale of one generation, while populations undergo evolutionary change, including by natural selection, over much longer timescales. On the film of evolution, each development lasts the time of a single photograph. Reciprocally, at the timescale of development, evolution looks still. In this view, there is no need to consider the dynamics of development when talking about evolution: a snapshot picture of development (e.g., the adaptive value of such and such life cycles) is enough. This is the view defended by Mayr (1961).

This view can be generalized to posthumous phenotypes. Say that a posthumous phenotype spans a few generations, but that selection is slow. Then, chances are that one be able to identify a selective process rescaled in time, where the timescale of the selected phenotype would span a few generations (Pocheville 2010). A number of examples will make this idea more intuitive.

Maternal effects occur when mothers transmit resources and information to their offspring independently of their genes (Mousseau and Fox 1998a). Proponents of niche construction count maternal effects as examples of niche construction (Odling-Smee, Laland, and Feldman 2003, 11, 125–127). A number of maternal effects can be seen, at least hypothetically, as adaptations, where the relevant timescale of the selected phenotype spans at least a pair of generations, from mother to offspring (Mousseau and Fox 1998a, 1998b; Marshall and Uller 2007, see also Duckworth, this volume). Under this view, the proximate source of offspring adaptation may be the mother, but the ultimate source lies in the evolution by natural selection of the mechanism of maternal transmission—much like an organ would adaptively respond to previous experiences in life, while at another timescale the adaptive response would be shaped by natural selection. This is not to say that maternal effects evolve like any other trait—they do not—or have no particular dynamical implications—they do—but that, *as far as adaptation goes*, the hypothetical source of adaptation is still long-term, slow natural selection.

The same holds for environmental inheritance, the inheritance of environments modified by the niche construction of relatives or ancestors.[15] Beaver dams are a paradigmatic example, some of them lasting for many beaver generations (Odling-Smee, Laland, and Feldman 2003, 226; Laland, Odling-Smee, and Feldman 2003, 119). Earthworms are a similar flagship example[16]: they modify the soil in a way which "enables them to survive in an otherwise forbidding environment" (Turner 2000, 6). As Laland, Odling-Smee, and Feldman (2005, 39) put it, "each worm directly benefits from its own activities" but "their impact on the soil accumulates over many generations." Lehmann (2008), however, has shown in a mathematical model how spatial viscosity (the fact that offspring dispersal and environmental modifications tend to remain local) can lead to the evolution by slow, natural selection of genetically determined posthumous extended phenotypes (see also Lehmann 2010). Insofar as beaver dams or soil modifications are beneficial to the beaver or worm offspring, a putative explanation for their durability is slow, kin selection in time on genes for dam building or soil modification—genes which are shared by the offspring inheriting the modified environment, genes, interestingly, the existence of which is classically hypothesized both by old "standardists" (Dawkins 1982, 209) and niche constructionists (Odling-Smee, Laland, and Feldman 2003, 334; see, however, Laland, Odling-Smee, and Feldman 2005, 44).

Nongenetic inheritance has often been invoked to argue for an extended view of evolution, especially—though not only—in that it seems to provide under certain conditions nonblind, adaptive variation (Jablonka and Lamb 1995; Danchin et al. 2011; Laland et al. 2015). These mechanisms may provide transgenerational plasticity,[17] enabling the accommodation of environmental challenges over longer timescales than the sole generation (Lamm and Jablonka 2008; Bonduriansky, Crean, and Day 2012). The "standard" argument is that nongenetic mechanisms of inheritance are themselves traits determined by genes subject to blind variation and natural selection over evolutionary timescales (Haig 2007; Scott-Phillips, Dickins, and West 2011; Dickins and Rahman 2012; but see Mesoudi et al. 2013). This can be seen for instance in a model by Lachmann and Jablonka (1996), where a gene determining the transition rate between two epigenetic states can evolve on the long term to track the fluctuation rates of a changing environment.

Considering posthumous phenotypes amounts to rescaling in time development and evolution by natural selection (see also Bouchard 2011). These processes can be rescaled in space or levels at the same time—say, for instance, that selection for dam building happens at the level of a family of beavers—but they need not be.[18] Importantly, rescaling may come with changing the focal trait featuring in an evolutionary explanation. With classical or transgenerational plasticity, rather than the current value of a trait, it is the capacity to accommodate (i.e., adaptively change value of the trait through lifetime or generations), which we now want to explain. Focusing on plasticity (intra- or inter-generational) amounts to discarding a "zeroth-order" trait (the current trait) to consider a "first order" one (the capacity to change that trait over time). This is, for instance, what is at stake when Odling-Smee, Laland, and Feldman (2003, 21) and Laland and Sterelny (2006, 1756) write: "the finch's capacity to use spines to grub for insects is not an adaptation. Rather, the finch … exploits a more general and flexible adaptation, namely the capacity to learn." To be sure, this change in order can be further recruited to argue for a gene-centered view of adaptation, where genes (or the genome architecture) would determine that higher order trait.

The view separating development (broadly construed) and evolution by natural selection can be inflected on a multiplicity of timescales, where adaptation would arise solely, ultimately, from the selection of blind variation (figures 13.3 and 13.4). For instance, at the developmental timescale, parts of the organism would accommodate to physiological challenges possibly partly by manifesting an already existing repertoire of responses,[19] and partly by blindly exploring diverse physiological responses—and stabilizing, if possible, the successful ones. Both the repertoire of responses, and the parameters of blind exploration, might be evolvable by natural selection at a longer timescale. This evolvability by natural selection may itself be subject to evolution by natural selection, at yet a longer timescale—and so on, in a possibly indefinite (but virtuous) regress across orders and timescales. This classical, multiscale (and multi-order) view of blind adaptation traces back to the early days of Darwinism, and is still alive today (Roux 1881, 90; Weismann 1904a, ix, 1904b, 119; Dawkins 1986; see Gould 2002, 223, quoting Weismann, cited

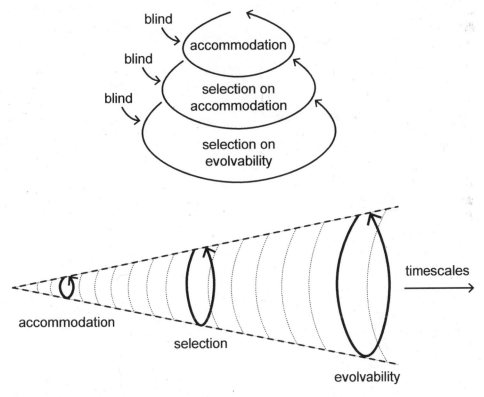

Figure 13.3
A (classical) multiscale, multi-order view of evolution, where the ultimate source of adaptation is natural selection of blind variation. Each looping arrow represents the causal iteration of a process, with potential variation. Here processes are each supposed timescale separated from the others. Causation occurs "within" a timescale.

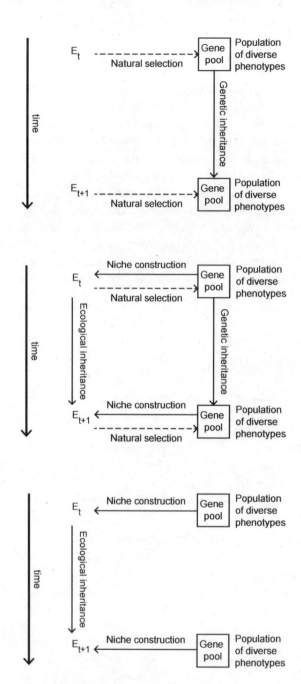

Figure 13.4
Timescale separation breaks reciprocal causation. The top and middle panels are redrawn from Odling-Smee, Laland, and Feldman (2003, 14), who give the following caption: "[Top]: Standard evolutionary perspective: Organisms transmit genes from generation t to generation t+1 with natural selection acting on phenotypes. [Middle]: With niche construction: Organisms also modify their local environment (E), as depicted by the arrow labeled "niche construction." Each generation inherits from ancestral organisms both genes and a legacy of modified selection pressures, described as "ecological inheritance." Such a proximate causal narrative, however, is insufficient to argue for reciprocal causation between proximate and ultimate processes. The narrative needs to be supplemented by showing that all causal arrows also consider long-term, commensurate timescales. [Bottom]: under the ("standard") hypothesis that niche construction and ecological inheritance consider much different timescales than natural selection and genetic inheritance, reciprocal causation disappears.

above; Hodgson 2005; Godfrey-Smith 2009, chap. 2; Kupiec 2012; Soen, Knafo, and Elgart 2015; Pocheville and Danchin 2017).

3.2.2 Slow development vs. fast evolution by natural selection

An inverse case is when development takes place on much longer timescales than its evolution by natural selection. Here again, there is no dynamical interplay between development and natural selection, but not because development looks instantaneous, rather, because it now looks still. I will illustrate how this idea makes sense with a well-known example, involving posthumous phenotypes:

When photosynthesis first evolved in bacteria, particularly in cyanobacteria, a novel form of oxygen production was created. The contribution of these ancestral organisms to the earth's 21% oxygen atmosphere must have occurred over billions of years, and it must have taken innumerable generations of photosynthesizing organisms to achieve. It is highly likely that modified natural selection pressures, stemming from the earth's changed atmosphere, played an enormous role in subsequent biological evolution. (Odling-Smee, Laland, and Feldman 2003, 12)

Here the posthumous development of each photosynthesizing organism is their own, individual contribution to the atmospheric dioxygen.[20] Its timescale is how long this contribution, however small, will last in the atmosphere, even after their death (it is commensurate with the residence time of dioxygen in the atmosphere). In this case, this posthumous development seems slow enough (ca. $> 10^6$ generations) to leave plenty of time for evolution by natural selection to possibly occur *within* a developmental timescale.[21]

If individual contributions are to make any difference at all, that is. In this example, individual contributions, relatively to the atmosphere, appear infinitesimal. To see some environmental effect one has to consider whole photosynthetic lineages. And here again, the environmental effect seems so slow (the atmosphere's evolution taking place over hundreds of millions of years), that it is virtually invisible for the evolution of these lineages on timescales smaller than billions of individual generations. This is a case of what has been named in the literature "niche alteration" or "mere effects" (resp. Dawkins 2004, 380; Sterelny 2005, 24; see Dutreuil and Pocheville 2015). This kind of niche construction is so slow that its dynamics seems negligible with respect to natural selection on these timescales. Here niche construction *constrains* selection: there is no reciprocal causation.[22]

Similar to slow processes, rare events can sometimes be ignored on relatively short timescales. Rare events may set initial conditions for a process, but they do not dynamically interact with it. For instance, in adaptive dynamics, selection is supposed to happen "in between" two mutation events. A putative similar example, in the niche construction literature, would be imprinting, the fact that an organism seeks a situation of which it has had an experience before (Odling-Smee, Laland, and Feldman 2003, 104). If changes in imprinting were rare events with respect to the evolution of a population (which I personally doubt, but this is independent from the theoretical argument), an evolutionary biologist might be justified in treating the behavior as a constant on their timescale of interest. This means that exhibiting cases of imprinting (or of other "single-event" niche construction or cultural changes) is not enough to argue for reciprocal causation: one has to specify empirically their rate of occurrence and compare it to the timescale of the evolutionary process of interest.

In a nutshell, constraints and initial conditions, as important as they are, do not exemplify reciprocal causation: the dynamics of interest conforms to the sole focal process.

3.2.3   Development and evolution by natural selection interacting on commensurate timescales ("niche interaction")

Now let us consider the case where development and evolution by natural selection are properly entangled—that is, interacting on a common timescale. Inflecting the niche construction terminology, I call this case niche interaction. I give here niche interaction as a theoretical possibility, although I think it is where the novelty—or one core novelty—of the extended views of evolution lies. Thus, empirical investigations aiming at justifying the importance of these frameworks may aim at documenting this case in priority—otherwise, empirical investigations, such as those proposed by Odling-Smee, Laland, and Feldman (2003, chap. 7, 292), would run the risk of being reducible to investigations of classical, natural selection-driven evolution of extended or posthumous phenotypes (Pocheville 2010, 88). Such an entanglement could be the case for the examples of supposedly fast posthumous development examined above, if we re-assess the common assumption that evolution by natural selection is slow, and consider that it can occur on relatively few generations (Thompson 1998; Hairston et al. 2005; Gingerich 2009; Richards, Bossdorf, and Pigliucci 2010; Shefferson and Salguero-Gomez 2015; Braun 2015). For instance, maybe sometimes evolution by natural selection occurs in worm populations *at the same speed* as the individual modifications of the soil happen and decay, leading to new evolutionary routes and dynamical behaviors (as exemplified in the models by Laland, Odling-Smee, and Feldman 1999). Gene-culture coevolution is another case where timescale nonseparability has been traditionally assumed. Culture (including individual learning) sets up selective pressures which evolve at a pace possibly commensurate with genetic evolution (Laland, Odling-Smee, and Myles 2010). This aspect has lead early theorists to propose to track cultural and genetic dynamics in parallel in dual-inheritance models (Boyd and Richerson 1985). Last, environmental induction of otherwise durably inherited epigenetic marks and the model of mutational assimilation offer similarly rich timescale structures, where brief developmental events can have long-term effects (Jablonka and Lamb 1995, 167–171; Danchin et al. 2018).

Niche interaction means that the dynamical aspects of each process are relevant for the dynamics of the other. In particular, evolution by natural selection is "dynamically insufficient" (Lewontin 1974, 6) to account for the evolution of the trait: the way posthumous phenotypes change over time in the population depends as much on their individual dynam-

Table 13.1.
Four possible ratios of the timescales of development (possibly posthumous) and evolution by natural selection (here abbreviated selection).

Timescale ratio	Examples	Consequences
development << selection	maternal effects, beaver dams, worm mucus, and other short-term consistent effects, which are good candidates of naturally selected posthumous phenotypes	selection constrains development; no reciprocal causation (classical view)
development >> selection	enrichment of atmospheric dioxygen by cyanobacteria over billions of generations	development constrains selection; no reciprocal causation
development ~ selection	possibly the posthumous phenotypes above if evolution by natural selection is fast; gene-culture coevolution; mutational assimilation	development and selection interact at the given timescale of study; reciprocal causation (called here "niche interaction")
no characteristic timescale	theoretical possibility	development and selection interact at all timescales; reciprocal causation of a probably new kind

ics as on the natural selection bearing on them. Of course, one might attempt to rescale the processes to separate their rescaled counterparts—maybe, at another timescale, natural selection is sufficient and adaptation emerges—but, on the focal timescale, one has to consider their interactions. It is unclear in which respect adaptation should emerge from such a dynamical interplay.

3.2.4 Development and evolution by natural selection with no characteristic timescale

A process can have no particular timescale, that is, continue to have significant manifestations (variations or fluctuations), or to "last," at all timescales. If (posthumous) development and evolution by natural selection have no particular timescale, rescaling as an attempt to separate them is not an option, because they will be intertwined at any considered timescale—even, theoretically, infinite ones. Either of them will (by hypothesis), taken on its own, always be dynamically insufficient to explain the change in phenotypes.

Although this is an exciting scenario, I mention it here as a theoretical speculation only. Let us say that the situation may not necessarily be unknown to biology. For instance, in humans, healthy hearts exhibits rate variations at all temporal scales, while unhealthy hearts do not (West 2006). Some results reviewed by Braun (2015) may also hint at such phenomena in experimental evolution. It is unclear whether any aspect of the Darwinian dream could be saved—there might still be natural selection, but its direction could change at any timescale—and whether this situation would be necessary for, or at least compatible with, adaptation.

4.   Why This Matters

I now wish to draw a few remarks and sketch some perspectives. The remarks respond to views often expressed orally, and offer a snapshot of current discussions.

4.1   Remarks

4.1.1   Reciprocal causation is neither reciprocal nor causation

Reciprocal causation is not a problem of infinite regress of prior causation in causal chains. An infinite regress of prior causation, even when empirically demonstrated, does not mean reciprocal, symmetrical explanation of one process by another. On the contrary, it is sometimes (if not often) possible to break an infinite regress in our explanations by timescale separating two processes which, when timescales are not taken into account, seem to exhibit intertwined causal chains. When processes are timescale separated, slow processes play the role of constraints for fast ones, and asymptotic values of variables in fast ones play the role of causal variables in slow ones.

Causation is more easily conceived as occurring at one timescale of study. Flows of causation can easily change direction when changing this timescale. The definition of a process is inseparable from the definition of a relevant timescale of study.

4.1.2   The question is not "How long ago?," but "How much slower?"

As Uller and Helanterä (2016) put it in their recent discussion of extensions to evolutionary theory, "[o]n the former account, process autonomy and strict genetic inheritance ensure that all adaptive directionality arising in development can be explained in terms of selection on genetic variation in the past. The same logic does not apply when the processes are intertwined" (see also Uller and Helanterä 2017). I feel of course very sympathetic to this view.[23] I would like to offer here a slightly different, complementary perspective on what the locus of the theoretical extension is.

Insofar as timescale separation is concerned, the question is not exactly whether one will be able to explain current directionality in development in terms of causation by past selection. Rather, it is whether one will be able to explain development in terms of the constraints imposed by another slower, wider in time, evolutionary process. Timescale separation puts an emphasis on constraints (local invariants), not causes,[24] as crucial features in our explanations (Pocheville 2010). This view is akin to how mathematical features can become explanatory by themselves, freeing our explanations from appeals to initial conditions (Huneman 2010).

A philosophical framework characterizing the structure of evolutionary theory in terms of process autonomy (or reciprocal causation) focusses on flows of causation, and on how our hypotheses on these will change. The "timescale" framework focusses on how our current view of the structure of evolutionary time may change. Whether timescales may

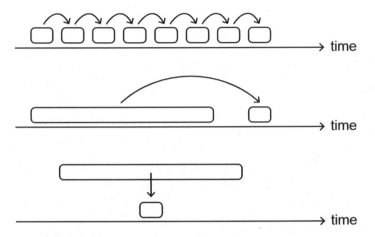

Figure 13.5
Three ways of telling causal narratives. Top: a chain of proximate events explains a current state. Middle: a current proximate process is explained by the action of another past ultimate process (e.g., natural selection). Bottom: a current proximate process is explained in terms of the constraints imposed by an ultimate process.

or may not be separable, embedded one in another like Russian dolls, not the procession of events in time, is the matter at stake (figure 13.5).

4.1.3 Externalism does mind

Timescale separation may affect whether our explanations will be "externalist," "internalist," or "interactionist"—that is, whether properties of the system will be explained by those of its environment, those of the system, or both (Godfrey-Smith 1996, 51). For instance, in the cyanobacteria case, an interactionist story (cyanobacteria have considerable influence on oxygen levels in the environment which also selects them) fades out into a classical externalist story (at nongeological timescales) when timescale separation is operated (cyanobacteria have negligible influence on oxygen levels in the environment which selects them). Similarly, interactionist stories of posthumous development at the few generations timescale (beavers build dams that influence the selection bearing on subsequent generations) can become externalist at a larger timescale (dam building is constrained by long-term natural selection stemming from properties of the landscape).

The question of externalism, however, is probably not the best way to frame extended views of evolution, such as niche construction theory (Pocheville 2010, 54–55, 95–100). While the environment is usually thought of as the source of natural selection, the constraints on a process need not be spatially located (e.g., physical laws). Constraints tell how the whole process is meant to happen—or rather, not meant to happen. They are invariant facts to which the process conforms, and do not necessarily coincide with physical objects in the external world. When we think of the environment as a constraint (to which organisms must adapt), it only means that we think that some relevant properties of the environment are stable enough for their dynamics to be separated from the activities of the organisms. How and why timescale separation may come to coincide with spatial delineation is an interesting question, but the spatial dimension, unlike time, is not vital for the definition of processes. This means that the (or one) core claim of, for instance, niche construction theory, is not about organisms modifying environments in any significant way (non-externalism), but that processes of development and evolution by natural selection are entangled (nonseparation). Externalism here is temporal, not spatial.[25]

4.1.4   Gene reductionism is a temporal holism

One frequent contention with respect to extended views of evolution is that, ultimately, the contention goes, everything must come from the genes. For instance:

"[U]ltimately the extra information (e.g., methylation) is provided by enzymes (methylases) encoded by genes in the genome. Epigenetics, per se, doesn't add any new information. It's just a consequence, or outcome, of the information already in the DNA." (Moran 2016, cited by Griffiths 2017)

We are now in a position to give a charitable interpretation of this view: it is (I contend) the view that evolutionary biology deals with evolution, and that evolution, in particular cumulative adaptation, is a long-term process, which operates solely over long-lasting, genetic materials (e.g., Haig 2007).[26] In the case above, as Griffiths (2017) replies, some information in the epigenetic patterns comes from the environment, so epigenetics *does* add new information. However, the charitable interpretation goes, that information that matters evolutionarily (and not just developmentally) must be long lasting, and it is uncertain whether epigenetics adds any new long-lasting information:

"If an epimutation is to have evolutionary importance, it must persist. Either epigenetic states must be highly stable or they have to be converted to some form of hard inheritance or, quite probably, both." (Wilkins 2011, 391, partly cited by Griffiths 2017)

The traditional view, in informal discussions, is sometimes thought of as a kind of reductionism (reducing organic evolution to the evolution of gene sequences). It is however less akin to a constitutive reductionism (that which happens at this level can be explained in terms of that which happens at a lower level) than to a "temporal holism": that which happens at this timescale is explained in terms of that which happens at a larger one.

4.1.5   A change in evolution is not a change in evolutionary theory

A frequent argument for theoretical change is that phenomena such as imprinting, maternal effects, environmental modification, and so on change the course of evolution.

Surely, they do. Any new trait can virtually change the course of evolution. Any new mode of locomotion, of digestion, and so on will enable new evolutionary outcomes. But it does not ensue that we must change evolutionary theory. It is the aim of evolutionary theory to accommodate new traits and treat them as specific features of a more generic framework.[27] Changing the course of evolution is not enough: possibly (this is the "standard" view), the theoretical change will be limited to the particulars of biological parameter implementations in specific models, leaving the general structure of our explanations untouched (especially in regard to adaptation). Timescale separation, as we have seen, is the royal way to accommodate niche construction and nongenetic inheritance as posthumous phenotypes, to save, despite their effects on evolution, the general structure of evolutionary theory.

4.1.6   Timescale separation has deep and subtle consequences
Timescale separation can have deep, subtle consequences. For instance, Griffiths (2017) has argued that nongenetic inheritance is significant in that it modifies the laws of heredity (an argument reminiscent of Odling-Smee 1988, 74):

"The primary significance of Mendelism for the theory of natural selection was that it specified the form of the transmission phase. Epigenetic and exogenetic heredity change this form." (Griffiths 2017)

This argument is correct. However, it does not guarantee that, still, the transmission of that which evolves on the long term is not Mendelian (or more generally genetic). Similarly, short-term nonblind variation is theoretically compatible with long-term blind variation. Short-term neo-Lamarckism is compatible with long-term neo-Darwinism.

4.1.7   Big, slow drivers can have virtually null short-term effects
Big, slow drivers need not have big, short-term effects. On the contrary, very small, almost undetectable effects can accumulate through time. This means that even with almost null heritability attributable to genetic variation at the generation timescale, an evolutionary process can still be completely driven by natural selection operating on genetic variation on the long term.

4.1.8   Empirical work must be theoretically informed
The theoretical considerations above put a strong constraint on empirical (and philosophical) work. If a biologist looks at empirical examples and finds cases which, on closer scrutiny, might turn out to be classical phenotypes, extended in space and (now) time, and yet evolved by natural selection, then the biologist might actually be reinforcing neo-Darwinism without noticing. For instance, it has often been argued that niche construction by organisms "can result in significant, consistent, and directed changes in their local environments" (Odling-Smee, Laland, and Feldman 2003, 5). This is not providing evidence for a neglected process in evolution but listing ideal conditions for (classical, neo-Darwinian) evolution by natural selection of extended and posthumous phenotypes. Generally speaking, to the extent that the case for an extended evolutionary synthesis is based on reciprocal causation, providing evidence of organisms modifying their environments (even in a way relevant to selection) or of a new inheritance route is not enough to argue that our view of long-term evolution is incomplete: one has also to provide (reasonable) evidence that the missing part is not timescale separable from the existing picture.

Process separation being the current default assumption, and a simplifying assumption (or parsimonious, if you will), the burden of proof seems to bear on those who argue for nonseparation.

4.1.9   The origin of timescales is not the origin of life

There is a temptation, in a multilevel view of evolution, to assimilate the origin of life with something akin to the lowest currently existing levels (e.g., cells, or nucleic acids, depending on the view). The temptation goes on that, intuitively, small materials should have short timescales, and that one could assimilate the origin of life with the smallest timescales.[28] However, timescales (and levels) are themselves evolvable and can emerge or change in an evolutionary process. Mechanisms stabilizing DNA lengthen the timescale of genetic evolution, the stability of epigenetic materials can evolve to match the lability of the environment (Lachmann and Jablonka 1996), and, most probably, a heritable material occupying a new "niche" on the axis of timescales enables or leads other substrates to move to other timescales (see Uller, English, and Pen 2015). What lies at the origin of timescales is what has a short timescale now, that is, the materials currently undergoing the fast processes. It is most probably a result, rather than the origin, of evolution.

4.1.10   Infinite regress in timescales is virtuous

Those who are presented rescaling techniques for the first time often worry that it may lead to an infinite regress. For instance, say that we observe adaptive physiological variation and wonder how to explain it. Say that we discover that it can be explained in terms of adaptive nongenetic inheritance. This is already a first explanatory move upward in orders and scales of evolution. Now this adaptive nongenetic inheritance itself seems to need some explanation. One can attempt to repeat the explanatory move: rescale again and try to explain it in terms of natural selection operating over genes. Now, say that we discover that the genetic variation involved was not exactly totally blind, and that some crucial loci were more evolvable than others, which turned out to be adaptive. What would prevent us from going one step further and hypothesize that the patterns of genetic variation were actually determined by something else, itself the true, ultimate target of natural selection? Here comes the worry of infinite regress: each time adaptive, nonblind variation appears, rescaling pops up.

This worry can be alleviated. First, rescaling or changing order (e.g., from first order to second order selection) does not mean going up to infinity. One can stop after just one iteration. After all, explanations have to start and end somewhere. It might also be a genuine feature of the biological world that there is no infinity of scales and orders of evolutionary processes—for instance, maybe there is no pattern of nonblind genetic variation, and rescaling ends there. In this case, even the best possible multiscale explanation will consider a finite number of orders and scales.

Moreover, were we to identify an infinite regress, it would be good news. The traditional philosophical wisdom assimilates infinite regress with a fallacy, but it is exactly that which forms, in science, the basis of asymptotic reasoning. In the case of scaling, asymptotic reasoning is instantiated in renormalization techniques, the most powerful techniques to date to deal with multiscale phenomena (Lesne 2003).[29]

4.1.11   Rescaling is but a tool in a research program

Rescaling, some worry, is ad hoc. This is something akin to the critique of Gould and Lewontin (1979, 586), that the adaptationist program is endless: "If one adaptive argument fails, try another." This would become, in our case, "If variation is adaptive at some timescale, try to explain it by natural selection operating on blind variation at another, longer timescale." Indeed, this is the selectionist program. Rescaling is but a tool in a research program. This program will be considered virtuous to some, problematic to others, and history will decide between them.

4.1.12   Extended views are renewed Darwinism

Extended views on evolution are Darwinian (often explicitly, e.g., Jablonka and Lamb 2005, 2). They have sometimes been claimed to differ from the "traditional Darwinian perspective" in that they seek the laws of variation (Laland et al. 2015, 7). By so doing however, they precisely renew Darwin's own care about variation (Darwin 1859, chap. V; Darwin 1863; Hoquet 2009).

Such a renewal does not necessarily mean challenging the current view. Variation and selection are two complementary parts of the evolutionary process, and, under the hypothesis that they are separable (which, arguably, Darwin doubted, as do many contributors to this volume), looking more closely into one leaves the other intact. I will illustrate this question with a last example, which will bring us back to the question of explaining adaptation.

Niche construction has been offered as a second route to adaptation (Odling-Smee, Laland, and Feldman 2003, 43). It is however unclear whether niche construction should not rather be considered as "blind" to adaptation—that is, the capacity to construct niches would explain the existence of beneficial (and detrimental) niche constructing variations in the space of possible variations, but it would not explain why niche construction should more often lead to beneficial or viable activities than classical phenotypes (Pocheville 2010, 90–95; Uller and Helanterä 2017).[30]

Maybe the claim that niche construction is second a route to adaptation should be rephrased as follows: "Niche construction is another word for the development of extended and posthumous phenotypes, and future theories of organization will tell us why there exist beneficial or viable variations in the space of possible developments. Thus we will have two explanations for adaptation: the existence of beneficial (or at least viable) variations, and the selection of these variations" (see, e.g., Laland and Sterelny 2006, 1759). Even so, however, the separability of variation and natural selection would, possibly, be intact.

4.2   Perspectives

4.2.1   Should we expect timescale separability?

It has been the subject of empirical investigation ever since Darwin whether much of the living world conforms to the Darwinian dream (of course phrased in different terms). Intuitively, adaptation is probably more easily understood when adaptive processes at different timescales are separated—the first intuition of an engineer would probably be to design an adaptive system in such a modular way (modular here in time). This might explain why time-rescaling as a way of restoring the prominence of blind variation is so appealing. Under this view, one might expect a temporal self-organization of adaptive processes (much like a niche differentiation in scales). For instance, different hereditary materials would evolve toward different, somewhat regularly spaced, timescales.

Another view would be that adaptability requires variation occurring at all timescales—and conversely that adaptation (by accommodation or natural selection) can lead to perturbations at all scales (in organisms or ecosystems). Under this view, living beings would be able to sustain themselves precisely insofar as the processes occurring at different timescales (and levels) would be nonseparable.[31] The model of mutational assimilation cited above, where fast physiological processes can inscribe long-term effects in the biological history—and which at the moment might look like a theoretical curiosity—would exemplify a fundamental and widespread feature of living beings (figure 13.6).

4.2.2   A Statement of incompleteness

While I have been using (mathematical) dynamical systems as a formal comparison to inform current informal debates, it is worth reemphasizing before concluding this contribution that this account is not meant to be complete. Recent—and less recent—empirical

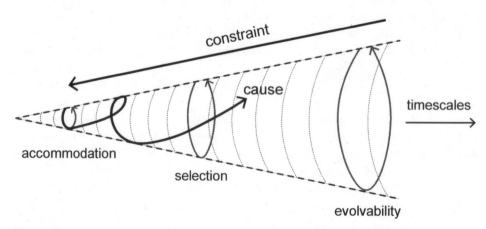

Figure 13.6
A somewhat ill-defined situation, where fast processes can cause long-term effects (as exemplified by mutational assimilation).

works call for theoretical extensions in many other ways than just a possible timescale entanglement of proximate and ultimate processes (pertaining, to name but a few, to the nature of novelty, of randomness, of time, etc.). A theoretical biologist would not make impossibility statements: she would be sure to be soon empirically disproved. Neither would she say that the current theoretical tools are complete. Nothing guarantees, for instance, that dynamical systems as we now know them can fully grasp the richness of timescale relationships in the living.

5. Conclusion

The line is subtle between the empirical cases that call for theoretical revisions and those that fit into our current scheme: it is the width of a timescale separation. Inter-generational proximate processes do not in principle threaten the distinction between proximate and ultimate causes, as formal models show. The "standard view," if it exists, is that the locus of evolutionary biology is a long-term process, separable from the short-term ones. It is reasonable. The extended view, in this regard, is that short-term processes may have long-term effects. It is exciting. Fleshing it out empirically, however, will require much more than a stall of cases exemplifying multiple inheritance systems or nonblind variation on short timescales. It will require a dive deep in time, to explore whether, and how, processes percolate along the scales.

Glossary

Adaptation   Property of a feature fit to a purpose, or the process leading to this fit (discussed in Burian 1983).

Blind   A process is blind (or epistemically blind) to something if this process (or explaining it) is independent from this thing (that is, the process is symmetrical through transformations of the thing).

Blind variation   Variation, the fitness effects of which are assumed irrelevant to computing its possibility or probability of occurrence (discussed in Pocheville and Danchin 2017).

Cause   A variable in a causal process, represented by a variable lying on the right-hand side of a causal equation, or by a variable in an algorithm.[32]

Constraint   Invariant in a dynamic, represented by the structure of a causal equation or of an algorithm, for instance the values of constant parameters (Pocheville 2010, 40; Montévil and Mossio 2015).

Development   Edification of a trait at the individual level. Includes niche construction and properties of hereditary materials originating in the focal individual and its experience of the environment. Development can start before conception or continue after death (Pocheville 2010, 84).

Evolution   Change at the population level in inherited characteristics. For instance, an environmentally induced epigenetic modification may constitute evolution of the epigenetic materials (similar to a directional mutation), but also development of a posthumous phenotype constrained by putative genes (similar to a cross-generational reaction norm).

Natural selection   Differential survival or multiplication of entities according to their intrinsic properties (Godfrey-Smith 2009, 53; Bouchard 2011; Bourrat 2015, 33).

Niche construction   The modification of the environment. Niche construction can lead to environmental inheritance when environments are differentially associated to offspring.[33]

Niche interaction   The modification of the environment modifying the effects of selection on a timescale commensurate with the selection bearing on the capacity to modify the environment (Pocheville 2010). A toy-comparison is as follows: consider beans in a sift, such that beans would get sifted according to their size (natural selection) but each bean would individually, dynamically vary in size (niche construction) at the same speed as the sifting would proceed (niche interaction).

Nongenetic inheritance   Inheritance of characteristics other than the DNA sequence.[34]

Posthumous phenotype   Effect of an organism (or gene, in a gene-centered view) extending beyond its life span. Posthumous phenotypes can be mere effects or evolve by kin-selection in time when they affect related individuals (Lehmann 2008).

Process   Phenomenon represented by a dynamical, causal equation, or a function in a causal algorithm. A timescale separation of processes is represented by the timescale separation of their respective equations.[35] In an algorithm, separated processes can be represented with different functions (one can call the other).

Proximate   Short term relatively to a longer timescale (after Mayr 1961). It is an empirical hypothesis, and not a conceptual necessity, that "proximate" coincides with "individual level." (The coincidence of proximate and individual level, and of ultimate and population level, however, is desirable for explanations of evolution in terms of natural selection.)

Proximate/ultimate cause   Strictly speaking, causal variable involved in a proximate (resp. ultimate) process. Here, however, it is more loosely used in the sense of an explanation framed in terms of this proximate (resp. ultimate) process. For instance, evolution by natural selection of genes is a putative ultimate process, usually referred to as an ultimate cause, explaining adaptation.

Reciprocal causation   Situation where two processes are nonseparable. Traditionally defined by intertwined causal chains: "Process A is a cause of process B and, subsequently, process B is a cause of process A, with this feedback potentially repeated in causal chains" (Laland et al. 2015, 6). I contend here that intertwined causal chains are not enough: one has to show that they consider similar timescales.

Timescale   Characteristic time of a phenomenon (e.g., inverse of a rate, average lifetime of an object). Also used here in the sense of a time span of description. The characteristic time (and the very aspect of the process one looks at) can depend on the timescale of description.

Timescale separation   Approximation technique to simplify the study of a process by ignoring the dynamics of otherwise possibly relevant processes, because they are either fast or slow enough.

Ultimate   Long term, relatively to a shorter timescale (after Mayr 1961).

Acknowledgments

I am grateful to Maël Montévil and Étienne Danchin for unconditional support and lively discussions, and to Tobias Uller, Kevin Laland, an anonymous referee, the participants of the workshop *Cause and Process in Evolution*, the members of the group Theory and Methods in Biosciences, especially Pierrick Bourrat and Elena Walsh, and Régis Ferrière, for kind and constructive feedback. My gratitude also goes to the many friends, colleagues, and mentors (including Régis Ferrière, Philippe Huneman, and Frédéric Bouchard) who greatly helped during my thesis, on which this chapter is based. This chapter was made possible through the support of a grant from the Templeton World

Charity Foundation, Inc. (TWCF0242). The opinions expressed in this chapter are those of the author and do not necessarily reflect the views of the Templeton World Charity Foundation, Inc.

Notes

1. Dawkins (2006, 19).

2. The theoretical argument is reminiscent of, but does not coincide with, Crick's Central Dogma (Crick 1958; see other historical references in Bonduriansky 2012, Box 3). More on this history can be found in Sapp (1987, chap. II), Gayon (1998), and Bowler (2001).

3. A view akin to what Godfrey-Smith (2001) would call "explanatory adaptationism," the view that "[t]he apparent design of organisms, and the relations of adaptedness between organisms and their environments, are the big questions, the amazing facts in biology ... [and that] [n]atural selection is the key to solving these problems."

4. I call it a "dream" to acknowledge well-grounded but somewhat unsubstantial arguments targeted against references to a fictitious "standard" evolutionary theory. While I think the idealization retains core aspects of the Darwinian explanation of adaptation, possible variations are innumerable, as future work will show.

5. In the case of frequency-dependence, similar idealizations can be used, as adaptive dynamics show (see below, and Dieckmann and Ferrière [2004]); however, the concept of adaptation itself must be revisited. For instance, the absolute fitness could stay constant over the selective trajectory (e.g., Sarkar 2014).

6. Darwin (e.g., 1859, 102) seems to have regarded large numbers and long timescales primarily as enablers of variation (more than of selection).

7. Genetic algorithms would be another such model, which I will leave aside here.

8. Arlin Stoltzfus (this volume) questions the idealization of nonlimiting mutation in population genetics.

9. In discrete time, the equivalent requirement is that the feature to be selected occurs before selection.

10. On levels, see also Walsh (this volume).

11. In this contribution, the vocable "to constrain," in contrast to "to determine," is meant to leave room for the possibility of indeterminism in the dynamics.

12. I take the word from Godfrey-Smith (2009, 9).

13. Two limit cases are when the phenotype is not extended at all (classical phenotype), or extended up to the population so that all individuals are affected in the same way (classical, nonspatial frequency-dependence). In both limit cases individuals are symmetrical with respect to the selective challenge.

14. Except when there is no characteristic space-scale, as in critical phenomena.

15. I prefer here the term "environmental" inheritance over that of "ecological" inheritance (Odling-Smee, Laland, and Feldman 2003, 419), to better specify the locus of what is inherited. "Ecology" is polysemous and most often embeds the idea of an "ecological" timescale, with respect to which I remain neutral here.

16. See, for example, Odling-Smee, Laland, and Feldman (2003, 11, 160, 291, 375); Laland (2004, 319–321); Laland, Odling-Smee, and Feldman (2005, 39); Laland and Brown (2006, 99); Laland and Sterelny (2006, 1754, 1758–1760); Laland, Odling-Smee, and Gilbert (2008, 552, 554, 560); and Laland et al. (2009, 199).

17. See Sultan (this volume) and Stotz (2017).

18. In biology, local processes need not have shorter timescales than larger ones (Lesne 2013, 9; see Duckworth, this volume, for a similar point). For instance, the mutation of a gene can be orders of magnitude slower than the firing of an action potential traveling along a giant calamar's neuron. Fast, far-reaching couplings might be a mark of biological phenomena (Longo, Montévil, and Pocheville 2012). The space-time relation has been discussed in field ecology (see, e.g., Clark 1985; Holling 1992; Peterson, Allen, and Holling 1998; and their discussion in Jeliazkov 2013).

19. On plasticity, see Sultan (this volume) and Dayan et al. (this volume).

20. I focus here on photosynthesizing organisms as a way to analyze reciprocal causation (without meaning to reduce, in general, niche construction theory to this sole aspect). If the selected organism is not the constructor,

there is no possible reciprocal causation between niche construction and natural selection (Pocheville 2010, 64–66).

21. The residence time of diogygen is estimated to be now within the 10^3–10^4 years range (Lenton 2001); the doubling-time for some current cyanobacterias lies in the half-day range (Mori, Binder, and Johnson 1996), yielding a posthumous phenotype of $> 10^6$ generations.

22. This timescale ratio can be reversed on timescales large enough to witness atmospheric evolution. Consider, for instance, that oxygen is a toxin. Say that the first photosynthesizing organisms came to be adapted to deal with oxygen to protect their cellular machinery during photosynthesis (this adaptation may have arisen through classical, individual natural selection [Kirschvink and Kopp 2008]). The increase in atmospheric oxygen may then lead to mass extinctions, wiping out whole anaerobic lineages, favoring photosynthesizing organisms—a kind of geological allelopathy, if you will (Dutreuil 2016, 276–279). One could speculate that geological allelopathy is an adaptation by natural selection over geological timescales: some lineages may fare better than others with respect to releasing and handling toxins (over millions of generations), thus getting more numerous (over billions of generations). This adaptionist reasoning would amount to rescaling in time (and levels) development and selection, arguably to an extreme degree, in an attempt to again separate them (Pocheville 2010, 83).

23. Process autonomy (locally) ensues from timescale separation; but not the other way around. Two processes can be autonomous and consider the same timescale of interest; only their putative effects on one another are supposed negligible on this timescale.

24. A causal explanation has two elements: initial conditions (the values of causal variables) and constraints (the local invariants) (Pocheville and Montévil n.d.).

25. Whether non-externalism is the relevant characterization for niche construction theory is related to the question tackled by Lynn Chiu (this volume), whether niche construction should be thought of as causal (modification of the relevant environment) or constitutive (definition of the relevant environment by the organism) (Lewontin 1983, 220; Godfrey-Smith 1996, 144–151). I have myself briefly discussed these issues in Pocheville (2010, 99–100).

26. I mean here genetic in the sense of DNA sequences. My colleagues Lu and Bourrat (2017) have proposed an alternative explication of the conservative view, arguing that evolutionary genes need not, from a theoretical point of view, be made of DNA sequences.

27. I am here stretching the "standard" view to its best. In another context, my colleagues and I have explored the idea that any new trait can virtually change evolutionary theory (Montévil et al. 2016).

28. On this subject, see Watson (this volume).

29. The (currently science-fictional) renormalization of natural selection through timescales would amount to something like: "Does our model of variation converge toward something when our rescaling operations tend to infinity?" Notice this is different from a regress in causal chains.

30. Indeed, a few pages later, the same authors consider that niche construction can be detrimental as well as beneficial (Odling-Smee, Laland, and Feldman 2003, 47).

31. This view is akin to putting an emphasis on critical phenomena in living systems (see the review by Longo and Montévil 2013).

32. The definition is not necessarily circular: the scientist can decide that an equation is causal by other means than just saying that it contains causes. For instance, the equation can sum up principles of conservation and results of experiments. More on the concept of causality used here can be found in Woodward (2003), Pearl (2009), Pocheville and Montévil (n.d.), and Bailly and Longo (2006). See also the contribution by Jun Otsuka (this volume).

33. This sense draws on Laland, Odling-Smee, and Feldman (1999), and departs from the original sense(s) in Odling-Smee, Laland, and Feldman (2003, 419), where niche construction is the modification of selective pressures (see also Odling-Smee 1988, 89–100). Niche construction is different from ecological engineering, which only considers nontrophic modifications of the physical environment (Jones, Lawton, and Shachak 1994). The relation to the extended phenotype notion is more ambiguous (Dawkins 1982, 2004; Laland, Odling-Smee, and Feldman 2005). The many meanings of niche construction are discussed in Pocheville (2010, 61–64, 109–110).

34. Nongenetic inheritance includes environmental inheritance by definition, but to avoid overlap with the discussion on niche construction I focus here on other characteristics such as epigenetic marks (see Griffiths and Stotz 2013, 110–124).

35. A similar ontology has been defended by DiFrisco (2017).

References

Avilés, L., and P. Tufino. 1998. "Colony Size and Individual Fitness in the Social Spider *Anelosimus eximius*." *The American Naturalist* 152 (3): 403–418.

Bailly, F., and G. Longo. 2006. *Mathématiques et sciences de la nature: La singularité physique du vivant*. Paris: Hermann

Bonduriansky, R. 2012. "Rethinking Heredity, Again." *Trends in Ecology and Evolution* 27 (6): 330–336.

Bonduriansky, R., A. J. Crean, and T. Day. 2012. "The Implications of Nongenetic Inheritance for Evolution in Changing Environments." *Evolutionary Applications* 5 (2): 192–201.

Bouchard, F. 2011. "Darwinism without Populations: A More Inclusive Understanding of the 'Survival of the Fittest.'" *Studies in History and Philosophy of Science Part C: Studies in History and Philosophy of Biological and Biomedical Sciences* 42 (1): 106–114. https://doi.org/10.1016/j.shpsc.2010.11.002.

Bourrat, P. J.-N. 2015. "Reconceptualising Evolution by Natural Selection." PhD thesis, Sydney, Australia, The University of Sydney.

Bowler, P. 2001. *The Mendelian Revolution: The Emergence of Hereditarian Concepts in Modern Science and Society*. London: Athlone Press.

Boyd, R., and P. J. Richerson. 1985. *Culture and the Evolutionary Process*. Chicago: University of Chicago Press.

Braun, E. 2015. "The Unforeseen Challenge: From Genotype-to-Phenotype in Cell Populations." *Reports on Progress in Physics* 78 (3): 1–51.

Burchfield, J. D. 1990. *Lord Kelvin and the Age of the Earth*. Chicago: University of Chicago Press.

Burian, R. M. 1983. "Adaptation." In *Dimensions of Darwinism*, edited by Marjorie Grene, 287–314. New York: Cambridge University Press.

Claidière, N., and J.-B. André. 2012. "The Transmission of Genes and Culture: A Questionable Analogy." *Evolutionary Biology* 39 (1): 12–24.

Clark, W. C. 1985. "Scales of Climate Impacts." *Climatic Change* 7 (1): 5–27.

Corsi, P. 2011. "Jean-Baptiste Lamarck: From Myth to History." In *Transformations of Lamarckism: From Subtle Fluids to Molecular Biology*, edited by S. B. Gissis and E. Jablonka. Cambridge, MA: MIT Press.

Crick, F. 1958. *On Protein Synthesis*. Symposium of the Society for Experimental Biology, XII. New York: Academic Press.

Crow, J. F., and M. Kimura. 1970. *An Introduction to Population Genetics Theory*. 2009 Reprint of the 1970 Edition by Harper and Row, Publishers, Inc. Caldwell, NJ: The Blackburn Press.

Danchin, É., A. Charmantier, F. A. Champagne, A. Mesoudi, B. Pujol, and S. Blanchet. 2011. "Beyond DNA: Integrating Inclusive Inheritance into an Extended Theory of Evolution." *Nature Reviews Genetics* 12 (7): 475–486.

Danchin, É., and A. Pocheville. 2014. "Inheritance Is Where Physiology Meets Evolution." *Journal of Physiology* 592 (11): 2307–2317.

Danchin, É., A. Pocheville, O. Rey, B. Pujol, and S. Blanchet. 2018. "Epigenetically Facilitated Mutational Assimilation: Epigenetics as a Hub within the Inclusive Evolutionary Synthesis." *Biological Reviews*. https://doi.org/10.1111/brv.12453.

Darwin, C. R. 1859. *On the Origin of Species by Means of Natural Selection, or the Preservation of Favoured Races in the Struggle for Life*. London: John Murray.

Darwin, C. R. 1863. To Asa Gray Darwin Correspondence Project, "Letter no. 4153." Accessed October 24, 2018. http://www.darwinproject.ac.uk/DCP-LETT-4153.

Dawkins, R. 1982. *The Extended Phenotype: The Gene as the Unit of Selection*. Oxford: Oxford University Press.

Dawkins, R. 1986. "Universal Darwinism." In *Evolution from Molecules to Men*, edited by Derek S. Bendall, 403–425. Cambridge: Cambridge University Press Archive.

Dawkins, R. 2004. "Extended Phenotype—But Not too extended. A Reply to Laland, Turner and Jablonka." *Biology and Philosophy* 19 (3): 377–396.

Dawkins, R. 2006. *The Selfish Gene*, 3rd ed. Oxford: Oxford University Press.

Day, R. L., K. N. Laland, and J. F. Odling-Smee. 2003. "Rethinking Adaptation: The Niche-Construction Perspective." *Perspectives in Biology and Medicine* 46 (1): 80–95.

Dickins, T. E., and Q. Rahman. 2012. "The Extended Evolutionary Synthesis and the Role of Soft Inheritance in Evolution." *Proceedings of the Royal Society B: Biological Sciences* 279 (1740): 2913–2921. doi: 10.1098/rspb.2012.0273.

Dieckmann, U., and R. Ferrière. 2004. *Adaptive Dynamics and Evolving Biodiversity.* Interim Report IR-04-063. International Institute for Applied Systems Analysis, Schlossplatz 1, A-2361. Laxenburg, Austria.

DiFrisco, J. 2017. "Time Scales and Levels of Organization." *Erkenntnis* 82 (4): 795–818.

Dutreuil, S. 2016. "Gaïa: Hypothèse, programme de recherche pour le système terre, ou philosophie de la nature?" Thèse de doctorat, Université Paris 1.

Dutreuil, S., and A. Pocheville. 2015. "Les organismes et leur environnement: la construction de niche, l'hypothèse Gaïa et la sélection naturelle." *Bulletin de la SHESVIE* 22 (1).

Ferrière, R., and S. Legendre. 2013. "Eco-evolutionary Feedbacks, Adaptive Dynamics and Evolutionary Rescue Theory." *Philosophical Transactions of the Royal Society B* 368 (1610): 20120081.

Gayon, J. 1998. *Darwinism's Struggle for Survival: Heredity and the Hypothesis of Natural Selection.* Cambridge: Cambridge University Press.

Gayon, J., and M. Montévil. 2017. "Repetition and Reversibility in Evolution: Theoretical Population Genetics." In *Time in Nature and the Nature of Time*, edited by C. Bouton and P. Huneman. Dordrecht: Springer.

Geritz, S. A. H., G. Mesze, and J. A. J. Metz. 1998. "Evolutionarily Singular Strategies and the Adaptive Growth and Branching of the Evolutionary Tree." *Evolutionary Ecology* 12 (1): 35–57.

Gingerich, P. D. 2009. "Rates of Evolution." *Annual Review of Ecology, Evolution, and Systematics* 40:657–675.

Gissis, S. B., and E. Jablonka. 2011. *Transformations of Lamarckism: From Subtle Fluids to Molecular Biology.* Cambridge, MA: MIT Press.

Godfrey-Smith, P. 1996. *Complexity and the Function of Mind in Nature.* Cambridge: Cambridge University Press.

Godfrey-Smith, P. 2001. "Three Kinds of Adaptationism." In *Adaptationism and Optimality*, edited by S. H. Orzack and E. Sober, 335–357. Cambridge: Cambridge University Press.

Godfrey-Smith, P. 2009. *Darwinian Populations and Natural Selection.* Oxford: Oxford University Press.

Gould, S. J. 2002. *Structure of Evolutionary Theory.* Cambridge, MA: Harvard University Press.

Gould, S. J., and R. C. Lewontin. 1979. "The Spandrels of San Marco and the Panglossian Paradigm: A Critique of the Adaptationist Programme." *Proceedings of the Royal Society of London. Series B. Biological Sciences* 205 (1161): 581–598.

Griffiths, P. E. 2017. "Genetic, Epigenetic and Exogenetic Information in Development and Evolution." *Royal Society Interface* 7:20160152.

Griffiths, P. E., and K. Stotz. 2013. *Genetics and Philosophy: An Introduction.* Cambridge: Cambridge University Press.

Haig, D. 2007. "Weismann Rules! OK? Epigenetics and the Lamarckian Temptation." *Biology and Philosophy* 22 (3): 415–428.

Hairston, N. G., S. P. Ellner, M. A. Geber, T. Yoshida, and J. A. Fox. 2005. "Rapid Evolution and the Convergence of Ecological and Evolutionary Time." *Ecology Letters* 8 (10): 1114–1127.

Haldane, J. B. S. 1954. "The Statics of Evolution." In *Evolution as a Process*, edited by Julian S. Huxley, A. C. Hardy, and Ford, 109–121. London: Allen/Unwin.

Hanahan, D., and R. A. Weinberg. 2000. "The Hallmarks of Cancer Review." *Cell* 100:57–70.

Hodgson, G. M. 2005. "Generalizing Darwinism to Social Evolution: Some Early Attempts." *Journal of Economic Issues* 39 (4): 899–914.

Holling, C. S. 1992. "Cross-Scale Morphology, Geometry, and Dynamics of Ecosystems." *Ecological Monographs* 62 (4): 447–502.

Hoquet, T. 2009. *Darwin contre Darwin: Comment lire L'Origine des espèces?* Paris: Seuil.

Huneman, P. 2010. "Topological Explanations and Robustness in Biological Sciences." *Synthese* 177 (2): 213–245.

Jablonka, E., and M. J. Lamb. 1995. *Epigenetic Inheritance and Evolution: The Lamarckian Dimension.* Oxford: Oxford University Press.

Jablonka, E., and M. J. Lamb. 2005. *Evolution in Four Dimensions: Genetic, Epigenetic, Behavioral, and Symbolic Variation in the History of Life.* Cambridge, MA: MIT Press.

Jeliazkov, A. 2013. "Effets d'échelles dans les relations agriculture-environnement-biodiversité." PhD diss., Université Pierre et Marie Curie.

Jones, C. G., J. H. Lawton, and M. Shachak. 1994. "Organisms as Ecosystem Engineers." *Oikos* 69 (3): 373–386.

Kirschvink, J. L., and R. E. Kopp. 2008. "Palaeoproterozoic Ice Houses and the Evolution of Oxygen-Mediating Enzymes: The Case for a Late Origin of Photosystem II." *Philosophical Transactions of the Royal Society of London B: Biological Sciences* 363 (1504): 2755–2765.

Kupiec, J.-J. 2012. *L'ontophylogenèse: évolution des espèces et développement de l'individu.* Versailles: Editions Quae.

Lachmann, M., and E. Jablonka. 1996. "The Inheritance of Phenotypes: An Adaptation to Fluctuating Environments." *Journal of Theoretical Biology* 181 (1): 1–9.

Laland, K. N. 2004. "Extending the Extended Phenotype." *Biology and Philosophy* 19 (3): 313–325.

Laland, K. N., and G. R. Brown. 2006. "Niche Construction, Human Behavior, and the Adaptive-Lag Hypothesis." *Evolutionary Anthropology: Issues, News, and Reviews* 15 (3): 95–104.

Laland, K. N., J. R. Kendal, and G. R Brown. 2007. "The Niche Construction Perspective: Implications for Evolution and Human Behaviour." *Journal of Evolutionary Psychology* 5 (1): 51–66.

Laland, K. N., J. F. Odling-Smee, and M. W. Feldman. 1999. "Evolutionary Consequences of Niche Construction and Their Implications for Ecology." *Proceedings of the National Academy of Sciences USA* 96 (18): 10242.

Laland, K. N., J. F. Odling-Smee, and M. W. Feldman. 2003. "Niche Construction, Ecological Inheritance, and Cycles of Contingency in Evolution." In *Cycles of Contingency: Developmental Systems and Evolution*, edited by S. Oyama, R. Gray, and P. Griffiths, 117–126. Cambridge, MA: MIT Press.

Laland, K. N., J. F. Odling-Smee, and M. W. Feldman. 2005. "On the Breadth and Significance of Niche Construction: A Reply to Griffiths, Okasha and Sterelny." *Biology and Philosophy* 20 (1): 37–55.

Laland, K. N., J. F. Odling-Smee, M. W. Feldman, and J. Kendal. 2009. "Conceptual Barriers to Progress within Evolutionary Biology." *Foundations of Science* 14 (3): 195–216.

Laland, K. N., J. F. Odling-Smee, and S. F. Gilbert. 2008. "EvoDevo and Niche Construction: Building Bridges." *Journal of Experimental Zoology Part B: Molecular and Developmental Evolution* 310 (7): 549–566.

Laland, K. N., J. F. Odling-Smee, and S. Myles. 2010. "How Culture Shaped the Human Genome: Bringing Genetics and the Human Sciences Together." *Nature Reviews Genetics* 11 (2): 137–148.

Laland, K. N., and K. Sterelny. 2006. "Perspective: Seven Reasons (Not) to Neglect Niche Construction." *Evolution* 60 (9): 1751–1762.

Laland, K. N., K. Sterelny, J. F. Odling-Smee, W. Hoppitt, and T. Uller. 2011. "Cause and Effect in Biology Revisited: Is Mayr's Proximate-Ultimate Dichotomy Still Useful?" *Science* 334 (6062): 1512–1516.

Laland, K. N., T. Uller, M. W. Feldman, K. Sterelny, G. B. Müller, A. Moczek, E. Jablonka, and J. F. Odling-Smee. 2015. "The Extended Evolutionary Synthesis: Its Structure, Assumptions and Predictions." *Proceedings of the Royal Society B: Biological Sciences* 282:20151019.

Lamarck, J.-B. de. *Philosophie zoologique ou exposition des considérations relatives à l'histoire naturelle des animaux.* 2 vols. Paris: Dentu, 1809. https://gallica.bnf.fr/ark:/12148/bpt6k5675762f/.

Lamm, E., and E. Jablonka. 2008. "The Nurture of Nature: Hereditary Plasticity in Evolution." *Philosophical Psychology* 21 (3): 305–319.

Lehmann, L. 2008. "The Adaptive Dynamics of Niche Constructing Traits in Spatially Subdivided Populations: Evolving Posthumous Extended Phenotypes." *Evolution* 62 (3): 549–566.

Lehmann, L. 2010. "Space-Time Relatedness and Hamilton's Rule for Long-Lasting Behaviors in Viscous Populations." *The American Naturalist* 175 (1): 136–143.

Lenton, T. M. 2001. "The Role of Land Plants, Phosphorus Weathering and Fire in the Rise and Regulation of Atmospheric Oxygen." *Global Change Biology* 7 (6): 613–629.

Lesne, A. 2003. "Approches Multi-Échelles En Physique et Biologie." Mémoire d'habilitation à diriger les recherches, Université Pierre et Marie Curie.

Lesne, A. 2013. "Multiscale Analysis of Biological Systems." *Acta Biotheoretica* 61 (1): 3–19.

Lewontin, R. C. 1974. *The Genetic Basis of Evolutionary Change.* New York: Columbia University Press.

Lewontin, R. C. 1983. "Gene, Organism and Environment." In *Evolution from Molecules to Men*, edited by D. S. Bendall. Cambridge: Cambridge University Press.

Longo, G., and M. Montévil. 2013. *Perspectives on Organisms: Biological Time, Symmetries and Singularities.* Dordrecht: Springer.

Longo, G., M. Montévil, and A. Pocheville. 2012. "From Bottom-up Approaches to Levels of Organization and Extended Critical Transitions." *Frontiers in Physiology* 3 (232): 1–10.

Lu, Q., and P. Bourrat. 2018. "The Evolutionary Gene and the Extended Evolutionary Synthesis." *British Journal for the Philosophy of Science* 69:775–800.

Marshall, D. J., and T. Uller. 2007. "When Is a Maternal Effect Adaptive?" *Oikos* 116 (12): 1957–1963.

Mayr, E. 1961. "Cause and Effect in Biology." *Science* 134:1501–1506.

Mayr, E., and W. B. Provine. 1998. *The Evolutionary Synthesis: Perspectives on the Unification of Biology, with a New Preface.* Cambridge, MA: Harvard University Press.

Mesoudi, A., S. Blanchet, A. Charmantier, É. Danchin, L. Fogarty, E. Jablonka, K. N. Laland, et al. 2013. "Is Non-genetic Inheritance Just a Proximate Mechanism? A Corroboration of the Extended Evolutionary Synthesis." *Biological Theory* 7 (3): 189–195.

Metz, J. A. J., S. A. H. Geritz, G. Meszéna, F. J. A. Jacobs, and J. S. Van Heerwaarden. 1996. "Adaptive Dynamics: A Geometrical Study of the Consequences of Nearly Faithful Reproduction." In *Stochastic and Spatial Structures of Dynamical Systems*, edited by S. J. van Strien and S. M. Verduyn Lunel, vol. 45. North-Holland: KNAW Verhandelingen, Afd. Natuurkunde, Eerste reeks.

Montévil, M., and M. Mossio. 2015. "Biological Organisation as Closure of Constraints." *Journal of Theoretical Biology* 372:179–191.

Montévil, M., M. Mossio, A. Pocheville, and G. Longo. 2016. "Theoretical Principles for Biology: Variation." *Progress in Biophysics and Molecular Biology* 122 (1): 36–50.

Moran, L. A. 2016. "Sandwalk: Extending Evolutionary Theory?—Paul E. Griffiths." *Sandwalk.* Accessed October 18, 2017. http://sandwalk.blogspot.com/2016/10/extending-evolutionary-theory-paul-e.html.

Mori, T., B. Binder, and C. H. Johnson. 1996. "Circadian Gating of Cell Division in Cyanobacteria Growing with Average Doubling Times of Less Than 24 Hours." *Proceedings of the National Academy of Sciences* 93 (19): 10183–10188.

Mousseau, T. A., and C. W. Fox. 1998a. *Maternal Effects as Adaptations.* New York: Oxford University Press.

Mousseau, T. A., and C. W. Fox. 1998b. "The Adaptive Significance of Maternal Effects." *Trends in Ecology and Evolution* 13 (10): 403–407.

Odling-Smee, J. F. 1988. "Niche-Constructing Phenotypes." In *The Role of Behavior in Evolution*, edited by H. C. Plotkin. Cambridge, MA: MIT Press.

Odling-Smee, J. F. 2007. "Niche Inheritance: A Possible Basis for Classifying Multiple Inheritance Systems in Evolution." *Biological Theory* 2 (3): 276–289.

Odling-Smee, J. F., K. N. Laland, and M. W. Feldman. 2003. *Niche Construction: The Neglected Process in Evolution.* Princeton, NJ: Princeton University Press.

Pearl, J. 2009. *Causality: Models, Reasoning, and Inference*, 2nd ed. New York: Cambridge University Press.

Peterson, G., C. R. Allen, and C. S. Holling. 1998. "Ecological Resilience, Biodiversity, and Scale." *Ecosystems* 1 (1): 6–18.

Pigliucci, M., and G. B. Müller. 2010. "Elements of an Extended Evolutionary Synthesis." In M. Pigliucci and G. B. Müller, *Evolution: The Extended Synthesis*, 3–17. Cambridge, MA: MIT Press.

Pocheville, A. 2010. "What Niche Construction Is (Not)." In "La Niche Ecologique: Concepts, Modèles, Applications. (Thèse de Doctorat)," 39–124. Paris: Ecole Normale Supérieure Paris.

Pocheville, A., and É. Danchin. 2017. "Genetic Assimilation and the Paradox of Blind Variation." In *Challenging the Modern Synthesis*, edited by D. M. Walsh and P. Huneman. Oxford: Oxford University Press.

Pocheville, A., and M. Montévil. n.d. "Intervening in a Continuous World: When Achilles Finds His Tortoise." Unpublished manuscript.

Richards, C. L., O. Bossdorf, and M. Pigliucci. 2010. "What Role Does Heritable Epigenetic Variation Play in Phenotypic Evolution?" *BioScience* 60 (3): 232–237.

Roux, W. 1881. *Der Kampf der Theile im Organismus*. 1. Auflage. Leipzig: Engelmann.

Ruse, M. 1992. "Darwinism." In *Keywords in Evolutionary Biology*, 4th ed., edited by E. F. Keller and E. A. Lloyd, 74–80. Cambridge, MA: Harvard University Press.

Sapp, J. 1987. *Beyond the Gene: Cytoplasmic Inheritance and the Struggle for Authority in Genetics*. New York: Oxford University Press.

Sarkar, S. 2014. "Formal Darwinism." *Biology and Philosophy* 29 (2): 249–257.

Scott-Phillips, T. C., T. E. Dickins, and S. A. West. 2011. "Evolutionary Theory and the Ultimate–Proximate Distinction in the Human Behavioral Sciences." *Perspectives on Psychological Science* 6 (1): 38–47.

Shefferson, R., and R. Salguero-Gomez. 2015. "Eco-evolutionary Dynamics in Plants: Interactive Processes at Overlapping Time-Scales and Their Implications." *Journal of Ecology* 103:789–798.

Sinervo, B., and C. M. Lively. 1996. "The Rock-Paper-Scissors Game and the Evolution of Alternative Male Strategies." *Nature* 380 (6571): 240–243.

Sober, E. 1983. "Equilibrium Explanation." *Philosophical Studies: An International Journal for Philosophy in the Analytic Tradition* 43 (2): 201–210.

Soen, Y., M. Knafo, and M. Elgart. 2015. "A Principle of Organization which Facilitates Broad Lamarckian-like Adaptations by Improvisation." *Biology Direct* 10 (1): 1–17.

Sonnenschein, C., and A. M. Soto. 1999. *The Society of Cells: Cancer and Control of Cell Proliferation*. Oxford: Bios Scientific Publishers.

Sterelny, K. 2005. "Made By Each Other: Organisms and Their Environment." *Biology and Philosophy* 20 (1): 21–36.

Stotz, K. 2017. "Why Developmental Niche Construction Is Not Selective Niche Construction: And Why It Matters." *Interface Focus* 7 (5): 1–10.

Thompson, J. N. 1998. "Rapid Evolution as an Ecological Process." *Trends in Ecology and Evolution* 13 (8): 329–332.

Turner, J. S. 2000. *The Extended Organism: The Physiology of Animal-built Structures*. Cambridge, MA: Harvard University Press.

Turner, P. E., E. S. C. P. Williams, C. Okeke, V. S. Cooper, S. Duffy, and J. E. Wertz. 2014. "Antibiotic Resistance Correlates with Transmission in Plasmid Evolution." *Evolution* 68 (12): 3368–3380.

Uller, T., S. English, and I. Pen. 2015. "When Is Incomplete Epigenetic Resetting in Germ Cells Favoured by Natural Selection?" *Proceedings of the Royal Society of London, B: Biological Sciences* 282 (1811): 20150682.

Uller, T., and H. Helanterä. 2016. "Heredity Concepts and Evolutionary Explanations." Paper presented at New Trends in Evolutionary Biology: Biological, Philosophical and Social Science Perspectives, joint Discussion Meeting of the British Academy and the Royal Society, November 7–9.

Uller, T., and H. Helanterä. 2017. "Niche Construction and Conceptual Change in Evolutionary Biology." *British Journal for the Philosophy of Science*: axx050.

Van Baalen, M., and D. A. Rand. 1998. "The Unit of Selection in Viscous Populations and the Evolution of Altruism." *Journal of Theoretical Biology* 193:631–648.

Waxman, D., and S. Gavrilets. 2005. "20 Questions on Adaptive Dynamics." *Journal of Evolutionary Biology* 18 (5): 1139–1154.

Weismann, A. 1891. *Essays Upon Heredity and Kindred Biological Problems*, Vol. II. Oxford: Clarendon Press.

Weismann, A. 1893a. "The All-sufficiency of Natural Selection: A Reply to Herbert Spencer." *The Contemporary Review* 64:309–338.

Weismann, A. 1893b. *The Germ-Plasm a Theory of Heredity.* Translated by Newton W. Parker and Harriet Rönnfeldt. New York: Charles Scribner's Sons.

Weismann, A. 1904a. *The Evolution Theory*, Vol. 1. Translated by Arthur J. Thomson and Margaret Thomson. London: Edward Arnold.

Weismann, A. 1904b. *The Evolution Theory*, Vol. 2. Whitefish, MT: Kessinger Publishing.

West, B. J. 2006. *Where Medicine Went Wrong: Rediscovering the Path to Complexity.* Singapore: World Scientific.

West-Eberhard, M. J. 2003. *Developmental Plasticity and Evolution.* New York: Oxford University Press.

Wilkins, A. 2011. "Epigenetic Inheritance: Where Does the Field Stand Today? What Do We Still Need to Know?" In *Transformations of Lamarckism: From Subtle Fluids to Molecular Biology*, edited by S. B. Gissis and E. Jablonka, 389–393. Cambridge, MA: MIT Press.

Woodward, J. 2003. *Making Things Happen: A Theory of Causal Explanation.* New York: Oxford University Press.

Wright, B. E. 2000. "A Biochemical Mechanism for Nonrandom Mutations and Evolution." *Journal of Bacteriology* 182 (11): 2993–3001.

14 Decoupling, Commingling, and the Evolutionary Significance of Experiential Niche Construction

Lynn Chiu

1. Introduction

Evolutionary theory is dominated by externalist models, theories, strategies, and metaphors that appeal to the environment to explain adaptive fit, or "good design" (Endler 1986; Godfrey-Smith 1996, 2001; Gould 1977, 2002; Lewontin 2000; Mayr 1982; Walsh 2015; Williams 1966). Externalist explanatory strategies presume that adaptations map onto environmental features. Externalist metaphors speak of traits as "solutions" to environmental "problems" or keys fitted to "fill" an environmental niche or lock.

Richard Lewontin, however, argued that externalism is "bad biology" (Levins and Lewontin 1985; Lewontin 1982, 1983, 2000). Organisms are not "adapted" to environments. Instead, they "construct every aspect of their environment themselves. They are not the passive objects of external forces, but the creators and modulators of these forces" (Levins and Lewontin 1985, 104). The better metaphor for evolution should be "construction," not adaptation.

Lewontin's constructivism has inspired a large and expanding literature on the causal significance of "niche construction" on evolution.[1] However, most followers reject Lewontin's radical ontology of the environment. According to Lewontin, organisms do not just alter the world they occupy, they also change how the world is *experienced*. The niche constructed by the organism is thus not entirely made up by the external world, but by the experiences of the organism as well. Many admirers of Lewontin find it difficult to comprehend or operationalize "experienced niches" as external environments and causes of evolution (Brandon and Antonovics 1995; Godfrey-Smith 1996, 2001; Odling-Smee 1988; Odling-Smee, Laland, and Feldman 2003; Sterelny and Griffiths 1999). They thus opt to identify constructed environments as the intrinsic properties of the external world, albeit those relevant to the organism. An evolutionary theory of niche construction concerns the evolutionary significance of constructed, intrinsically defined environments.

In this chapter, I make the case for the evolutionary significance of "experiential niche construction" (coined by Sultan 2015). I start by arguing that recent analyses that draw on the agency and plasticity of organisms (Sultan 2015; Walsh 2015) can address a major objection against experiential niche construction (Godfrey-Smith 1996, 2001). I then propose a way experiential niche construction is evolutionarily significant. Theories of niche construction that leave out the experiential variety tend to maintain an externalist characterization of natural selection and argue that niche construction feeds into the environmental causes of natural selection. Natural selection, however, does not adapt a population to its environment when different organisms of a population construct and experience different environments. Instead, in these scenarios, the causes of selection are spread across varying organisms and their varying constructed environments. I argue that experiential niche construction helps maintain the spread of selective causes across organism and environment interactions. It thus creates the conditions for a kind of natural selection that is not "externalist."

My approach is pluralistic. Sometimes, natural selection can be heuristically approximated as environmental selection and niche construction contributes to selection's environmental sources. In these scenarios, organisms and environments are "decoupled" causes. Other times, however, natural selection cannot be heuristically treated as environmental selection. This occurs when organisms and environments "commingle" and niche construction *constitutes* the conditions of natural selection. I propose a decoupling/commingling framework that specifies when it is and is not appropriate to heuristically assume that natural selection explanations are externalist.

In the following, I will refer to the external world surrounding the organism, characterized by its intrinsic properties, as the external or physical world or environment. When I am referring to the environment experienced by the organism (which will be determined in part by the properties of the organism), I will qualify the world or environment with terms such as "experienced" or "experiential."

2.   Experiential Niche Construction and Its Discontents

According to Lewontin, mainstream evolutionary theory assumes that the environment presents well-defined problems for organisms to solve.

The word "adaptation" reflects this point of view, implying that the organism is molded and shaped to fit into a preexistent niche, given by the autonomous forces of the environment, just as a key is cut and filed to fit into a lock. (Levins and Lewontin 1985, 98)

He argues that there are two problems with this problem-solution metaphor. The conceptual problem is that organisms do not "fit into a preexistent niche" as the niches of a species come to exist through interactions between organisms and their environments. It is conceptually impossible, then, for a niche to preexist and select organisms. The empirical problem is that niches are not "given by the autonomous forces of the environment" as they are instead determined by the biology of the organism. Organisms create niches by determining what's relevant, by altering properties of the world, by transducing external signals (into different types of signals), or by transforming environmental patterns (Levins and Lewontin 1985, 98–106; Lewontin 2000, 55–68). The properties of environments emerge from interactions with organisms.

The inter-relation between organism and environment is dynamic and dialectical. Lewontin asks us to consider the way plant engineers attempt to improve crop yield by designing (through artificial selection or genetic engineering) leaf phenotypes optimized to a measured microenvironment, for example, the temperature, light exposure, humidity, and oxygen and carbon dioxide concentration around the plant. The problem is that the newly selected leaf morphologies tend to alter the humidity, light, carbon dioxide, and other distribution and create a different, less optimal microenvironment. The plant engineers can intervene again, but only to have the plants change the environment once more. "The plant engineers are chasing not only a moving target but a target whose motion is impelled by their own activities," states Lewontin. "This process is a model for a more realistic understanding of evolution by natural selection" (Lewontin 2001, 57). In the wild, plants are constantly changing the environments as they develop and evolve in response to them, which in turn results in further change to their environments. The properties of organism and environment that emerge through their interactions can propel their future change.

Mediational Niche Construction

Lewontin's examples of niche construction fall under two main categories: physical and experiential niche construction.

Physical Niche Construction (also known as perturbational niche construction or habitat construction) is the causal manipulation of the external world by the organism, changing the environment's intrinsic properties.

Experiential Niche Construction is changes in the environment experienced by the organism without changes to the intrinsic properties of the external world.

There are two types of experiential niche construction:

Relocational Niche Construction (also habitat choice) determines which intrinsic properties of the external world surround an individual.

Mediational Niche Construction determines the relevance, impact, and significance of the external world for the organism. It determines *how* the intrinsic properties of the environment is experienced by the organism.

What I coin "mediational niche construction" is currently an underexplored category.[2] The core question of this paper is whether *mediational* niche construction is a type of niche construction, and if so, whether it has evolutionary significance *qua* niche construction.

Even though relocational and mediational niche construction are both cases of experiential niche construction (they do not change the intrinsic properties of the environment), I single out mediational niche construction as the central concern. Relocational and physical niche construction both determine *which* intrinsic properties surround an organism (the first by choosing an environment, the second by altering an environment). However, mediational niche construction does not alter nor determine which intrinsic properties are around an organism. Instead, it changes the way an organism *experiences* them.[3] Is this niche construction?

Lewontin raises several examples of mediational niche construction. One example is the transduction of temperature into the biochemical signals of organisms such that a one-degree drop in the outside world is experienced as a smaller difference for one organism but a greater difference for another. Another example is the way an organism perceives its environment as resource rich or poor. The perceived scarcity of an environment is relative to the organism's level of fat storage. Yet another example is when the physiology of an organism incorporates rates of change of environmental factors into its experienced environment, thus perceiving and reacting to sudden changes instead of absolute levels.

Mediational niche construction occurs because the organism stands between itself and the world. Through this type of niche construction, states Lewontin, "the common external phenomena of the physical and biotic world pass through a transforming filter created by the peculiar biology of each species, and it is the output of this transformation that reaches the organism and is relevant to it" (Lewontin 2000, 64). It is as if the organism is residing inside a Plato's cave of its own making, "determined by the shadows on the wall, passed through a transforming medium of its own creation" (Lewontin 2000, 64). An organism in its self-created "bubble" is still affected by the physical world, but the effects of the world are distorted and transformed by organismal activities and physiology.

Mediational niche construction was dismissed by philosophers (Godfrey-Smith 1996, 2001) and left out by proponents of NCT (Odling-Smee, Laland, and Feldman 1996; Odling-Smee et al. 2003). In the following, I consider Godfrey-Smith's objections and argue that new developments on mediational niche construction (Sultan 2015; Walsh 2015) can address his concerns.

Godfrey-Smith against Mediational Niche Construction

It is not obvious that mediational niche construction is a type of niche construction. How can *internal* changes within an organism count as changes to the *environment*? Internal changes are usually considered phenotypes under selection, not determinates of selective pressures. That is why some argue that relocational niche construction is actually a phenotype for habitat choice while mediational niche construction is a phenotypic response to environmental pressures (Brandon 1990; Godfrey-Smith 1996, 2001).

Godfrey-Smith suggests that mediational niche construction *as* niche construction might make sense if we adopt one of Lewontin's dialectical principles (found in the conclusion chapter of *The Dialectical Biologist*). Lewontin and colleague Richard Levins have long advocated for a dialectical biology against the "Cartesian" decoupling of causes and effects, parts and wholes, and insides and outsides (Levins and Lewontin 1985; Lewontin and Levins 2007). Organism and environment do not exist independently of each other as causes and effects, they argue, but "interpenetrate," or "commingle." Godfrey-Smith suggests that the dialectical principle that parts do not exist independently of each other can help make a case for mediational niche construction. If we assume that organisms and environments are two parts of a whole and parts do not independently exist, then a change to any part would *logically* (not causally) entail a change in the other parts. The part-whole principle thus explains how an internal change to organisms is a change in their environments.

Applying this dialectical principle to organisms and environments, however, seems to create an undesirable mix of antirealism and intractable holism. The "environment" or "niche" of an organism is not the objectively measurable environment, but something (in part) subjectively constructed by the organism. Furthermore, any change to the organism, the environment, or their relation will count as niche construction, thus trivializing the concept. An all-inclusive notion of niche construction fails to capture the complex and varied relations between organisms and environments that are important for empirical study.

Godfrey-Smith proposes that it is fruitful to just acknowledge physical niche construction as an evolutionarily relevant process and treat mediational niche construction as mere traits undergoing selection. To illustrate, consider two species of bacteria in an environment of toxic molecules. One evolves a different internal physiology such that the chemical is no longer toxic. For instance, the organisms may no longer have the receptors or signaling pathways that react to the chemical in a self-destructive way. The other evolves a mechanism that excretes toxin-degrading enzymes. These are two distinct evolutionary responses to an environmental challenge. The first is a selected internal accommodation to the environment, whereas the second is a selected trait that *also* alters the environment. The second trait changes the environment for future generations to come whereas the descendents of the first are still living with the toxin. Yet for Lewontin, both count as niche construction as the organisms all end up experiencing a nontoxic environment. This coarse-grained dialectical perspective glosses over important differences and outcomes of distinct evolutionary responses.

There are many advantages to a narrower scope of niche construction. First, it gives us a sharper boundary of what counts as niche construction and what does not. Not any change to organism or environment is niche construction. It is clear that the line is drawn at whether organisms alter the intrinsic properties of the external world. Second, we retain a commonsensical notion of the environment as an objective, physical phenomenon instead of constructed experiences. Third, it distinguishes between mere adaptations from those that also alter the external, selective environment.

There is a final reason for discounting mediational niche construction. This phenomenon is not even needed to reject externalism. Godfrey-Smith argues that there are two types of externalist explanations: symmetric and asymmetric. The problem with adaptationist thinking is *asymmetric externalism*, that is, the position that while the environment accounts for organismal evolution, organisms cannot account for changes in the selective environment. Externalism can also be of the symmetric kind, which permits the evolutionary effects of niche construction. As physical niche construction alone rejects asymmetric externalism, leaving out experiential niche construction does not diminish the Lewontinian challenge against what's wrong with externalist thinking.

In sum, contrary to Lewontin's call to replace adaptation with construction, Godfrey-Smith insists that there is a complementary coexistence between the two. "Rather than a replacement, there should be a supplementation. Both adaptation and construction are real relationships that organisms have, in particular instances, to environmental conditions" (Godfrey-Smith 2001, 263). Physical niche construction provides "constructivist" explanations that explain the intrinsic properties of the environment by properties of organisms, whereas natural selection supplies "externalist" explanations that explain the properties of organisms in terms of the intrinsic properties of the environment. Together, they give a fuller picture of how evolution works.

3.   New Support for Mediational Niche Construction

Godfrey-Smith's major concern was that mediational niche construction implies holistic intangibility and antirealism about the environment. I argue that recent developments, in particular Denis Walsh's affordance theory of niches and Sonia Sultan's mechanistic account of plastic cue and response systems, provide rich and testable characterizations of mediational niche construction that address each of the challenges. They offer clear-cut distinctions between the various ways affordances or experienced environments can change and explain why experienced environments *are* constructed environments.

Walsh's Affordance Framework

Philosopher Denis Walsh (2012, 2014, 2015) argues that adaptive traits are not evolutionary responses to the external world per se, but responses to opportunities in the environment that appear to the organism *as* opportunities for action. Organisms, as purposeful agents, perceive and respond to opportunities that appear because of what the organism can do and what it aims to do.[4]

An example from Walsh can help clarify how opportunities depend on the goals and capacities of organisms. A stick does not become a tool just because sticks can solve a problem and there are sticks lying around. The problem-solving agent needs to have the capacity to perceive the stick *as* a usable tool that can potentially solve a problem, that is, to experience it as affording a particular type of action that can fulfill a goal. For a hominoid, opposable thumbs and "precision grip" are necessary for the manipulation of hand-held tools such as sticks. Without these hand structures, objects in the environment would not seem "grip-able" or "grasp-able" for use. Yet the appearance of precision grip is possibly not the result of direct selection for tool use, and an evolutionary byproduct of bipedalism. The structural changes in feet that enabled hominoids to stand up and run also altered the structures of hands (Rolian, Lieberman, and Hallgrìmsson 2010). These new hand structures opened new possibilities: objects previously inconceivable as graspable are now within "reach." The evolution of precision-grip is a nice example whereby organisms "can make a change in its own form, without affecting the environment, which in turn alters the affordances provided to the organism" (Walsh 2015, 181–182).

According to Walsh, environmental opportunities are *affordances*[5] that appear only when there is a purposeful agent. An affordance exists when the environment is experienced by the agent as having the sorts of properties that can either enable or disrupt it from achieving its goals. An agent responds to these affordances by interacting with them in ways that help attain its goals, either through the exploitation of facilitators or the elimination of obstructions. Affordances are emergent properties of a purposeful agent and its environment. Agents and environment both constitute[6] an agent's affordances, that is, their properties and relations underlie what an environment can afford to an agent.

Walsh thus presents an alternative version of adaptive evolution whereby organisms adapt to affordances: "adaptation is not the process in which the external environment moulds passive form. Rather it is the process by which organisms respond to, and in the process create, their own system of affordances" (Walsh 2015, 164). He proposes a "situated adaptationism" of organisms adapting to affordances. In contrast to a standard evolutionary theory that treats the environment supplying a fixed adaptive landscape that populations climb as they evolve, Walsh proposes a dynamic "affordance landscape" that changes its shape as organisms move across it. A stick that can't be gripped might be an obstacle, but once an organism is able to grip sticks they become tools. The internal change of an organism thus constructed a new environment filled with tools instead of obstacles even though the intrinsic nature of the environment remained unaltered.

Sultan's Cue and Response System

While Walsh provides a conceptual framework for niches as constructed experiences, biologist Sonia Sultan offers an underlying mechanistic framework. Sultan (2015) coined the term "experiential niche construction" and has documented a rich and diverse set of examples from the microbial world to animals and plants. When organisms change, their experience of the environment changes even though the external world remains the same. Some of these changes are achieved by moving around, for instance, to spots where resources are more densely packed (i.e., relocational niche construction). Others are achieved merely by altering how the organism mediates environmental stimuli and signals (i.e., what I refer to as "mediational niche construction").

Cases of experiential niche construction include the construction of perceived resource environments, predator environments, competition environments, and temperature environments. For instance, a resource can be experienced as more or less plentiful depending on the organism's behavior or sensory faculties. The morphology of leaves can alter the perceived density of a plant's photon environment. Root spread and morphology can determine the experienced humidity of the soil.

Phenotypic plasticity, a property of organisms, is a key mechanism of experiential niche construction. Organisms are plastic when they can develop different phenotypes in different environments.[7] More specifically, organisms are *passively* plastic when it is an inevitable result that different conditions (e.g., more nutrition) contribute to different phenotypes (e.g., taller or larger body size). They are *actively* plastic when they have specific mechanisms that allow them to switch to different, and typically adaptive, phenotypes depending on how they receive and process environmental signals. These mechanistic pathways of phenotypic plasticity constitute "cue and response systems," that is, systems that enable the organism to developmentally, physiologically, or behaviorally adjust to perceived environmental information (i.e., cues).

The experienced environment is a combination of "cues" and responses to these cues. What becomes a cue depends on the sensory system and the way the organism conducts its behavior to choose and sample environmental bits and pieces. In these ways, "usable cues are specifically 'constructed' by each organism from the myriad variables in its habitat" (Sultan 2015, 52). Organisms can also sense environmental cues directly from the environment or from themselves (e.g., the chemistry of the tissue, the growth rate of the body, the levels of various internal activities), which often involves feedback loops that further remove the cue from the environmental source. A cue could furthermore be an anticipatory mark of a future condition that has yet to exist, and thus the reaction has little to do with the immediate properties of the current environment. The transduction pathways that lead to responses are also mediated by the organism. Environmental signals are transduced into chemical and cellular interactions. They are mediated by the organism's physiology and behavior at the cellular, tissue, and organ levels, triggering downstream effects that regulate gene expressions or determine physiological responses or social behaviors.

In sum, adaptive interpretations need to distinguish "between traits that comprise functionally adaptive responses to a given environmental challenge and those that instead remediate that challenge" (Sultan 2015, 165). The latter is the effect of experiential niche construction. Experiential niche construction occurs when there are changes in the cues or responses. Cue and response systems "mediate the organism's experience of its external environment. In this sense, plasticity (i.e., aspects of trait expression that change from one environment to another) can be seen as a mode of niche construction that allows the organism to *experience* a different and often more favorable physical and biotic environment, even without altering external conditions as such" (Sultan 2015, 71, original italics).

A New Case for Mediational Niche Construction

These new conceptual and mechanistic developments of mediational niche construction do not suffer from Godfrey-Smith's objections. Godfrey-Smith reconstructed and then rejected Lewontin's argument based on a dialectical principle about parts and wholes. Yet the part-whole principle is not the best dialectical rule to appeal to. Lewontin did not use it to support his claims but instead referred to the separation of causes and effects in his section about niche construction. To Lewontin, the problem with externalism is not that it prohibits organismal sources of environmental change (asymmetric externalism), but that environments are used to explain evolution. Niche construction is supposed to show that environments cannot be prior, explanatory causes of organismal change, not that organisms and environments are both parts of some larger whole.

Walsh and Sultan have each developed a framework that explains why environments cannot be the type of prior causes assumed in externalist thinking. Walsh argued that organisms adapt and respond to organism-mediated affordances. Sultan argued that organisms oftentimes respond to organism-mediated cues through organism-mediated signaling. The external environment underdetermines what would count as an organismal response. Therefore, external environments oftentimes do not serve as a major cause of organismal evolution. Evolution by natural selection is not always a process that adapts organisms *to* environments.

Godfrey-Smith objected that mediational niche construction suffers from holism. This is not the case for Walsh's and Sultan's views. They both clearly dissect the various components of the commingled organism and environment. To Walsh, the organism–environment pair consists of the agent's goals and capacities, the external environment, and the affordances that emerge from the former two. To Sultan, the organism–environment pair consists of the external environment and the organism-mediated cues, transduction pathways, and response. Changes to each component makes different predictions about future evolution. Therefore, it is not the case that any change to any part of the organism–environment system are glossed over as the same type of change.

Finally, Godfrey-Smith worries that mediational niche construction implies an antirealistic position on an organism's "niche" or environment. To Walsh, affordances are not subjective constructs but emergent properties of the organism and its environment. To Sultan, the mechanistic involvement of each aspect of the cue and response system dispels any worries that the experiences of organisms are not objectively measurable properties. The experiential properties of "threatening," "resource-poor," "warm," and so on, that can change when the cue and response system changes are operational because they are features of cues and signals that determine how an organism will react.

There is one more obstacle to overcome. Godfrey-Smith argues that mediational niche construction is just a trait while physical niche construction can also be a cause of natural selection. How can experiential niche construction be a cause of natural selection when there are no changes made to the external environment?

Stotz (2017) argues that *developmental* niche construction should not be conflated with *selective* niche construction. The former occurs when organisms construct their developmental niches, which is "a multi-dimensional space of environmentally induced and developmentally regulated, heritable resources that scaffold development" (Stotz 2017, 2). Developmental niche construction creates new variants. The latter occurs when organisms construct their selective niche, which "is defined by the parameters that determine the relative fitness of competing types in a population" (Stotz 2017, 3). Selective niche construction alters the sources of natural selection.

The types of phenomena described by Sultan and Walsh might seem like developmental niche construction. For instance, one of Sultan's examples is the water flea (*Daphnia sp.*) that plastically morphs into a "defensive" phenotype when exposed to predator cues. This is a classic case instance of a cue-and-response system in action. However, Stotz argues that while the cue is part of the developmental niche, it is not part of the selective niche. Only the predator, characterized by the intrinsic properties of the external world, is part of the selective niche.[8] Contrary to Sultan, cues are not part of the experienced, selective environment.

In response, I argue that a distinction should be made between the causes that make a difference between phenotypic variants (for instance, the causes of developmental niches) and the causes that make a difference to the fitness variables of preexisting variants. The fitness of a trait is context dependent. Cumbersome long claws are fitter when a nutrition source is hidden deep inside a trunk, but they become less fit when the food (e.g., worms) now tends to stay at the surface. A variant can thus have a different fitness value because its context has changed, even though the variant itself is the same. A behavior that alters the context of fitness is selective niche construction because it changes a component of natural selection—differential fitness.

Sultan and Walsh are concerned with cases where the organism constructs its fitness context. For instance, when water fleas are exposed to predator cues, their offspring tend to adopt the acquired "defensive" phenotype even though there are no predators nor cues in the offspring environment. Sultan considers this a case of experiential niche construction, where the parent creates a *less threatening environment* due to the transmission of defensive phenotypes. *If* predators were to appear (suppose the probability of predator encounter is unchanged), the offspring are less vulnerable to their attack (if there is any attack at all, as the predators may be deterred by the defense). While Stotz argues that this is an even stronger mismatch between a developmental response and the selective environment, to Sultan, the developmental response creates a differently experienced environment and thus a different selective environment.

4. Commingling and Decoupling Strategies

In this section, I argue that niche construction theories can differ in their conceptual models about the relation between niche construction and natural selection processes. I will present a heuristic framework that recommends either "decoupling" or "commingling" explanatory strategies depending on the conceptual model. Under this framework, I make a new case for the evolutionary significance of mediational niche construction, one that complements the proposals from Walsh and Sultan.

Niche Construction Theory and Decoupling

Consider the causal commitments of Niche Construction Theory (NCT). NCT is a concrete elaboration of Lewontin's vision that organisms have an active agency in their own evolution (Day, Laland, and Odling-Smee 2003; Laland 2015; Laland and Sterelny 2006; Odling-Smee 1988; Odling-Smee et al. 1996, 2003). Niche construction, according to NCT, is defined as

the process whereby organisms, through their metabolism, their activities, and their choices, modify their own and/or each other's niches. Niche construction may result in changes in one or more natural selection pressures in the external environments of populations. (Odling-Smee et al. 2003, 419)

According to proponents of NCT, organisms construct their niches either by changing the properties of the environment surrounding them (perturbation niche construction, or physical niche construction) or by moving to environments with different intrinsic properties (relocation niche construction, the first type of experiential niche construction).

Evolutionary theory is meant to explain the extraordinary match between organism and environment. NCT[9] models niche construction and natural selection as distinct evolutionary processes that can shape this match. These causal processes point in opposite directions: niche construction is a causal process from organism to environment, whereas natural selection points from environment to organism. Separately, niche construction and natural selection can each adjust the complementarity between organisms and environment: one by changing the organisms (natural selection), the other by changing the environment (niche construction). Jointly, niche construction can either reinforce or alter the direction of natural selection through changes to the external environment.

Organisms are not just evolutionary products but also causes of their evolution. A co-evolution between organism and environment is driven by their reciprocal causation (Laland et al. 2011; Laland et al. 2012). When organisms change the environment, the altered environment can be experienced by future generations as an ecological inheritance, either because offspring tend to construct similar environments or inherit a changed environment.

NCT supporters take for granted the assumption that natural selection is a process that adapts organisms to environments.[10] The theory concerns the way intrinsic properties of

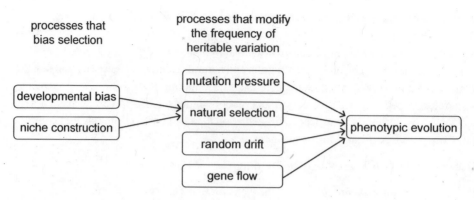

Figure 14.1
Relation between developmental bias, niche construction, and natural selection recreated from Laland et al.
(2015).

the environment are manipulated or chosen by organisms, such that niche construction can impose direction and order on these selective pressures.

To further illustrate the NCT conceptual model of niche construction and natural selection, consider the way Laland and colleagues take niche construction and developmental bias as evolutionary processes that bias natural selection (Laland 2015; Laland et al. 2015) (figure 14.1).

The relation between developmental bias and natural selection is usually framed as an *internal versus external* tug-o-war between the relative strength of internalist (developmental bias) and externalist (natural selection) processes in the evolutionary origin of adaptive traits (Amundson 2005; Sansom 2009). Some argue that internal processes constrain the production of possible variants, thus impeding the capacity of selection to build new traits (Gould and Lewontin 1979). Others argue that internal processes generate novel and adaptive variants, thus rivaling the creativity of selection (Gerhart and Kirschner 2007; Müller and Newman 2003; Walsh 2007). The NCT frames the relation between niche construction and natural selection in a similar way. Proponents argue that niche construction can act against or amplify natural selection. By itself, niche construction can also rival the ability of natural selection to create complementarity between organism and environment.

The conceptual model underlying NCT lines up nicely with Godfrey-Smith's position about the relation between niche construction and natural selection. Both focus on the types of niche construction that pertain to the *intrinsic* properties of their environments, and both endorse an externalist interpretation of natural selection as environmental selection. Even though NCT does not explicitly endorse nor reject mediational niche construction,[11] whereas Godfrey-Smith explicitly dismisses it, both approaches share a similar causal schema of evolutionary causes and processes: organism and environment

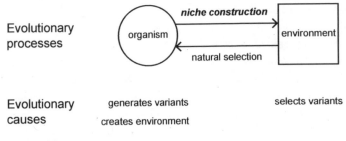

Figure 14.2
Causal schema of Godfrey-Smith and NCT.

are decoupled as cause and effect; niche construction and natural selection are decoupled as distinct causal processes (figure 14.2).

Lewontin and Commingling

In contrast, in a new introduction for his 1983 paper on organisms and environments, Lewontin (2001) proposed a different type of conceptual relation between niche construction and natural selection. He argued that a consequence of niche construction is that *different* types of organisms within the same population will determine *different* types of environments. When organisms have different niches, the selective environment of the entire population is dependent on the specific mixture of constructed environments by actual niche constructing phenotypes. Therefore, as the population evolves, the causes of natural selection simultaneously change.

To Lewontin, when there is intra-population variation in niche construction, natural selection operates in a frequency-dependent-like[12] manner: "the fitness of a genotype is dependent upon the mixture of other genotypes in the population. ... It is hopeless to measure the net fitnesses of many genotypes in an immense array of different frequency combinations" (Lewontin 2001, 57). As a consequence, "realistically, selection coefficients are frequency-dependent, and theoretical modeling of the effect of natural selection must use frequency-dependent formulations." In contrast, the standard model of evolution "assumes constant fitnesses or, at the most, fitnesses that vary with some autonomous extrinsic force or stochastically" (Lewontin 2001, 57).

Lewontin's conceptual breakthrough is that natural selection *by nature* should be conceptually understood as a type of evolutionary process whereby the fitness differences of phenotypes are in part the result of variation in constructed environments. The central takeaway is that *variation* in niche construction matters for the nature of natural selection. The implications challenge many philosophical characterizations of natural selection, such as the Principle of Natural Selection defended by Brandon (1990). Brandon argues that the selective environment presents a common background condition that supports but does not make a causal difference to variation in fitness. Natural selection occurs only if fitness differences between organismal types are completely accounted for by differences between their internal, intrinsic properties. However, when organisms vary in their constructed niches, differences in fitness are accounted for by both variation in organisms and variation in their constructed niches as well as the interaction between the two.

Intra-population variation in niche construction and plastic response to varying niches have been shown to have many consequences for evolutionary parameters (Saltz and Nuzhdin 2014). An example is social niche construction, where individuals of the same population create different types of social subgroups. Saltz and Foley (2011) studied the role aggressive behaviors play in constructing social groups of fruit flies, and the consequences of these social groups for fitness. Highly aggressive genotypes tend to displace other males, forming a smaller social group with fewer males. The other social groups come to consist of a larger number of less aggressive males. Interestingly, males with genotypes that have higher mating rates after winning a fight thrive only when they are part of the first type of group. The males with genotypes whereby winning or losing does not make a difference thrive when they are in the second type of group. When males of each genotype find themselves in the other group, however, they have lower fitness. The fitness of these fight-related genes are thus dependent on the social context, which in turn are environments constructed by the ration of aggression-related traits. The fight-related genes are not selected against a uniform environment but are against a mixture of environments that depend on the specific composition of the population.[13]

In sum, the conceptual model underlying Lewontin's theory of niche construction does not treat niche construction and natural selection as distinct, decoupled processes. Niche construction is not a prior cause that can incur changes to the environmental causes of natural selection. Instead, variation in niche construction is part of the conditions of natural selection. Niche construction and natural selection are *commingled processes.*

Odling-Smee, Decoupling, and Commingling

Even though NCT assumes the decoupling of niche construction and natural selection, the first model of NCT is compatible with both decoupled and commingled relations between the two processes.

In 1988, John Odling-Smee presented the first model for NCT. His quest was to find a way to measure evolution of the environment so that he could track the coevolution between organism and environment. The Modern Synthesis does not provide such a measure for two reasons: it defines evolution (of organisms) in terms of gene frequencies, and it uses the environment as a "reference device." A reference device is "the final source of the theory's explanations." When the reference device is the external environment, the environment is the foundational explanation that explains all evolutionary change, including organism–environment coevolution. In a sense, by explaining coevolution in terms of the environment (e.g., organisms are selected to respond to and change the environment in a certain way), "co"-evolution is explained away.

Odling-Smee proposed a new reference device to handle the currency of coevolutionary change—the organism–environment relationship (O_iE_i) or (OE). Coevolution and evolution of the organism or the environment, respectively, are explained in terms of changes in the (OE) relation. E_i (individual level) or E (population level) is the part of the environment that is relevant to organism perception and action.[14] In Odling-Smee's setup, organisms are purposive agents with the aim to persist and reproduce. They have predictive and decision-making capacities to decide whether to act on themselves, the environments, or on the relation between organisms and environments. The environments, on the other hand, do not have agency, but they can passively change themselves, the organisms, or the relation between organisms and environments. The passive and active actions of organisms and environments are framed in terms of how they contribute to changes in two features of (OE): organism–environment adaptedness and spatio-temporal properties.

Odling-Smee's model is compatible with both decoupling and commingling situations. He identifies the strong causal influence of environments on organisms via changes in (OE)s as natural selection. The strong causal influence of organisms on environments via changes in (OE)s is instead niche construction. The coevolution between organisms and environments can occur when there is strong bi-directional interaction between natural selection and niche construction via changes to (OE)s. These types of interactions became the sole focus of NCT in later publications with Kevin Laland, Marcus Feldman, and others. A decoupling strategy is appropriate in this case, as the focus is on the reciprocal feedback between the intrinsic properties of organisms and environments.

Odling-Smee's framework also includes the many other types of changes to organisms, environments, and organism–environment relations as a result of changes in (OE). Since Odling-Smee considers (OE) at multiple hierarchical levels from the individual (O_iE_i) to the populational (OE), *prima facie*, a distribution of changes to both Os and Es can create commingling conditions. When some of these varying organism–environment relations determine differences in fitness, then a commingling strategy is more suitable and natural selection is not environmental selection.

The Decoupling/Commingling Framework

To Lewontin, organism and environment do not exist without each other. They "interpenetrate," to use Lewontin's terminology, or in this chapter, "commingle." Instead of treating these relations as ontological properties, I interpret "decouple" and "commingle" as evolutionary scenarios and explanatory strategies. A *decoupling strategy* treats the intrinsic properties of organisms and environments as distinct explanatory entities. It is suitable only if local environments can be reified into a single, causally effective entity that explains organismal evolution, *a decoupled evolutionary scenario*. Natural selection is then an externally driven process that alters organisms. *Commingling strategies* are instead called for when the causes of natural selection are distributed across varying organisms

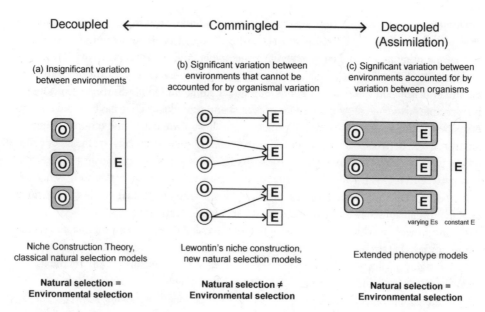

Figure 14.3
The decoupling/commingling framework.

and their varying environments, a *commingled evolutionary scenario*. In these cases, the environment cannot be reified into a single explanatory entity. When organisms and environments are commingled, natural selection is not environmental selection (see figure 14.1).

An external environment is a selective environment, that is, the causal source of natural selection, when it is possible to reify a single environment from the local environments different organisms. Emphasis on the intrinsic properties of the world around an organism promotes the reification of the external environment as the selective environment. Physical niche construction and relocational niche construction both determine the intrinsic properties of the external world surrounding an organism or a population of organisms. When the population constructs an environment that is experienced in roughly the same way[15] by all its members, slight variation in local environments is not significant for evolutionary change (figure 14.3a). The selective environment can thus be abstracted into a single causal entity that interacts with the population.

Another way organisms and environments can be "decoupled" is when each type of organism constructs a unique environment that is only experienced by its own type, this is when environmental differences can be entirely accounted for by variation between organismal types (figure 14.3c, "assimilated" into the population). In this, niche construction is an extended phenotype that is subject to natural selection by a wider environmental context. The organisms are not decoupled from their constructed environments, but they are decoupled from the wider selective environment they're embedded in.

However, niche construction can also lead to complex variations in the experienced environments of organisms, preventing the abstraction of local environments into a single, common environment. When variation in constructed environments is significant yet not fully accounted for by differences between individuals (figure 14.3b), the external environment, as a whole, does not have a single explanatory role in natural selection explanations. Instead, the causes of selection are no longer just in the external environment. They are distributed across organisms and environments. When the causes of natural selection are distributed across organism–environment relations, organism and environment are said to be "commingled." In these cases, the difference makers of fitness are not just the varying organisms but the variation between constructed environments as well. Since the causes of selection are not in the external environment, a "decoupling" heuristic does not capture the causes of selection in these situations. Under a "commingling" scenario, natural selection is the type of process sensitive to the evolving frequencies of constructed environments.

A Different Case for Mediational Niche Construction

Mediational niche construction makes a stronger case for commingling strategies. When organisms vary in mediational niche construction, organism and environment are commingled, yet the commingling conditions of variation between constructed niches are more difficult to cancel out. Contrast the situation with physical niche construction, which can also result in variation in constructed environments. When organisms alter the intrinsic properties of their external world, those properties are *intrinsic* to the environment despite having an organismal origin. That means they are separate from the properties of the organisms. So given enough time, organisms can move about and come to experience the environments constructed by others. The selective environment thus becomes insensitive to population composition as the varying phenotypes eventually experience the same constructed environment.

The different environments caused by mediational niche construction, however, are not easily cancelled out. Mediational niche construction does not leave a mark in the world, and thus variation in created environments is not easily accessible by conspecifics. When two individuals experience the significance of the same environment differently, these differences are not canceled out even if they exchange their spatial locations. Variation in experiential niche construction is more likely to be preserved by the idiosyncratic experiences of individuals. Through mediational niche construction, the same physical world can form different experienced environments for different variants and hence different patterns of selection. The greater the variation in constructed environments, the less impact of the physical world *as is* on relative fitness.

The decoupling/commingling framework is a pluralistic approach that allows both strategies to operate. Which one is appropriate depends on the evolutionary scenarios.

When a decoupling strategy is appropriate, natural selection can be treated as environmental selection and niche construction as the causal contributor to the sources of selection. However, when a commingling strategy is applicable, natural selection supervenes on the varying niche constructing activities and outcomes of organisms. Niche construction is constitutive of natural selection. Mediational niche construction can help retain variation in constructed environments, and thus provides the strongest support for commingling strategies.

5.   Conclusion

A common criticism against NCT is that it fails to offer anything new (Gupta et al. 2017; Scott-Phillips et al. 2014). One of the reasons NCT is seen as rehashing old phenomena is its adoption of decoupling strategies. If environmental selection is still an important actor of adaptive evolution, why would adding organisms as a causal source of the selective environment be considered a conceptual breakthrough? Standard evolutionary thinking already recognizes that environments can be caused by prior organismal activities or that organisms can buffer or strengthen selective pressure by changing their environments (Fisher 1930; see for instance, Mayr 1960).

Mediational niche construction and commingling strategies are some of the unique perspectives NCT can bring to evolutionary thinking (see also Laland et al., this volume). The decoupling/commingling framework proposed in this chapter is an attempt to expand the explanatory strategies that can fall under Niche Construction Theory 2.0. Niche Construction Theory (NCT) adopted decoupling strategies due to their focus on the ways organisms determine the intrinsic properties of their surroundings, either by perturbing the environment or relocating to other environments. But NCT can also take on commingling strategies.

By connecting early theories of niche construction with recent developments on mediational niche construction under a heuristic framework, I have shown that the strongest case for commingling strategies comes from a type of niche construction that occurs without any changes in the intrinsic properties of the environment. When a population of organisms *varies* in their experiences of the environment, the same environment has different relevance, impact, and significance to the organisms, thus creating different patterns of selection. In these cases, organism and environment commingle and the causes of selection are distributed across organism–environment relations. Natural selection is not environmental selection.

Despite Lewontin's original arguments for both, mediational niche construction has been largely neglected in debates over NCT. Perhaps the neglect is part of a heuristic "screening off" strategy to enable productive research (Laland et al. 2011), as many neglected scientific theories and concepts are not empirically refuted but set aside due to pragmatic or conceptual limitations. I argued that Lewontin's niche construction as well as the earliest model of NCT (Odling-Smee 1988) already contain the conceptual resources for commingling strategies. Proponents of NCT may find that revisiting neglected aspects of earlier proposals can gain new currency and viability, affording new avenues for future development and renewed defenses. There is a vast and exciting middle ground of commingled natural selection scenarios to explore.

Acknowledgments

I would like to thank Kevin Laland and Tobias Uller for the opportunity to develop this chapter at the EES workshop at the KLI in Klosterneuberg, Austria. Special thanks to André Ariew, Denis Walsh, Sonia Sultan, John Odling-Smee, Kevin Laland, Tobias Uller, and Jean-François Moreau for valuable feedback and comments on the various versions. Previous versions of this work benefited from the audience of ISHPSSB 2013 at Montpellier, the KLI (in particular, the 2013 cohort) in Austria, the 2nd Toronto Annual Roundtable Meeting 2015 (especially Rasmus Grønfeldt Winther), and members of the ERC-IDEM group at Bordeaux (led by Thomas Pradeu). Parts of the chapter have been developed from the second chapter of my PhD dissertation (University of Missouri). Copyeditor Scott Dyson provided an invaluable service during the initial writing stages. Detailed comments from the three reviewers contributed greatly to the final product, especially the last section. Finally, many thanks to Greg Dupuy for final stylistic comments. This project has received funding from the European Research Council (ERC) under the European Union's Horizon 2020 research and innovation program—grant agreement no. 637647—IDEM.

Notes

1. "Niche construction" was coined by Odling-Smee (1988). The majority of theoretical and empirical work on niche construction is currently developed under the banner of "Niche Construction Theory," spearheaded by John Odling-Smee, Kevin Laland, and Marcus Feldman. It includes hundreds of researchers in a wide range of disciplines. For a list of researchers under NCT, please consult the St. Andrews website: https://synergy.st-andrews.ac.uk/niche/others-working-with-niche-construction/.

2. In the immediate aftermath of the publication of *Niche Construction: A Neglected Evolutionary Process*, much hair has been split over which phenomena should count niche construction, driven by the worry that the concept would be trivialized if defined too broadly (Archetti 2015; see for instance, Dawkins 2004; Griffiths 2005; Laland 2004; Laland, Odling-Smee, and Feldman 2005; Okasha 2005; Sterelny 2005). However, very few analyses cover the full extent of Lewontin's experiential niche construction. Most leave out mediational niche construction.

3. It is important to clarify that even though mediational niche construction involves the physiological faculties of the organism, it is not *developmental* niche construction: the construction of a developmental environment. Following (Stotz 2017), our concern here is selective niche construction, the construction of the selective environment.

4. The goal-directedness of organisms does not entail high capacities of cognition or consciousness. Organisms can possess basic goals related to survival and reproduction without consciousness, cognition, or self-awareness, and can select from a repertoire of possible actions without conscious deliberation.

5. "Affordance," or "ecological affordance," is a concept developed by ecological psychologist J. J. Gibson. "The affordances of the environment are what it offers the animal, what it provides or furnishes, for good or ill ... I mean by it something that refers to both the environment and the animal.... It implies the complementarity of the animal and the environment" (Gibson 1986, 127). There are many opposing interpretations of Gibsonian affordances. Walsh's interpretation, an agency account, is that affordances are emergent properties of the organism–environment system.

6. Walsh distinguishes between a causal and a constitutive relation between organisms and environments. Churchill once remarked "we shape our buildings; thereafter they shape us." This type of niche construction is reciprocal causation between organisms and their environments. Marx, on the other hand, stated "the animal is immediately one with its life activity. It does not distinguish itself from it." This second type of niche construction is instead reciprocal constitution.

7. Plasticity also plays a powerful role in determining the evolutionary outcome of physical niche construction. Organisms have "norms of reaction" that represent their phenotypic responses to different environmental conditions. When organisms alter the properties of their external environment, they supply new environmental cues that can trigger different phenotypic responses. Sultan concludes with at least four evolutionary outcomes of plasticity in physical and experiential niche construction: (1) changed trait expressions in subsequent selection, (2) changed selective pressures on the population, (3) intra-population variation in niche construction, with consequences for evolutionary parameters such as heritability, epistasis, and pleiotropy, and (4) selection of traits that favor the constructed environments (see also Donohue 2005; Saltz and Nuzhdin 2014).

8. "... the cue of the developmental niche to induce the predator-protected morph is not the predator itself but chemicals released by the predator. It is, however, not these so-called kairomones, but the predator itself that is the defining parameter of the selective niche" (Stotz 2017, 3).

9. The primary focus of this chapter is NCT as described in Odling-Smee, Laland, and Feldman 1996, 2003; Laland 2015; Laland et al. 2015.

10. From the very beginning, Odling-Smee was skeptical of Lewontin's claim that adaptation metaphors should be replaced with construction metaphors. "Lewontin is overstating the case slightly. The idea that active organisms construct their own environments does not replace the idea of adaptation. It merely proposes that the adaptive fit which organisms regularly express in their environments could be arrived at by a different route from the one described by the Modern Synthesis" (Odling-Smee 1988, 77).

11. Personal communication with Kevin Laland and John Odling-Smee revealed that the decision to leave out mediational niche construction was largely pragmatic. Doing so allowed them to capture the agency of organisms and implement environmental modification in both experimental and theoretical work. Incidentally, Kevin also revealed that they corresponded frequently with Peter Godfrey-Smith and Richard Lewontin while developing NCT. Peter probably played an active role in their decision to remove mediational niche construction from their analyses.

12. To clarify, Lewontin does not mean to state that all cases of natural selection are frequency dependent or that frequency-dependent selection is natural selection. Nor is he saying that some forms of selection are dependent on the frequency of conspecifics or the environment (e.g., prey, predator, resources). Nor is the point of bringing organisms into selection that selection standards are always relative to a population (i.e., soft selection). Many such forms of frequency-dependent natural selection have long been recognized under standard evolutionary theory, and they do not challenge the conceptual structure of evolution. Many thanks to Kevin Laland and Tobias Uller for pushing me to clarify the difference between Lewontin's notion of frequency-dependent-like natural selection and the more familiar types of frequency-dependent selection.

13. In the wild, the scenario is even more complex. There are multiple dynamic and transient social environments within the same population depending on the composition of aggression-related genotypes and their likelihood of meeting each other. Flies are constantly and freely forming or dissolving social groups. The constructed environments are not entirely unique to the genotypes, as different genotypes can share the same social environment. Furthermore, the non-aggression social groups are quite heterogeneous.

14. The environment variable E is defined in terms of the "observational horizon" (the environment the organism can monitor at a particular space and time) and "action horizon" (the environment the organism can influence at a particular space and time) of an organism.

15. Mediational niche construction does not create commingling between organisms and environments just because it distorts environmental signals. Instead, it creates commingling scenarios because of the variation between the ways the same environment is experienced. Distortion itself is not a sufficient reason to adopt a commingling model. When the constructed experiential environments of all individuals in a population "distort" physical properties in the same way (the physical world has the same relevance, impact, and significance to all individuals regardless of phenotype), there is still a one-to-one mapping between the physical world and the constructed environments (figure 14.3c). In these cases, a decoupling strategy is still apt. Relative fitness can be accounted for in terms of varying phenotypes in a physical world, not an experienced environment.

References

Amundson, R. 2005. *The Changing Role of the Embryo in Evolutionary Thought: Roots of Evo-Devo*. New York: Cambridge University Press.

Archetti, E. 2015. "Three Kinds of Constructionism: The Role of Metaphor in the Debate over Niche Constructionism." *Biological Theory* 10 (2): 103–115. https://doi.org/10.1007/s13752-015-0205-3.

Brandon, R. N. 1990. *Adaptation and Environment*. Princeton, NJ: Princeton University Press.

Brandon, R. N., and J. Antonovics. 1995. "The Coevolution of Organism and Environment." In *Concepts, Theories, and Rationality in the Biological Sciences*, edited by G. Wolters, J. G. Lennox, and in collaboration with Peter McLaughlin, 211–232. The second Pittsburgh-Konstanz Colloquium in the Philosophy of Science. University of Pittsburgh, October 1–4, 1993. Pittsburgh: University of Pittsburgh Press.

Dawkins, R. 2004. "Extended Phenotype—But Not Too Extended. A Reply to Laland, Turner and Jablonka." *Biology and Philosophy* 19 (3): 377–396. https://doi.org/10.1023/B:BIPH.0000036180.14904.96.

Day, R. L., K. N. Laland, and F. J. Odling-Smee. 2003. "Rethinking Adaptation: The Niche-Construction Perspective." *Perspectives in Biology and Medicine* 46 (1): 80–95.

Donohue, K. 2005. "Niche Construction through Phenological Plasticity: Life History Dynamics and Ecological Consequences." *New Phytologist* 166 (1): 83–92.

Endler, J. A. 1986. *Natural Selection in the Wild*. Princeton, NJ: Princeton University Press.

Fisher, R. A. 1930. *The Genetical Theory of Natural Selection*. Oxford: Clarendon Press.

Gerhart, J., and M. Kirschner. 2007. "The Theory of Facilitated Variation." *Proceedings of the National Academy of Sciences USA* 104 (Suppl 1): 8582–8589. https://doi.org/10.1073/pnas.0701035104.

Gibson, J. J. 1986. *The Ecological Approach to Visual Perception*. New York: Taylor & Francis.

Godfrey-Smith, P. 1996. *Complexity and the Function of Mind in Nature*. New York: Cambridge University Press.

Godfrey-Smith, P. 2001. "Organism, Environment, and Dialectics." In *Thinking About Evolution: Historical, Philosophical, and Political Perspectives*, edited by R. S. Singh, C. B. Krimbas, D. B. Paul, and J. Beatty, 253–266. New York: Cambridge University Press.

Gould, S. J. 1977. "Eternal Metaphors of Paleontology." In *Patterns of Evolution as Illustrated by the Fossil Record*, edited by A. Hallam, 1–26. Amsterdam: Elsevier.

Gould, S. J. 2002. *The Structure of Evolutionary Theory*. Cambridge, MA: Belknap Press of Harvard University Press.

Gould, S. J., and R. Lewontin. 1979. "The Spandrels of San Marco and the Panglossian Paradigm: A Critique of the Adaptationist Programme." *Proceedings of the Royal Society of London. Series B. Biological Sciences* 205 (1161): 581–598.

Griffiths, P. 2005. "Review of 'Niche Construction.'" *Biology and Philosophy* 20 (1): 11–20.

Gupta, M., N. G. Prasad, S. Dey, A. Joshi, and T. N. C. Vidya. 2017. "Niche Construction in Evolutionary Theory: The Construction of an Academic Niche?" *Journal of Genetics* 96 (3): 491–504.

Laland, K. N. 2004. "Extending the Extended Phenotype." *Biology and Philosophy* 19 (3): 313–325.

Laland, K. N. 2015. "On Evolutionary Causes and Evolutionary Processes." *Behavioural Processes* 117:97–104. https://doi.org/10.1016/j.beproc.2014.05.008.

Laland, K. N., F. J. Odling-Smee, and M. W. Feldman. 2005. "On the Breadth and Significance of Niche Construction: A Reply to Griffiths, Okasha and Sterelny." *Biology and Philosophy* 20 (1): 37–55. https://doi.org/10.1007/s10539-004-6834-8.

Laland, K. N., F. J. Odling-Smee, W. Hoppitt, and T. Uller. 2012. "More on How and Why: Cause and Effect in Biology Revisited." *Biology and Philosophy* 28 (5): 719–745. https://doi.org/10.1007/s10539-012-9335-1.

Laland, K. N., and K. Sterelny. 2006. "Seven Reasons (Not) to Neglect Niche Construction." *Evolution* 60 (9): 1751–1762. https://doi.org/10.1111/j.0014-3820.2006.tb00520.x.

Laland, K. N., K. Sterelny, F. J. Odling-Smee, W. Hoppitt, and T. Uller. 2011. "Cause and Effect in Biology Revisited: Is Mayr's Proximate-Ultimate Dichotomy Still Useful?" *Science* 334 (6062): 1512–1516. https://doi.org/10.1126/science.1210879.

Laland, K. N., T. Uller, M. W. Feldman, K. Sterelny, G. B. Müller, A. Moczek, E. Jablonka, and F. J. Odling-Smee. 2015. "The Extended Evolutionary Synthesis: Its Structure, Assumptions and Predictions." *Proceedings of the Royal Society B* 282 (1813): 20151019. https://doi.org/10.1098/rspb.2015.1019.

Levins, R., and R. Lewontin. 1985. *The Dialectical Biologist*. Cambridge, MA: Harvard University Press.

Lewontin, R. 1982. "Organism and Environment." In *Learning, Development and Culture*, edited by H. C. Plotkin, 151–170. Chichester: Wiley.

Lewontin, R. 1983. "The Organism as the Subject and Object of Evolution." *Scientia* 118:63–82.

Lewontin, R. 2000. *The Triple Helix: Gene, Organism and Environment*. Cambridge, MA: Harvard University Press.

Lewontin, R. 2001. "Gene, Organism, and Environment: A New Introduction." In *Cycles of Contingency: Developmental Systems and Evolution*, edited by S. Oyama, P. E. Griffiths, and R. D. Gray, 55–57. Cambridge, MA: MIT Press.

Lewontin, R., and R. Levins. 2007. *Biology under the Influence: Dialectical Essays on Ecology, Agriculture, and Health*. New York: Monthly Review Press.

Mayr, E. 1960. "The Emergence of Evolutionary Novelties." In *Evolution after Darwin*, Vol. 1, edited by S. Tax. Chicago: University of Chicago Press.

Mayr, E. 1982. *The Growth of Biological Thought: Diversity, Evolution, and Inheritance*. Cambridge, MA: Harvard University Press.

Müller, G. B., and S. A. Newman. 2003. *Origination of Organismal Form: Beyond the Gene in Developmental and Evolutionary Biology*. Cambridge, MA: MIT Press.

Odling-Smee, F. J. 1988. "Niche-constructing Phenotypes." In *The Role of Behavior in Evolution*, edited by H. C. Plotkin, 73–132. Cambridge, MA: MIT Press.

Odling-Smee, F. J., K. N. Laland, and M. W. Feldman. 1996. "Niche Construction." *The American Naturalist* 147 (4): 641–648. https://doi.org/10.2307/2463239.

Odling-Smee, F. J., K. N. Laland, and M. W. Feldman. 2003. *Niche Construction: The Neglected Process in Evolution*. Princeton, NJ: Princeton University Press.

Okasha, S. 2005. "On Niche Construction and Extended Evolutionary Theory." *Biology and Philosophy* 20 (1): 1–10.

Rolian, C., D. E. Lieberman, and B. Hallgrimsson. 2010. "The Co-Evolution of Hands and Feet." *Evolution* 64:1558–1568.

Saltz, J. B., and B. R. Foley. 2011. "Natural Genetic Variation in Social Niche Construction: Social Effects of Aggression Drive Disruptive Sexual Selection in Drosophila Melanogaster." *The American Naturalist* 177 (5): 645–654.

Saltz, J. B., and S. V. Nuzhdin. 2014. "Genetic Variation in Niche Construction: Implications for Development and Evolutionary Genetics." *Trends in Ecology and Evolution* 29 (1): 8–14.

Sansom, R. 2009. "The Nature of Developmental Constraints and the Difference-Maker Argument for Externalism." *Biology and Philosophy* 24 (4): 441–459.

Scott-Phillips, T. C., K. N. Laland, D. M. Shuker, T. E. Dickins, and S. A. West. 2014. "The Niche Construction Perspective: A Critical Appraisal." *Evolution* 68 (5): 1231–1243.

Sterelny, K. 2005. "Made by Each Other: Organisms and Their Environment." *Biology and Philosophy* 20 (1): 21–36.

Sterelny, K., and P. Griffiths. 1999. *Sex and Death: An Introduction to Philosophy of Biology*. Chicago: University of Chicago Press.

Stotz, K. 2017. "Why Developmental Niche Construction Is Not Selective Niche Construction: And Why It Matters." *Interface Focus* 7 (5): 20160157. https://doi.org/10.1098/rsfs.2016.0157.

Sultan, S. E. 2015. *Organism and Environment: Ecological Development, Niche Construction, and Adaptation*. New York: Oxford University Press.

Walsh, D. M. 2007. "Development: Three Grades of Ontogenetic Involvement." In *Philosophy of Biology*, edited by M. Matthen, C. Stephens, P. Thagard, and J. Woods, 179–200. Amsterdam: Elsevier.

Walsh, D. M. 2012. "Situated Adaptationism." In *The Environment: Philosophy, Science, and Ethics*, edited by W. P. Kabasenche, M. O'Rourke, and M. H. Slater, 89–116. Cambridge, MA: MIT Press. Retrieved from http://www.scopus.com/inward/record.url?eid=2-s2.0-84894865948&partnerID=40&md5=56c1db99e4546bf89 d52c3f003353aec.

Walsh, D. M. 2014. "The Affordance Landscape: The Spatial Metaphors of Evolution." In *Entangled Life*, edited by G. Barker, E. Desjardins, and T. Pearce, 213–236. Dordrecht: Springer. https://doi.org/10.1007/978-94-007 -7067-6_11.

Walsh, D. M. 2015. *Organisms, Agency, and Evolution*. Cambridge: Cambridge University Press.

Williams, G. C. 1966. *Adaptation and Natural Selection: A Critique of Some Current Evolutionary Thought*. Princeton, NJ: Princeton University Press.

15 Biological Information in Developmental and Evolutionary Systems

Karola Stotz

1. Introduction: The Problem of Information in Biology

There is a widespread use of informational language to describe and explain all kinds of processes in biology. For example, most people agree that *development* is the expression of information accumulated during *evolution* and that *heredity* is the transgenerational transmission of this information. This idea, however, is not only surprisingly hard to cash out in scientific terms, it also turns out to be a misleading description of development, heredity and evolution (Oyama, Griffiths, and Gray 2001; Laland et al. 2011). Nevertheless, there is a widely held belief that biological systems are distinctive among physical systems in being informed systems. Hence the lack of a rigorous account of biological information in both developmental and evolutionary systems is a striking gap in the scientific worldview. This chapter is part of a project that aims to fill that gap by grounding the idea of biological information in a contemporary philosophical account of causation, and showing that *biological information is a substantive causal factor in living systems*.

This chapter argues that biological systems are distinctive among physical systems in being an informed system, without subscribing to a narrow gene-centric view of biological information. It stands in strong opposition to the idea that development is best understood as the expression of information accumulated during evolution, and that heredity is the transmission of just this information. Instead of seeing evolution as the only information-accumulating process in biology, this chapter paints a picture of development as the acquisition, expression, and storage of information that is variably sourced from the environment, from parents through a range of inheritance channels, or emerges anew during self-organizing processes of development. Heredity—by whatever channel—is the transgenerational transmission or reconstruction of some of this information used to both reproduce and modify the developmental system's life cycle. It is the process that provides the heritable variation necessary for evolution to take place, which is one of the ways in which proximate factors affect this ultimate process. One important question is how alternative ways to represent and understand heredity affect our interpretation of causation in an evolving system (Helanterä and Uller 2010).

This chapter has the main goal of providing a general account of biological information as a causal factor in both development and evolution and furnishing strategies for how information from different sources can be quantified and compared. To this end, section 2 elucidates how biologists use and understand biological information within developmental or proximate biology to elucidate the nature of key biological processes. This will relate information talk in biology to the transfer of causal specificity. It further shows that causal specificity is distributed between genetic (both coding and noncoding) and nongenetic (epi- and exogenetic) sources. In section 3, I present an information-theoretic approach of causation to distinguish between two conceptions of causal specificity—fine-grained control and limited but precise control (aka one-to-one conception)—that are employed by biological systems to fine-tune and control different biological processes. This information-theoretic treatment allows elucidating a competitive approach to proximate specificity. Section 4 argues for the relevance of proximate sources of specificity as a cause in evolution. After a short, critical discussion of this distinction and potential conflations of the terms, section 4.2 goes on to dissect the main ultimate cause, natural selection, to expose its proximate basis. This sets the stage for explaining the ultimate relevance of proximate, exogenetically inherited information in evolution (section 4.3), and to discuss the quantification of the transgenerational transmission of specificity (section 5).

2.   Biological Information in Developmental Systems

2.1   Crick Information

The concept of information is widely used in areas of biology that are concerned with how organisms work. Most notably, when information entered biology in the 1950s, it was applied to describe the action and transfer of genetic information in protein synthesis, which by extension included its intergenerational transfer. Crick famously expressed this idea as the "Central Dogma of Molecular Biology" in his paper "On Protein Synthesis" (1958):

The Central Dogma ... In more detail, the transfer of information from nucleic acid to protein may be possible, but transfer from protein to protein, or from protein to nucleic acid is impossible. *Information means here the precise determination of sequence*, either of bases in the nucleic acid or of amino-acid residues in the protein. (Crick 1958, 152–153, italics added)

Crick's key insight was that specificity can be carried by and transferred to a linear sequence, a process that precisely specifies this sequence and ultimately its three-dimensional structure. He therefore supplemented the idea of stereochemical specificity, underlying the well-known lock-and-key model of interaction between enzymes and their substrates, with informational specificity. The latter expressed how fine-grained changes in the DNA/RNA sequence *cause* fine-grained changes in protein structure. DNA carries information because changes to DNA can specifically change—and hence control—the resultant proteins. Crick understood information twofold, as (1) precise determination, and (2) the transfer of this biological specificity from one molecule to another.

Crick distinguished "the flow of information" from "the flow of energy" and "the flow of matter" within the process of protein synthesis (Crick 1958, 144). While the "exact chemical steps" were clearly important, he regarded as "the essence of the problem" how to join the amino acids in the right order in the crucial act of "sequentialization" (Crick 1958, 143–144). Distinguishing "matter and energy" from "information" corresponds to the distinction between the *efficiency* (e.g., yield) and *specificity* (fine-grained or precise control) of a molecular process (Griffiths et al. 2015). While the enzyme polymerase has minimal specificity for its substrate (i.e., coding DNA), it mainly delivers the efficiency of the chemical reaction involved in transcribing protein-coding DNA into RNA. It does not, however, impose any specificity on the resulting sequence of the transcribed product.

More than a decade later Crick restated his main achievement as solving "the formulation of the general rules for information transfer from one polymer with a defined alphabet to another" (Crick 1970, 561). So the essential idea to take home is that information in biology is a way to talk about a specific kind of causal relationship, that of specificity: DNA molecules carry information in virtue of their ability to specify precisely the linear structure of the resultant proteins (Stotz and Griffiths 2017a).

2.2 Co- and Post-Transcriptional Regulation

The molecular biology of protein synthesis is just one important area in biology that has employed the term "information." Immediately after the publications of Crick's new idea of specificity, David L. Nanney (1958) pointed out that regulating the expression of genetic information in development calls for "two types of cellular control systems."

On the one hand, the maintenance of a "library of specificities," both expressed and unexpressed, is accomplished by a template replicating mechanism. On the other hand, auxiliary mechanisms with different principles of operation are involved in determining which specificities are to be expressed in any particular cell.... To simplify the discussion of these two types of systems, they will be referred to as "genetic systems" and "epigenetic systems." The term "epigenetic" is chosen... to underscore their significance in developmental processes. (Nanney 1958, 712)

The auxiliary mechanisms, of which Nanney had little information, include co- and post-transcriptional processes, some of which share informational specificity with the underlying coding DNA sequence, as Griffiths and Stotz (2013) have recorded extensively. For example, splicing factors are responsible for selecting among a range of available coding sequences during the process of alternative splicing, while editing factors are involved in even partially creating a new sequence through the specific substitution, deletion or insertion of bases from the underlying RNA in the process of RNA editing. These factors—the environmental signals that recruit them—therefore also carry "Crick information" (Griffiths and Stotz 2013).

In line with Nanney's original insight, biologists extended the use of the term information beyond protein synthesis to describe the processes that regulate the time- and tissue-specific expression of genes.

Clearly, in concentrating on this aspect of informational transfer [Crick] was setting aside two questions about the control of gene expression—when in the life of a cell the gene is expressed and where in the organism. But these are also *questions of an informational nature*, although not falling within Crick's definition. (Olby 2009, 251, italics added)

2.3 Gene-Regulatory Networks

The regulatory architecture of cells that underlies the regulation of the conditional expression of genes is referred to as Gene Regulatory Networks (GRNs). It is formed by the complex regulatory connections between the molecular factors involved, coding sequences and their regulatory products binding to a diverse set of cis-regulatory DNA modules. GRNs provide us with a novel way of talking about genetic information: the production and action of TFs is referred to as the flow of information within a gene network. Eric Davidson speaks about the "high specificity binding interaction between cis-regulatory DNA and nuclear proteins" (Davidson 2001, 27). These activators and repressors are located downstream from intracellular signaling pathways acting as mediators to bring "situational biological information to the gene," while the modular cis-regulatory sequences fulfill the function of "information processing" (Davidson 2001, 10). Ultimately, regulatory proteins relay environmental information to the genome. GRNs are often likened to internal signaling systems between external inputs, components of the GRN, and final outputs (Calcott 2014; Calcott, Pocheville, and Griffiths 2018). They transform the informational input into developmental outcomes, that is, causing a precise change in the state of a subsystem due to its interaction with another subsystem (Jablonka 2002). However, this is not Crick information defined as the fine-grained control of the sequence of a molecule.

2.4 Developmental Induction

A final example of an area of information use is induction, a process in which developmental biologists commonly distinguish between instructive and permissive interaction. Induction means that signaling between neighboring cells or tissues can lead to further differentiation in the responder tissue as a result of its interaction with the inducer tissue.

The notion of the specificity of interaction is closely associated with the terms "instructive" and "permissive" interaction. When the action system is largely responsible for the specificity of the interaction through the transfer of a specific message, to which the reaction system responds by entering into a particular pathway of differentiation, we speak of an instructive action. When, on the other hand, the specificity of a reaction is largely due to the state of the competence of the reaction system, so that even rather unspecific messages can serve as signals to open up new developmental pathways, we speak of a permissive action. (Nieuwkoop, Johnsen, and Albers 1985, 9)

The specificity of permissive versus instructive actions thus refers to the question of "how much information is provided by the stimulus, the seeming cause, to produce an effect?" The two notions are meant to provide "a measure of exactly what each agent contributes to the response" (Kirschner and Gerhart 2005, 125). Hence, the distinction divides causal responsibility in development.

In each of the examples above, biologists have struggled to identify where the specificity actually comes from. For example, according to Kirschner and Gerhard (2005, 128) "inducers merely release the innate but self-inhibited capacity to develop these structures." Similarly Holtzer, who introduced the distinction between instructive and permissive in the first place, expressed great doubt about "an inducer molecule instructing naïve ... cells" because "there is no such thing as an undifferentiated, uninstructed cell." He goes on to specify that "they are uniquely primed by their previous history to respond to the inducers," or "at all times engaged in synthetic activities regulated by nuclear 'messages'" (Holtzer 1968, 154). He admits that this leaves the void of answering the problem of great importance that the concept of instructive induction was meant to address: "If there is no exogenous didactic inducer molecule, what mechanism orders the emergence of distinct cell types? How are progenitor cells, the covertly differentiated cells, programmed in the first place?" (Holtzer 1968, 162). His solution was to make induction permit "those ... cells programmed by earlier events to express their ... genetic information." So both genetic information plus the history of the cell, and hence earlier external processes, seem to be responsible for the instructive programming of cell.

The question of where the specificity comes from almost always turns on the distinction between genetic and nongenetic resources, something also discussed at length by David Nanney in his 1958 paper. Scott Gilbert notes that the specificity of a reaction "has to come from somewhere, and that is often a property of the genome" (Gilbert 2003, 349). Cells, however, start with the same genetic "library of specificities." Nanney describes the question of how "cells with identical genetic composition acquire adaptive differences capable of being maintained in clonal heredity" as a developmental paradox (Nanney 1989). Giving causal responsibility solely to the genome neglects the cause acting as the *difference maker*, and dismisses that the genome's action—as Gilbert acknowledges—"depends upon its context. There are times where the environment gets to provide the specificity of developmental interaction" (2003, 350). How could it do that via "a signal of little complexity" (Kirschner and Gerhart 2005, 126)?

To answer this without assuming from the outset that genes are the only, or a special kind of, information in biology, we need to translate information into causation.

3. An Information-Theoretic Account of Biological Causation

How does information in biology exert its causal influence? Causal specificity, both understood as informational specificity of exactly the kind expressed in Crick's central dogma, as well of the older idea of stereochemical specificity from the pre-information era, form an extension of a widely supported interventionist approach to causation that grounds causal relationships in ideas about manipulability and control. By introducing the term "causal specificity" into James Woodward's account of causation (Woodward 2003), it is possible to distinguish between different kinds of causes in biological systems (Woodward 2010).

3.1 Causal Specificity

The phenomenon of biological specificity is explained by the existence of causes through which organisms exercise precise determination of outcomes. Following Woodward (2010), this can be cashed out in two ways.

A. Causal specificity 1: Fine-grained influence conception

Interventions on a *highly specific* causal variable C can be used to produce any one of a large number of values of an effect variable E. The ideal limit of this relationship would be that every value of E is produced by only one value of C.[1]

B. Causal specificity 2: One-to-one conception

One cause has exactly one effect, as can be seen in the precise affinity between molecules to cause specific reaction: enzyme-substrate, receptor-substrate, allosteric protein-effector-substrate, sperm-egg.[2]

Both of these conceptions meant to denote a high amount of specificity, contrasted with a "switch-like" cause that can only determine the outcome between "on" and "off"—like the radio switch rather than the station dial—allegedly carries very little specificity. Coding DNA and RNA that precisely determines sequences of amino acids provides a paradigm case of fine-grained control. The one-to-one conception instead figures in the lock-and-key model of "biological specificity" introduced in 1894 by Emil Fischer, and which figured hugely in pre-informational biology. There are a myriad of causal factors in biological systems that exhibit switch-like behavior, which Kirschner and Gerhard (2005, 126 and 111) call "a signal of little complexity," and section 3.4 gives an example and alternative interpretation of how to understand such signals.

3.2 An Information-Theoretic Framework

The problem with Woodward's and other philosophers' work on causal specificity (Stotz 2006; Weber 2006; Waters 2007) is that it relies too much on intuition. A recent information-theoretic approach was therefore developed to clarify and quantify the fine-grained specificity of causal relationships within the interventionist framework (see Griffiths et al. 2015).

This body of work formalizes the idea that, other things being equal, the more a cause specifies a given effect, the more that knowledge of how we have intervened on the cause variable will inform us about the value of the effect variable. This idea can be quantified using Shannon information theory with the addition of an intervention operator that allows us to isolate the causal component of the correlation between variables:

SPEC: the specificity of a causal variable is obtained by measuring how much mutual information interventions on that variable carry about the effect variable. (Griffiths et al. 2015, 538)

A more intuitive way to think about the specificity measure is that it measures the extent to which an agent can reduce its uncertainty about the value of the effect variable if it can change the value of the cause, that is, the extent to which the agent can precisely determine the value of E by intervening on C.

3.3 A Competitive Approach

SPEC can be used to measure either how specifically two variables are connected (*potential* causal influence or difference maker) or how much of the actual variation in E in some data is causally explained by variation in C (*actual* causal influence or difference maker). It can therefore elucidate the difference in specificity between different sources, such as coding sequences of DNA, on the one hand, and RNA polymerase, on the other. The main aim, however, was to generalize this approach to Crick information in a wider sense in order to compare the fine-grained specificity of genetic and non-genetic causal factors (Griffiths et al. 2015). It therefore provides a versatile tool for asking general questions concerning the respective specificity of different causes for the same effect, which includes accounting for the role of epigenetic and exogenetic factors in development, and to incorporate the emergence of *de novo* information (Stotz and Griffiths 2017a).

Griffiths and collaborators constructed a toy causal model of transcription and splicing to show how coding and noncoding regions in the genome can be compared quantitatively in terms of their respective contribution to the precise determination of the sequence of a biomolecule. For example, the outcome of a splicing process is influenced by the conserved core splicing regulatory sequences, intronic and exonic splicing enhancer and silencer regions that contribute to the probability by which exons will be removed from the resulting transcript. Hence there is Crick information in noncoding regions of the genome.

One can extend this measure to variables representing epigenetic modifications (The ENCODE Project Consortium 2007; Djebali et al. 2012; Griffiths 2017). Originally thought to just up- or down-regulate the transcription of certain genes, we now know that epigenetic modification can also make a difference to the precise sequence of biomolecules through their role in the regulation of transcriptional and post-transcriptional processes such as splicing and editing. For instance, histone modifications can influence the inclusion and exclusion of exons in the final transcript. So, by a direct application of Crick's original reasoning, epigenetic modifications—both intra- and transgenerational—can contain Crick information.

As pointed out above, the specificity contributed by epigenetic processes are often considered to be spurious, and in effect attributed to the genome. That is, epigenetic modifications are genetically controlled (Rosenberg 2006; Haig 2007, 423). One argument refers to the inherent ability of specific DNA sequence to be epigenetically modified as a result of natural selection (Haig 2007, 423). While it is undisputed that some sequences may be more prone to epigenetic modifications than others, there is nothing to suggest a complete genetic mechanism controlling epigenetics in general. Others may argue that the *trans*-regulatory factors involved in establishing epigenetic modifications are products of the genome and bring epigenetic states under genetic control via this route. But as we have shown in section 2.3, trans-regulatory factors that are involved in the regulation of time- and tissue-specific expression do not carry sequence-specific Crick information. A third argument refers to parent-of-origin imprinting of certain genes, which very likely is under genetic control, but should not be referred to as epigenetic inheritance (Rosenberg 2006).

Many of these arguments fail to clearly distinguish the material from the informational side of either genetic or epigenetic resources. Identifying the sources of biological specificity requires measuring causal control, not material contributions. Although some epigenetic modifications are specified by genomes while others are not, this distinction does not in itself address the question where the control originates. While the *general* ability for epigenetic regulation—just as for genetic inheritance—clearly is a product of natural selection, the *specific* information contained in an experience-dependent modification is clearly not.[3] Similarly, trans-acting factors are gene products, yet their transcription and activation is often under environmental control. Lastly, genomic imprinting is a special case of epigenetically mediated, genetic parental effects that derive from the parent's—not the offspring's—genome. Many cases of epigenetic modifications that contribute to the precise determination of phenotype are influenced by the parent's environment (in the form of an inborn environmental effect) or the offspring's environment. It seems reasonable to describe this as a mechanism for conveying environmental information to the genome, so that genome expression can be correctly matched to the environment, something Sonia Sultan has labeled "functionally appropriate trait adjustments" (Sultan 2017, 1).

The real issue is where causal control is being exercised over the transcription, production, and processing of those sequences. In the latter case, evolution has designed a mechanism that detects and responds to information from the environment. Because mechanisms, processes, or factors contain "Crick information" when they contribute substantially to the precise determination of the linear structure of biomolecules, this means that exogenetic factors in these examples contain Crick information.

3.4   A Hierarchical Approach to Causal Specificity

Brett Calcott has applied the measure of Crick Information to the permissive versus instructive distinction in developmental biology (Calcott 2017), which will help us to better understand two central notions introduced earlier, namely "control" and "switch." As a result, he modified Woodward's one-to-one conception of specificity as one of "precise but limited control." Permissive delineates a "canalized cause" that produces "one definite end-result" (e.g., a signal activating downstream event, such as transcription, splicing, differentiation).

The permissive signal gains this kind of control due to its specific linkage to an instructive cause. One example is the conserved core processes of biology—metabolism, gene expression, cell signaling—that can be reused in different circumstances by connecting them via weak regulatory linkage through signal transduction that employs simple signals. Kirschner and Gerhard call this weak linkage "indirect, undemanding, low-information kind of regulatory connection" (Kirschner and Gerhard 2005, 111). The signal can simply activate biological process, while the core process produces a final developmental effect, like the growth of a trait. In other words, the signal is a foreground cause controlling the relationship between the instructive background cause—the conserved process—and the final effect. Because both causes have different effects (the activation of the process versus the production of a developmental outcome) their specificity shouldn't be compared competitively.

In short, a permissive cause has limited causal capacity but nevertheless exhibits precise control over an effect whose range of possible states exceeds this capacity. Because this only occurs under some specific background conditions we look to those conditions for an explanation. We seek an instructive cause. (Calcott 2017, 496)

Understanding where the control for either the foreground or the background cause lies, we have to look at two different time dimensions. The control dimensions of the instructive cause having to do with the general programming of the GRNs lies at the phylogenetic level; a more fine-grained epigenetic programming occurs at the development level, which directs potential and long-term downstream events (genetic down-regulation or silencing, canalizing splicing choices, etc.). The control dimension of the permissive cause having to do with connecting parts of the gene regulatory networks via weak linkages lies at the phylogenetic level (see English et al. 2015 for a model of the evolution of a developmental switch); the switches that are connected to part of the network can then be activated at the developmental level.

Distinguishing between the "fine-grained" and the "precise but limited" control, and extending Woodward's original twofold distinction, elucidates a whole range of topics that are central to my argument.

1. Both kinds of specificity play a different explanatory role in biology: the fine-grained control characterizes the specification of a gene product; the limited but precise control characterizes the regulation or control of biochemical reactions or the conditional expression of genes (see sections 2 and 3).

2. This distinction elucidates the complexity of causal relationships in biological systems, which can be interrogated via different approaches: a competitive approach that compares the fine-grained control of two competing causes (section 3.3), and a hierarchical approach that looks at the interaction of connected (i.e., instructive versus permissive) causes (section 3.4).

3. Both conceptions of specificity are distributed among genetic and nongenetic causes (sections 2.2–2.4).

4. Both kinds are controlled at both developmental (sections 2 and 3 and evolutionary dimensions (sections 4 and 5).

4. From Proximate Information to Ultimate Explanations

The argument up to now implies that genetic, epigenetic, and exogenetic factors can be sources of information in the same sense that was introduced by Francis Crick for genetic variables: they contain Crick information. At other times they are sources of information in a different sense, like those involved in regulatory processes. To show that genetic, epigenetic, and exogenetic sources of information are "equally" important, we need to measure their contribution in empirically well-grounded models of development.

The phenomenon of biological specificity is explained by the existence of causes through which organisms exercise precise determination of outcomes. Development can be understood as the expression of biological specificity that can be measured using causal information theory, and extended heredity is the trans-generational transfer of this biological specificity. The functional expression of this specificity is usually explained by natural selection acting on those causes, but we will see that due to exogenetic inheritance this is not necessarily so (e.g., Helänterä and Uller 2010; Laland et al. 2015; Stotz 2017). The existence of biological information in epigenetic and exogenetic factors is relevant to evolution as well as to development because they can be passed on from previous generations through a range of inheritance mechanisms (Jablonka and Lamb 2005; Badyaev and Uller 2009; Shea, Pen, and Uller 2011; Stotz and Griffiths 2016; Stotz and Griffiths 2017b).

The relevance of this exogenetic information for evolution depends on a range of issues, including (1) the extent to which environmentally acquired and developmentally regulated information is trans-generationally transmitted; (2) how stable this transmission turns out to be; and (3) to what extend this information produces specifically adaptive adjustment to the phenotype, and hence shows some sensitivity toward the requirement of organisms in specific environmental conditions. But before these can be addressed, exogenetic information needs to pass the hurdle constructed by the proximate-ultimate distinction.

4.1 Conflating Proximate and Ultimate Causation

The last few years have seen a proliferation of arguments that call for an extension of the Modern Synthesis (MS) into an Extended Evolutionary Synthesis (EES) (e.g., Pigliucci 2007, 2009; Pigliucci and Müller 2010a, 2010b; Laland et al. 2015). One notable feature of these calls is their suggestion to rethink Mayr's (1961) distinction between proximate and ultimate causes (Laland et al. 2011, 2013). The proximate-ultimate distinction remains hugely influential to this day and hence this criticism has not been left unchallenged. For example, Dickins and Barton counter-argue that attempts to replace the proximate-ultimate distinction with reciprocal causation still rely on Mayr's distinction (Dickins and Barton 2013). Dickins and other collaborators also accuse the above proposals of a conflation between proximate and ultimate explanations (Scott-Phillips, Dickins, and West 2011; Dickins and Rahman 2012; see also Haig 2013). All three papers use a similar strategy of arguing that the phenomena, processes, and causes that are being proposed to extend the MS are all proximate mechanism; hence, they cannot have any relevance for evolution, and are therefore of no concern to the MS.

These counter-arguments rest on a misunderstanding of the critique of Mayr's distinction. Neither is it an attempt to get rid of the distinction between proximate and ultimate causes, nor is it a proposal to reclassify proximate into ultimate causes. Second, the counter-arguments are based on and perpetuate a flawed implication of Mayr's distinction. So let us have a closer look at how Mayr introduced the distinction:

The functional biologist is vitally concerned with the operation and interaction of structural elements. His ever-repeated question is "How?" ... The evolutionary biologist differs in his methods and in the problems in which he is interested. His basic question is "Why?" ... The functional biologist deals with all aspects of the decoding of the programed information contained in the DNA of the fertilized zygote. The evolutionary biologist, on the other hand, is interested in the history of the programs of information and the laws that control the changes of these programs from generation to generation. ... It is evident that the functional biologist would be concerned with analysis of the proximate causes, while the evolutionary biologist would be concerned with analysis of the ultimate causes. (Mayr 1988, 25–26)

Mayr's distinction is threefold: it is based on the existence of (a) two questions—how and why questions, (b) two disciplines—developmental and evolutionary biology, and (c) two kinds of causes—proximate and ultimate. The three distinctions are then assembled into two package deals: developmental "how" questions yield proximate explanations, while "why" questions concerning evolution yield ultimate explanations. This implies that both kinds of causes must be irrelevant to the other questions or disciplines.[4]

The existence of different kinds of questions, however, should not automatically imply that a given causal factor could only figure in the answer to one of the questions; a factor may play distinctive roles in different processes. Proximate and ultimate causes should be defined without reference to developmental and evolutionary processes—or related distinctions—to avoid vacuous or circular definitions. Predetermining what is the single ultimate cause able to answer evolutionary questions begs the question and precludes the discovery of other evolutionary causes. Accordingly, Laland and colleagues want to break "Mayr's association of the proximate with ontogeny and the ultimate with phylogeny" and instead ask for a "more nuanced conception of biological causation" (Laland et al. 2011, 1512).

4.2 The Proximate Foundation of Natural Selection

The defendants of the proximate-ultimate distinction cited above only accept natural selection as an ultimate cause:

Natural selection is the outcome of trait variation, inheritance and competition…Any trait variation that increases the differential in accessing those resources will be selected and the genes underpinning this will increase in relative frequency. This is the essence of ultimate causation; how genes work is not. (Dickins and Rahman 2012, 2916)

This seems to assume that extended inheritance only provides accounts of how exogenetic resources work, rather than how genetic and exogenetic information account for heritable trait variation. Beside this, all of natural selection's underlying conditions—trait variation, inheritance, and differential fitness—can be given a causal-mechanistic account of proximate processes. It is only through their role in the selection process that they are rendered as ultimate causes. Genetic mutations, epimutations, and environmentally or developmentally induced, but heritable, phenotypic changes are all causes of trait variation, and genetic, epigenetic, and exogenetic inheritance are the mechanisms of transmission that ensure their heritability. So why should we not accept them as part of an ultimate explanation?

It pays to have a closer look at this ultimate explanation in more detail. Following Matthen and Ariew (2002), Pigliucci and Kaplan (2006) distinguish between the informal (1) and formal (2) sense of fitness and natural selection:

(1) As the result of individual interactions between organism with particular traits and the physical processes that impinge on them, and (2) as a statistical distribution of changes in the makeup of populations. (Pigliucci and Kaplan 2006, 64)

They argue that describing natural selection as a *force* resulting from fitness differences between members of a population is a misleading metaphor. It distracts from understanding the origin of these differences as the result of some proximate or "physical instantiation of the organism involved, including the systems of inheritance available and the developmental systems that are reliably replicated through those inheritance systems" (Pigliucci and Kaplan 2006, 124). So the shape of particular traits and how they interact with particular processes in the environment influence the survival and reproduction of organisms and hence make natural selection possible (see Walsh, this issue).

The idea that a theory of forces can explain evolutionary change confuses what is in need of an explanation. Insofar as we are trying to explain changes in the distribution of heritable traits in a single population, the only "forces" at issue are straightforwardly physical ones. Individual populations change over time because of the way physical processes interact with particular organisms (Pigliucci and Kaplan 2006, 31).

So we can describe the ultimate cause of natural selection as a proximate mechanism, if mechanism is understood in the informal sense as a phenomenon at the individual level where the variation in a trait interacts with a particular physical process in a discriminate way. It is *only the informal sense* that explains fitness differences—even if the underlying stochastic nature of the physical processes involved makes it impossible to predict the outcome—because the formal level is strictly speaking not a process or a cause but a statistical distribution. There are ultimately two proximate processes that substantially impinge on the outcome of selection: the origin of heritable phenotypic variation among individuals and the ecological processes with which individuals interact to produce fitness differences.

While the modern synthesis explains these by the origin of variation (explained by stochastic genetic mutations) and an externally determined ecological niche, both are influenced in substantial ways by developmental processes. *Developmental niche construction* focuses on the influence of exogenetic resources (the developmental niche) on developmentally induced traits, which then become the targets of selection (Stotz 2017). *Niche construction theory* is concerned with how certain traits of the organism constructs the physical processes that impinge on the organisms (the selective niche; Odling-Smee, Laland, and Feldman 2003).

For Mayr's proximate-ultimate distinction and its current-day defenders an ultimate cause is typically what explains adaptations. They commonly see natural selection as the single ultimate cause, which is only true if it is solely responsible for the adaptive match between organism and environment. If, however, environmental components of development can lead to non-genetically caused adaptation because they alter the organism's response to their environmental conditions in a functionally appropriate way, then the creation of heritable variation would be rendered an ultimate cause in its own right. In the end "the efficacy of selection would depend on the nature of phenotypic variation" (Kirschner and Gerhart 2005, 13).

The mapping between parent and offspring phenotype can be significantly changed in two generations in cases where such variation of hereditary information plays a substantial role in development (Griffiths and Stotz 2013). Sonia Sultan and Mary-Jane West-Eberhard have long argued that even transient factors on development and on fitness will have an evolutionary impact. "… [I]nherited and immediate effects on phenotypes require that we decouple allelic change as a record of microevolutionary history from the much more complex, environmentally mediated causes of fitness variation that contributed to that history" (Sultan 2015, 145–146).

4.3   Rendering Proximate Mechanisms Evolutionary

Dickins and Rahman (2012) claim that understanding extended inheritance as introducing new heritable variation that changes frequencies at the population level conflates ultimate and proximate causation. This is not because they deny the inheritance of epigenetic variation, but because it is an inheritance system and as such based on proximate mechanisms, just like the genetic inheritance system:

The MS of Mendelian genetics and Darwinian evolution was precisely a synthesis of proximate genetic concerns with ultimate evolutionary ones—the how of genetic transmission with the why of evolution ... to simply outline other inheritance systems and describe how they work is not to provide an account at the ultimate level. (Dickins and Rahman 2012, 2917)

Just pointing out the existence of exogenetic inheritance systems may indeed commit what could be termed an ontogenetic fallacy: explaining an evolutionary outcome through causes of development (Hochman 2013). Investigating the role of heritable, exogenetic information on future phenotype, however, has impact on the evolutionary dynamics of the population (Helanterä and Uller 2010; Danchin et al. 2011; Otsuka 2015). Dickins and Rahman's intended conclusion is to deny epigenetic inheritance an ultimate status, however, their comparison of epigenetic inheritance with Mendelian inheritance establishes exactly the opposite. It shows that in contrast to the treatment of epigenetics by today's proponents of the MS, the founders of the MS did not dismiss Mendelian genetics as merely a proximate mechanism; they used it to derive the form of the transmission phase underlying heritable variation on which the process of natural selection relies. Quantitative genetics already accounts for nongenetic heredity based on epigenetic and exogenetic mechanisms as parental effects. But the common assumption that these are under "genetic control" (see above) obscures the fact that the incorporation of parental effects into evolutionary models significantly changes the evolutionary dynamics and hence contributes to explanations for the distribution of phenotypes over evolutionary time. In fact, any form of heredity, including epi- and exogenetic, that is a significant source of biological information in the sense defined above will be relevant in the same way because it substantially alters the mapping from parent phenotype to offspring phenotype (see figure 15.1). If Mendelian models of heredity were thought to be significant for evolution—even if the way genes work is not—we need to grant the same reasoning to epigenetic and exogenetic heredity. Both Mendelian and nongenetic inheritance systems are part of an ultimate explanation. Of course, not all genetic, epigenetic and exogenetic heredity has further consequences for the evolutionary process. However, if trans-generationally transmitted, it changes the form of the transmission phase, and that is substantially impacting on the dynamics of natural selection (Griffiths 2016, 2017).

5.   Quantifying the Transgenerational Transmission of Specificity

In the context of this paper we are most concerned with understanding the pattern of entanglement of hereditary causes, and the possibility of quantifying their influence. It was thought not possible to quantify the relative contributions of different causes to the production of a trait (see Tal, Kisdi, and Jablonka 2010 for a quantitative genetic/epigenetic model for teasing apart these contributions). Acknowledging that, the evolutionary biologist Ronald Fischer introduced the analysis of variance (ANOVA) for investigating the relative contributions of genotypic and environmental *variation* to total phenotypic *variation* in a population. The information-theoretic approach introduced in section 4 took advantage of Woodward's account of causation in terms of actual or potential "difference makers" (rather than total contribution to trait development) in order to develop a versatile tool for quantifying the different causal contribution to trait development using information theory. The concern of this section is interpreting causation in evolving systems, and to this end it becomes paramount to understand heredity.

The main thing left to reconsider here is how to most effectively demonstrate that it is worthwhile to compare the specificity of genetic versus inborn environmental variation in a population by effectively pointing out that the latter exist to a substantial amount.

As the original debate about the utility of heritability versus reaction norms did reference information, and these measures can be put into information-theoretic terms, a more formal approach to this idea should in principle be possible. The problem is that the different sources of variation (genetic, phenotypic, environmental and parental effects, or the inborn environmental variation, V_G, V_P, V_E, *and* V_{PE}) are typically not all either discrete or continuous variables from which follows that some different kind of mathematics might be needed from the entropy-based measures that have been used in the previous work described in section 3. One possible approach that could be used instead is to just discretize the parameters artificially. But that will remain a project for the future.

For now, it has to suffice to visualize the extent of exogenetic versus genetic factors. While there are limits to the reaction norm perspective (this is nicely covered by Sultan, this volume), some of the limitations could be solved by representing reaction norms in more than two dimensions. Since I want to identify the specificity of exogenetic inheritance, one needs to add inborn environmental information taken from the parental generation to environmental information taken from the current generation in order to show its influence on the phenotype.

For classical quantitative genetics, one complication in measuring heritability is gene-environment interaction—the extent to which the effect of the environment on the phenotype depends on the genotype, and the extent to which the effect of the genotype depends on the environment in which they are expressed. Another complication is the confounding of a number of factors in the broad heritability measure of genotypic variation (V_G); most relevant for our discussion is its inclusion of parental effects that influences the measurement for (V_P). The existence of parental effects (represented below as the inborn environmental effect on the parental generation) increases the complexity of heritability measures drastically.

In order to compare the contribution of genetically and exogenetically inherited resources to trait variation in a population, one needs to dissociate genetic from inborn environmental information and add this to the usual comparison between genetic and environmental information for phenotype. Unfortunately, a norm of reaction graph, which is useful because it offers more information than the standard outcome of a heritability measure, can only capture two inputs. If we have three inputs influencing the phenotype, it will depend on the question asked of how to represent the same data.

Figure 15.1 shows the results from a transgenerational plasticity experiment, shown in a multigenerational norm of reaction, which illuminates the complex situation nicely (Sultan 2017, 5–6).

The outcome of the same experiment can be represented to answer a range of different questions:

(1) Comparing the two norms of reactions in one single panel in figure 15.1 highlights the relative size of the influence of the inborn environmental effect and the environmental effect on the phenotype for each genotype. (2) One can see the genetic influence on the size of the transgenerational effect by comparing the three panels. There also exist important parent–offspring interaction effects: because the norms of reactions are not perfectly parallel, it shows that the different effects are not additive; in other words, the effect of offspring genotype on offspring phenotype differs depending on the offspring environment in which they are expressed, and the effect of the parental phenotype—and the developmental environment it produces—on offspring phenotype differs depending on offspring genotype. (3) Comparing the norms of reactions of all three genotypes from just one parental growth condition—thereby holding constant the effect of parental environment on offspring phenotype—measures the relative contribution of genotypic and current environmental variation to the total variation in a population (see figures 15.2a or 15.2b). (4) Comparing the norm of reactions under both parental growth conditions ($E_{P-shade}$ and E_{P-sun}) side by side is another way to highlight the influence of the inborn environment on the slope of the reaction norm of each genotype (redrawn in figure 15.2a and b).

(5) Finally and most important for our discussion, in order to estimate the component of nongenetic information on heritability one would need to compare the relative contribution of the genetic and the inborn environmental effect to the total variation. This requires holding the environment constant and drawing the reaction norm of each genotype against the parental growth conditions. This can be drawn as a norm of reaction of either the genotypes against the parental environments (see redrawn figure 15.3) or the different parental environments against the different genotypes (see redrawn figure 15.4).

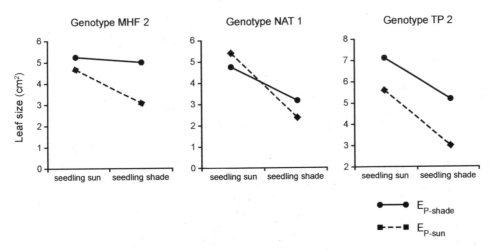

Figure 15.1
The figure shows three panels, each representing two norms of reaction from a single genotype of the species *Polygonum persicaria* affected by different parental growth conditions (E_p), sun and shade. The solid reaction norm shows the size of leaves in sun and shade from seedlings produced by parents that were grown in shade, the dotted one represents the reaction norm of seedlings by parents grown in sun. Redrawn with permission of the author from Sultan (2017).

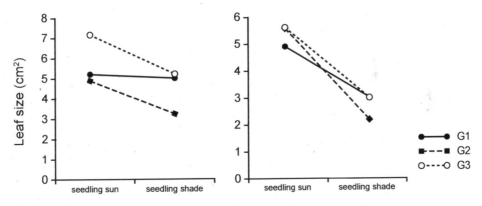

Figure 15.2
Each panel shows the relative contribution of genes and environment to the phenotype when parental environment (E_p) is held constant (left $E_{P\text{-shade}}$, right $E_{P\text{-sun}}$). Comparing both panels highlights the influence of the inborn environment on the slope of each genotype's reaction norm.

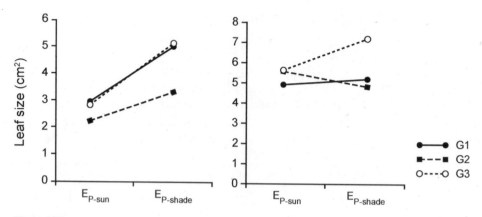

Figure 15.3
The figure plots the relative contribution of the genetic and the inborn environmental effect to the total variation under two different growing conditions (left seedling shade, right seedling sun).

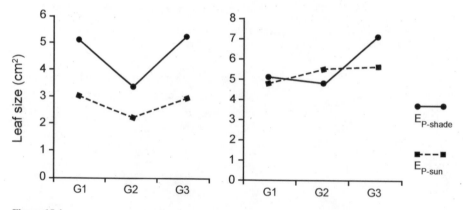

Figure 15.4
Genes as background in shade (left) and in sun (right) seedling growth conditions.

Obviously, having some data on the fitness differences due to the inborn environmental effect would make the whole result even more relevant to the debate, because one of the outstanding disputes with respect to the relevance of exogenetic variants for evolution is the possibility that they may produce adaptive adjustments of the phenotype, rather than being "blind" or insensitive to the demands created by the external environment. "The challenge to neo-Darwinism comes, not from the existence of epigenetic inheritance, but from the possibility that epimutations are directed rather than random" (Haig 2007, 421).

Extending inheritance with exogenetic inheritance mechanisms is not pitched against and in competition with genetic inheritance, but it is complementary to genetic inheritance serving a different, more ecologically responsive, role (Stotz and Griffiths 2016). Dickins and Rahman hypothesize "about a possible role for natural selection in building the proximate machinery for epigenetics and learning" (Dickins and Rahman 2012, 2917). But they are mistaken in thereby dismissing them as irrelevant for evolution. Instead, we should accept that selection has fashioned different inheritance channels serving distinctive roles (Shea et al. 2011): while genetic inheritance is the proximate mechanism for producing stable adaptations in slow-changing environments, exogenetic inheritance systems are proximate mechanisms with the ultimate function of calibrating organisms to environmental stochasticity.

6. Conclusion

It can be considered a main aim of an Extended Evolutionary Synthesis to provide *causal-mechanistic explanations* of the complex processes that generate phenotypes—a kind of theory of form—and their developmental and selective niches, and how those processes affect the traditional population-level mechanisms of evolutionary change (Laland et al. 2015). Underlying this is a profound shift from a programmed to a constructive view of development that recognizes the central role of the active organism in the evolutionary process.

Proponents of the existence and evolutionary significance of extended inheritance, developmental bias, self-organization, and developmental niche construction have long argued that an important difference between different heritable sources of information may be that some lead to functional and internally coherent phenotypic variants. These mechanisms could therefore cause some organism–environment complementarity, and therefore directionality in evolution, that is not due to natural selection.

The most striking and contentious difference from the original MS concerns the relative significance of natural selection versus generative variation in evolution, one of the oldest controversies in evolutionary biology. (Laland et al. 2015, 8; see also Stotz 2017)

The present chapter was mainly concerned in showing the significance of nongenetic, heritable variation both in development and in evolution. In order to evaluate the relative contribution of the different kinds of heritable information during development, we have introduced an information-theoretic treatment that may allow that. To get at the relative weight of the difference sources of information in populations would require the manipulation of these sources. Parental effects, other than some other sources of exogenetic information, strictly follow the same path, and hence masquerade, as genetic information, if no great care has been taken to distinguish between them.

Acknowledgments

This chapter was made possible through the support of a grant from the Templeton World Charity Foundation, Causal Foundations of Biological Information TWCF0063/AB37. The opinions expressed in this chapter are those of the author and do not necessarily reflect the views of the Templeton World Charity Foundation. The author also wants to thank Tobias Uller, Sonia Sultan, and Eva Jablonka for useful comments, and acknowledge the other collaborators on the grant, Paul E. Griffiths, Brett Calcott, and Arnaud Pocheville.

Notes

1. Woodward introduces the concept of "stability" to determine to what extent E is affected by interventions on other variables than C.

2. These examples don't perfectly illustrates Woodward's idea because the perfect fit is actually between reactants who together cause the reaction and hence the effect.

3. Similarly, while the ability of a songbird to learn a song has a genetic basis, the song it learns has not.

4. As an aside, Mayr's insistence on keeping the causes of development and evolution strictly separate looks rather ominous in light of the fact that Mayr himself was guilty of conflating them. He used to explain development with reference to the existence of an evolved genetic program. According to comparative psychologist Daniel Lehrman, this gives rise to what has been termed the "phylogeny fallacy": "although the idea that behavior patterns are 'blueprinted' or 'encoded' in the genome is a perfectly appropriate and instructive way of talking about certain problems in genetics and evolution, it does not in any way deal with the kinds of questions about behavioral development to which it is so often applied" (Lehrman 1970, 35). Referring to the genetic program or programmed information, which is hardly the same as providing "the mechanism that underpins the trait" talked about by Scott-Phillips et al. (2011) doesn't contribute in the least to our understanding of development (Griffiths and Tabery 2013).

References

Badyaev, A. V., and T. Uller. 2009. "Parental Effects in Ecology and Evolution: Mechanisms, Processes, and Implications." *Philosophical Transactions of the Royal Society, Biological Sciences* 364:1169–1177.

Calcott, B. 2014. "The Creation and Reuse of Information in Gene Regulatory Networks." *Philosophy of Science* 81 (5): 879–890.

Calcott, B. 2017. "Causal Specificity and the Instructive-Permissive Distinction." *Biology and Philosophy* 32 (4): 481–505.

Calcott, B., A. Pocheville, and P. E. Griffiths. 2018. "Signals That Make a Difference." *British Journal for the Philosophy of Science.* doi.org/10.1093/bjps/axx022.

Crick, F. H. C. 1958. "On Protein Synthesis." *Symposia of the Society for Experimental Biology* 12:138–163.

Crick, F. H. C. 1970. "Central Dogma of Molecular Biology." *Nature* 227:561–563.

Danchin, É., A. Charmentier, F. A. Champagne, A. Mesoudi, B. Pujol, and S. Blanchet. 2011. "Beyond DNA: Integrating Inclusive Inheritance into an Extended Theory of Evolution." *Nature Reviews Genetics* 12 (7): 475–486.

Davidson, E. R. 2001. *Genomic Regulatory Systems: Development and Evolution.* San Diego: Academic Press.

Dickins, T. E., and R. A. Barton. 2013. "Reciprocal Causation and the Proximate–Ultimate Distinction." *Biology and Philosophy* 28 (5): 747–756.

Dickins, T. E., and Q. Rahman. 2012. "The Extended Evolutionary Synthesis and the Role of Soft Inheritance in Evolution." *Proceedings of the Royal Society B* 279 (1740): 2913–2921.

Djebali, S., C. A. Davis, A. Merkel, A. Dobin, T. Lassmann, A. Mortazavi, A. Tanzer, et al. 2012. "Landscape of Transcription in Human Cells." *Nature* 489 (7414): 101–108.

The ENCODE Project Consortium. 2007. "Identification and Analysis of Functional Elements in 1% of the Human Genome by the ENCODE Pilot Project." *Nature* 447:799–816.

English, S., I. Pen, N. Shea, and T. Uller. 2015. "The Information Value of Non-Genetic Inheritance in Plants and Animals." *PLoS ONE* 10 (1): e0116996.

Gilbert, S. F. 2003. "The Reactive Genome." In *Origination of Organismal Form: Beyond the Gene in Developmental and Evolutionary Biology*, edited by G. B. Müller and S. A. Newman, 87–101. Cambridge, MA: MIT Press.

Griffiths, P. E. 2016. "Proximate and Ultimate Information in Biology." In *The Philosophy of Philip Kitcher*, edited by M. Couch and J. Pfeiffer. New York: Oxford University Press.

Griffiths, P. E. 2017. "Genetic, Epigenetic and Exogenetic Information in Development and Evolution." *Interface Focus* 7:20160152. doi: 10.1098/rsfs.2016.0152.

Griffiths, P. E., A. Pocheville, C. Calcott, K. Stotz, H. Kim, and R. Knight. 2015. "Measuring Causal Specificity." *Philosophy of Science* 82 (4): 529–555.

Griffiths, P. E., and K. Stotz. 2013. *Genetics and Philosophy: An Introduction*. Cambridge: Cambridge University Press.

Griffiths, P. E., and J. G. Tabery. 2013. "Developmental Systems Theory: What Does It Explain, and How Does It Explain It?" *Advances in Child Development and Behavior* 44:65–94.

Haig, D. 2007. "Weismann Rules! OK? Epigenetics and the Lamarckian Temptation." *Biology and Philosophy* 22 (3): 415–428.

Haig, D. 2013. "Proximate and Ultimate Causes: How Come? and What For?" *Biology and Philosophy* 28 (5): 781–786.

Helanterä, H., and T. Uller. 2010. "The Price Equation and Extended Inheritance." *Philosophy and Theory in Biology* 2:201306.

Hochman, A. 2013. "The Phylogeny Fallacy and the Ontogeny Fallacy." *Biology and Philosophy* 28 (4): 593–612.

Holtzer, H. 1968. "Induction of Chondrogenesis: A Concept in Terms of Mechanisms." In *Epithelial-Mesenchymal Interactions*, edited by R. Gleischmajer and R. E. Billingham, 152–164. Baltimore, MD: Williams & Wilkins.

Jablonka, E. 2002. "Information Interpretation, Inheritance, and Sharing." *Philosophy of Science* 69 (4): 578–605.

Jablonka, E., and M. J. Lamb. 2005. *Evolution in Four Dimensions: Genetic, Epigenetic, Behavioral, and Symbolic Variation in the History of Life*. Cambridge, MA: MIT Press.

Kirschner, M. W., and J. C. Gerhart. 2005. *The Plausibility of Life: Resolving Darwin's Dilemma*. London: Yale University Press.

Laland, K. N., J. Odling-Smee, W. Hoppitt, and T. Uller. 2013. "More on How and Why: Cause and Effect in Biology Revisited." *Biology and Philosophy* 28 (5): 719–745.

Laland, K. N., K. Sterelny, J. Odling-Smee, W. Hoppitt, and T. Uller. 2011. "Cause and Effect in Biology Revisited: Is Mayr's Proximate-Ultimate Dichotomy Still Useful?" *Science* 334 (6062): 1512–1516.

Laland, K. N., T. Uller, M. Feldman, K. Sterelny, G. B. Müller, A. Moczek, E. Jablonka, and F. J. Odling-Smee. 2015. "The Extended Evolutionary Synthesis: Its Structure, Assumptions and Predictions." *Proceedings of the Royal Society B* 282:20151019.

Lehrman, D. S. 1970. "Semantic & Conceptual Issues in the Nature-Nurture Problem." In *Development and Evolution of Behaviour*, edited by D. S. Lehrman, 17–52. San Francisco: W. H. Freeman.

Matthen, M., and A. Ariew. 2002. "Two Ways of Thinking about Fitness and Natural Selection." *Journal of Philosophy* 49 (2): 53–83.

Mayr, E. 1961. "Cause and Effect in Biology." *Science* 134 (3489): 1501–1506.

Mayr, E. 1988. *Toward a New Philosophy of Biology: Observations of an Evolutionist*. Cambridge, MA: Harvard University Press.

Nanney, D. L. 1958. "Epigenetic Control Systems." *Proceedings of the National Academy of Sciences USA* 44:712.

Nanney, D. L. 1989. "Metaphor and Mechanism: 'Epigenetic Control Systems' Reconsidered." *The Epigenetics of Cell Transformation and Tumor Development*. San Francisco, CA, Eightieth Annual Meeting of the American Association for Cancer Research, http://www.life.illinois.edu/nanney/epigenetic/sanfrancisco.html.

Nieuwkoop, P. D., A. G. Johnsen, and B. Albers. 1985. *The Epigenetic Nature of Early Chordate Development: Inductive Interaction and Competence*. Cambridge: Cambridge University Press.

Odling-Smee, F. J., K. N. Laland, and M. W. Feldman. 2003. *Niche Construction: The Neglected Process in Evolution*. Princeton, NJ: Princeton University Press.

Olby, R. C. 2009. *Francis Crick: Hunter of Life's Secret*. Cold Spring Harbor, NY: Cold Spring Harbor Laboratory Press.

Otsuka, J. 2015. "Using Causal Models to Integrate Proximate and Ultimate Causation." *Biology and Philosophy* 30:19–37.

Oyama, S., P. E. Griffiths, and R. D. Gray, eds. 2001. *Cycles of Contingency: Developmental Systems and Evolution*. Cambridge, MA: MIT Press.

Pigliucci, M. 2007. "Do We need an Extended Evolutionary Synthesis?" *Evolution* 61 (12): 2743–2749.

Pigliucci, M. 2009. "An Extended Synthesis for Evolutionary Biology." *Annals of the New York Academy of Sciences* 1168:218–228.

Pigliucci, M., and J. Kaplan. 2006. *Making Sense of Evolution: The Conceptual Foundations of Evolutionary Biology*. Chicago: University of Chicago Press.

Pigliucci, M., and G. B. Müller. 2010a. *Evolution—The Extended Synthesis*. Cambridge, MA: MIT Press.

Pigliucci, M., and G. B. Müller. 2010b. "Elements of an Extended Evolutionary Synthesis." In *Evolution: The Extended Synthesis*, edited by M. Pigliucci and G. B. Müller, 3–17. Cambridge, MA: MIT Press.

Rosenberg, A. 2006. "Is Epigeneis a Counterexample to the Central Dogma?" *History and Philosophy of the Life Sciences* 28:509–526.

Scott-Phillips, T. C., T .E. Dickins, and S. A. West. 2011. "Evolutionary Theory and the Ultimate–Proximate Distinction in the Human Behavioral Sciences." *Perspectives on Psychological Science* 6 (1): 38–47.

Shea, N., I. Pen, and T. Uller. 2011. "Three Epigenetic Information Channels and Their Different Roles in Evolution." *Journal of Evolutionary Biology* 24 (6): 1178–1187.

Stotz, K. 2006. "Molecular Epigenesis: Distributed Specificity as a Break in the Central Dogma." *History and Philosophy of the Life Sciences* 28 (4): 533–548.

Stotz, K. 2017. "Why Developmental Niche Construction Is Not Selective Niche Construction—And Why It Matters." *Interface Focus* 7:20160157.

Stotz, K., and P. E. Griffiths. 2016. "Epigenetics: Ambiguities and implicationssee for instance." *History and Philosophy of the Life Sciences* 38 (4): 1–20.

Stotz, K., and P. E. Griffiths. 2017a. "Biological Information, Causality and Specificity—An Intimate Relationship." In *Information and Causality: From Matter to Life*, edited by S. I. Walker, P. C. W. Davies, and G. F. R. Ellis, 366–390. Cambridge: Cambridge University Press.

Stotz, K., and P. E. Griffiths. 2017b. "Genetic, Epigenetic and Exogenetic Information." *Routledge Handbook of Evolution and Philosophy*. New York: Routledge.

Sultan, S. E. 2015. *Organism and Environment: Ecological Development, Niche Construction, and Adaptation*. London: Oxford University Press.

Sultan, S. E. 2017. "Developmental Plasticity: Re-conceiving the Genotype." *Interface Focus* 7:20170009.

Tal, O., E. Kisdi, and E. Jablonka. 2010. "Epigenetic Contribution to Covariance between Relatives." *Genetics* 184 (4): 1037–1050.

Waters, C. K. 2007. "Causes That Make a Difference." *Journal of Philosophy* 104 (11): 551–579.

Weber, M. 2006. "The Central Dogma as a Thesis of Causal Specificity." *History and Philosophy of the Life Sciences* 28 (4): 595–609.

Woodward, J. 2003. *Making Things Happen: A Theory of Causal Explanation*. Oxford: Oxford University Press.

Woodward, J. 2010. "Causation in Biology: Stability, Specificity, and the Choice of Levels of Explanation." *Biology and Philosophy* 25 (3): 287–318.

Contributor List

John A. Baker
Clark University

Lynn Chiu
University of Bordeaux

David I. Dayan
Clark University

Renée A. Duckworth
University of Arizona

Marcus W. Feldman
Stanford University

Susan A. Foster
Clark University

Melissa A. Graham
Clark University

Heikki Helanterä
University of Helsinki

Kevin N. Laland
University of St. Andrews

Armin P. Moczek
Indiana University

John Odling-Smee
Mansfield College

Jun Otsuka
Kyoto University

Massimo Pigliucci
City College of New York

Arnaud Pocheville
University of Sydney

Arlin Stoltzfus
Institute for Bioscience and Biotechnology
Research

Karola Stotz
Macquarie University

Sonia E. Sultan
Wesleyan University

Christoph Thies
University of Southampton

Tobias Uller
Lund University

Denis M. Walsh
University of Toronto

Richard A. Watson
University of Southampton

Index